SALMON FARMING

SALMON FARMING

Odd-Ivar Lekang

5m Books

First published 2022

Published by
5M Books Ltd,
Lings, Great Easton,
Essex CM6 2HH, UK,
Tel: +44 (0)330 1333 580
www.5mbooks.com

A Catalogue record for this book is available from the British Library

ISBN 9781789181814
eISBN 9781789182118
DOI 10.52517/9781789182118

Book layout by Cheshire Typesetting Ltd, Cuddington, Cheshire
Printed by CPI Anthony Rowe Ltd, UK
Photos by the author unless otherwise indicated
Illustrations by Tore Ensby

Contents

Preface

The salmon farming industry has been one of the fastest growing industries in the food sector since its start in the late 1960s. Today it is a multibillion-dollar industry with a yearly production above 2 million tonnes. It can rightly be called an industrial adventure. Atlantic salmon is by far the predominant salmon species used in salmon farming. Until recently, countries that have the natural conditions for sea cage farming have dominated the production of ongrowing salmon. However, the new technology of land-based farms, closed sea cages and large offshore cages is opening up the areas in the world that can be used for salmon production. Owing to increased demand and with little increase in production, the profitability of salmon farming has been very high in recent years. The future for salmon farming seems to be bright; even if prices drop, the market is growing. Atlantic salmon is a species known worldwide. It is healthy and can be used and processed in different ways.

This book gives an overview of salmon production, from eggs to slaughtering size, starting at the broodstock farm, continuing through the hatchery and smolt farm and ending up in the ongrowing sea cage farm until the salmon are ready for harvesting. Planning and forecasting of production on the farm, including the design of production plans, is also an important part of the salmon farming process and is covered in detail with worked examples. Supplementary chapters on salmon biology, feeding, diseases and breeding are also included. These are less substantial as these subjects lie outside the author's scientific background and are included as a introduction and guide to more detailed resources.

This book is based on material successfully used for BSc and MSc courses in intensive aquaculture given at the Norwegian University of Life Sciences (NMBU). The university has included courses in aquaculture since 1973. From 1990, it has provided special masters courses in aquaculture engineering, and from 2000, courses in optimalization of Atlantic salmon production. This textbook was written for a basic course in salmon production, but the university also has its own courses focusing on anatomy and physiology, feeding, breeding, diseases and harvesting and processing. The courses are parts of masters programmes in aquaculture and aquatic food production (for details, see www.nmbu.no).

During the compilation of material for this book many people have given useful advice. I especially thank my colleagues, friends and students including, among others, S.O. Fjæra, B.F. Eriksen, B.R. Hansen, H.H. Olsen, P.O. Skjervold, A. Bergheim, M. Brkic and S. Monteith.

Tore Ensby has drawn the majority of the line illustrations, and all photographs have been taken by the author.

Abbreviations

ADC	apparent digestibility coefficient
ADG	average daily growth
AGD	amoebic gill disease
AOP	apparent optical properties
ATU	accumulated temperature units
BFCR	biological feed conversion rate
BKD	bacterial kidney disease
CaSR	calcium sensing receptors
CCS	closed containment systems
CF	condition factor
CFTR	cystic fibrosis transmembrane conductance regulator
CV	coefficient of variation
DAA	dispensable amino acids
DE	digestible energy
DEN	digestible energy need
DM	dry matter
DP	digestible protein
EGI	Ewos growth index
EPA	eicosapentaenoic acid
FA	feed allowance
FAD	flavin adenine dinucleotide
FCR	feed conversion ratio
FE	feed efficiency
FF	feed factor
FFDR	forage fish dependency ratio
FIFO	fish in fish out
FSH	follicle-stimulating hormone
FT	flow through
GE	gross energy
GF	growth factor
GI	growth indexes
GIS	geographical information system
GM	genetically modified
GMO	genetically modified organism

GSI	gonad somatic index
HAB	harmful algal blooms
HI	heart index
HOG	head on gutted
HSMI	heart and skeletal muscle inflammation
HSS	haemorrhagic smolt syndrome
HUFA	highly unsaturated fatty acids
IAA	indispensable amino acids
IHN	infectious haematopoietic necrosis
IPN	infectious pancreatic necrosis
ISA	infectious salmon anaemia
LAP	land animal by-products
LED	light emitting diodes
LH	luteinizing hormone
LTLS	low trophic level species
MAB	maximum allowed biomass
MAS	marker-assisted selection
ME	metabolic energy
MR	mitochondria-rich
NALO	National Aquaculture Legislation Overview
NASCO	North Atlantic Salmon Conservation Organization
NE	net energy
NINA	Norsk Institutt for Naturforskning
NSP	non-starch part
OWI	operational welfare indicators
PAP	processed animal protein
PAR	photosynthetically active radiation
PD	pancreas disease
PGC	primordial germ cell
QTL	quantitative trait locus
RAS	recycling aquaculture system
RE	retained energy
RGL	relative gut length
ROMS	regional ocean modelling systems
ROV	remotely operated vehicles
SCM	single celled microorganisms
SCO	single cell oil
SCP	single cell protein
SGR	specific growth rate
SNP	single nucleotide polymorphisms
SPC	soy protein concentrate
SPDV	salmon pancreas disease virus
SRS	salmon rickettsial septicaemia
TAN	total ammonia nitrogen

TDC	true digestibility coefficient
TDER	total digestible energy requirement
TER	theoretical energy requirement
TGC	thermal growth coefficient
TU	thermal units
VHS	viral haemorrhagic septicaemia
WFE	whole fish equivalent

CHAPTER 1

Introduction

The Earth's growing population requires an ever increasing volume of food, including the provision of protein. The ocean makes up almost 71% of our planet's surface. By including the depth of the ocean, it is possible to grow and harvest in three dimensions, unlike on land where most food production is limited to two. The ocean therefore offers far greater potential for biological growth than dry land. To achieve future food security the ocean must factor as an important contributor [1–3].

Currently, only about 7% of protein for human consumption comes from fish [4]. It is therefore reasonable to expect that this sector has the potential for significant growth. Of the world's animal protein consumption, 17% comes from the ocean; the rest is represented by traditional livestock, such as poultry, pork and beef [5].

Food from the sea either comes from the capture of wild fish stock or from aquaculture. Wild catches have been quite stable during the last decades, at around 90 million tonnes. Most wild fish stocks are either fully exploited (61%) or overexploited (29%) [5]. The possibilities for increasing the tonnage of wild catches are very limited. The only way that the harvest from the ocean could increase appreciably is therefore through aquaculture. This has been recognized for some time; from the 1950s until today the aquaculture sector has been the fastest growing food production sector. In the early 1950s production was less than 1 million tonnes while production in 2018 was reported to be 81 million tonnes [6]. In the period from 2000 to 2018 there was a yearly increase in world aquaculture production of almost 6%.

Comparing the relationship between fishing and aquaculture to that between hunting and farming on land strengthens the claim that aquaculture will grow. Agriculture has almost totally taken over from the hunting of animals and gathering of edible plants in the wild in order to provide food for the growing human population. Through this shift, total harvested volumes have increased tremendously compared to those for hunting and gathering. It is obvious that traditional agriculture and fishing cannot support the growing population and this further indicates that the aquaculture sector must grow and develop in sophistication.

The aquaculture sector can be divided into shellfish (molluscs and crustacea), plants (seaweed and algae) and finfish. Finfish aquaculture (salmonids included) is of particular interest when it comes to satisfying the world's protein production needs. To be able to start sustainable aquaculture production of a species, the entire production cycle must be known and controlled, including spawning, hatching and start feeding. This means that the life cycle can be managed and manipulated artificially. Breeding programmes can then be established in which the best strains and individuals can be picked out as broodfish for the following generations. Through breeding programmes the fish will become better adapted to aquaculture production by selection of those best suited and, as a result, increased domestication will occur. Of course, also of major importance when starting aquaculture or farming of a new species is that it is a known species in the consumer market and that the price is high enough to cover the production cost related to farming.

Farming of Atlantic salmon (*Salmo salar*) has been a great success from the start, partly because it successfully fulfils the above-mentioned criteria, and the species has become a reference when it comes to aquaculture success. The salmon farming industry of today has a high level of industrialization compared to other farmed species within aquaculture, but salmonids still only account for 4.4% of the global seafood supply [4]. When commercial farming of Atlantic salmon started in the early 1970s the species was well known in the market and was a high-cost product; the market price then was higher than today's price (even taking account of inflation). The large-scale commercial fishing of Pacific salmon, another genus of salmon species quite equal in look and taste, was therefore an important door opener to the market. This was typically a seasonal fishery, so by use of aquaculture the market could be supplied with salmon all year round; this was successful in practice as the large growth in volume was consumed by the market. The biology and life cycle of Atlantic salmon are well known. Salmonids, especially Atlantic salmon, are highly appreciated by recreational anglers, which has driven an interest in enhancing wild stocks. From the 1950s a large number of hatcheries have been established worldwide for wild salmonid stock enhancement. There were even private initiatives undertaking private salmon ranching, through which knowledge about how to farm salmon was accumulated. Salmon ranching includes the set out of salmon in freshwater (or seawater) locations and letting them migrate to the sea where they find sufficient food to reach adult size, so there are no feed costs. When the salmon return to the place they were released as juveniles, they can be caught. However, the recapture percentage of this type of private ranching has been too low to achieve private economical sustainability.

To begin farming of a fish species domestication is a requirement. Domestication is necessary to ensure that the fish can perform optimally under farming conditions so that economic sustainability and profitability can be achieved. Atlantic salmon responded well to domestication and showed great potential for improvement by breeding. A large variation between families was documented and a large genetic gain has been attained on factors important to economy in farming.

Purely biologically, the salmonids, including Atlantic salmon, have a great advantage compared to other sea- and freshwater species because the eggs are large and result in larger fry. In addition, the hatched fry are equipped with a yolk sac, which supports growth in the first period of life. As a result, when the fry start to feed they are already well developed, and possess a functional digestive system that can digest dry formulated feed. Feeding can therefore be done directly with dry feed with a reasonable particle size. Other freshwater and seawater species are, however, so small and little developed after hatching that it will typically be necessary to use live feed in the first period of feeding. This is much more tricky and reduced survival rates should be expected. Compared to other salmonid species, including hybrids, Atlantic salmon offer a great advantage when it comes to growth in seawater and increased size at first maturation [7].

What makes modern salmon farming so sustainable is the good artificial control of the entire production cycle that can be obtained. Broodstock salmon are kept in sea- and freshwater. There is good knowledge of how to strip the brood fish and good fertilization rates of the eggs are achieved. The hatching process is known, and yolk sac fry can be relatively easily kept in good condition. Adapted feed for start feeding has been developed and low mortality occurs in this phase. Ongrowing in freshwater typically goes well, as does salmon smoltification before transport to sea cage farms for ongrowing until the required size for slaughter for human consumption or transfer to a broodstock farm is reached. The fish are then around 3 years old; brood fish approach maturity from the summer of their third or fourth year.

The success of Atlantic salmon aquaculture is evident when compared to other finfish species in aquaculture. During recent decades no other fish species have come close to the same growth in tonnage, and Atlantic salmon still achieve reasonably high prices.

Compared to traditional animal livestock, such as beef, pork, chicken and lamb, farming of Atlantic salmon has a number of advantages. The production represents a low overall carbon footprint, a high feed conversion ratio (FCR) is achieved, and less freshwater is used for production [8]. For the production of 1 kg edible product the release of carbon dioxide is estimated to be 7.9 kg, while for chicken, pork and beef similar values are estimated to be 6.2 kg, 12.2 kg and 39 kg, respectively [4,9]. Freshwater requirements are also favourable with 2000 l/kg for salmon (farmed salmon fillet in Scotland) compared to 4300 l/kg, 6000 l/kg and 15,400 l/kg, respectively, for poultry, pork and beef [4].

Atlantic salmon has good nutritional value; the protein content is high, it is rich in vitamins and minerals and it contains the healthy omega-3 fatty acids, which gives an advantage over traditional animal livestock. The edible meat produced per kilogram of food supplied is also much higher than for traditional animal livestock. To provide for the protein requirements of the growing human population a high FCR is required, which is another advantage of salmon farming. Atlantic salmon are also cold blooded and therefore do not need to spend energy on heating up their bodies as land animals do. They are weightless in water, meaning no energy is needed to stay in an upright position. Also, the release of metabolic waste products (ammonia) is more effective.

Today, farming of Atlantic salmon has become a large worldwide industry creating many jobs both directly related to the farming process and indirectly in the supplier industries, related to the processing of salmon and logistic services. Taking Norway as an example, over 8000 people are directly employed in the aquaculture industry in work related to salmon production [10] and more than twice as many are employed if activities connected to the business are included, meaning close to 30,000 full-time jobs either directly or indirectly. Similar numbers are reported from the other salmon producing countries, such as Scotland with over 8000, Canada with over 10,000 and Chile (including processing) with over 30,000 employed [11]. Taking the entire salmon farming industry, including processing and services related to the industry, it can be estimated that more than 100,000 jobs are sustained worldwide. Furthermore, it is also of major importance that many of the jobs are based in rural areas and in areas with limited employment, such as in small communities in coastal areas. In both the salmon production and processing industries both genders are employed, which is of additional importance to maintain settlement in rural areas. In many rural districts salmon aquaculture has also been a major driver for improving infrastructure, such as roads, bridges, electricity and the internet, and salmon farms and salmon slaughterhouses have become cornerstone companies in many rural communities.

However, only a limited number of countries have the natural conditions suited to production of Atlantic salmon in sea cages, which until today has been the method used to farm salmon for the consumer market (excluding a low tonnage produced on land or in closed cages) (Figure 1.1). The ideal natural conditions for sea cage farming include an acceptable seawater temperature and a sheltered coastline with fjords and bays. Such conditions feature in a latitude belt in both the northern and southern hemispheres. Today, Norway, Chile and the UK are top among a short list of producers (see Figure 2.3). Atlantic salmon is only a native species in the countries with a coastline on the Atlantic Ocean in the northern hemisphere; in all other salmon producing countries it is a non-native species. The introduction of land-based farms and the use of recycling aquaculture system (RAS) technology will allow other countries to efficiently farm salmon. These production method developments are already evident in the planning and establishment of a number of land-based farms, several outside the native area of the species. Also, the introduction of offshore cages that can tolerate higher waves will increase the number of countries that can start salmon production, because this removes the need for a sheltered coastline. Global heating will also be an important regulator for where it is possible to grow Atlantic salmon in the future. The rise in seawater temperature due to climate change will push ongrowing sea cage farms further north and south.

Figure 1.1 Ongrowing sea cage farm, with a well boat collecting fish for slaughtering.

The aim of this textbook is to supply information on how to produce Atlantic salmon under aquaculture conditions in an effective way. It is divided into chapters covering the operation of broodstock farms (5), hatcheries and smolt farms (6 and 7), ongrowing in sea cage farms (8) and production planning in salmon farms (9). Chapters 2 and 3 give an overview of the Atlantic salmon farming industry and the history of the salmon industry. Chapters 4 and 10–12 are simpler supplementary chapters covering salmon biology, salmon feeding, salmon breeding and the diseases encountered in salmon farming. The book concludes with Chapter 13 which discusses future perspectives.

References

1. Schubel, J.R. and Thompson, K. Farming the sea: the only way to meet humanity's future food needs. *GeoHealth*. 2019. 3(9): 238–244.
2. Duarte, C.M. et al. Will the oceans help feed humanity? *BioScience*. 2009. 59(11): 967–976.
3. Costello, C. et al. The future of food from the sea. *Nature*. 2020: 1–6.
4. MOWI. *Salmon Industry Handbook*. 2020. https://corpsite.azureedge.net/corpsite/wp-content/uploads/2021/05/Salmon-Industry-Handbook-2021.pdf.
5. Kituyi, M. and Thomson, P. 90% of fish stocks are used up – fisheries subsidies must stop. 2018. https://www.weforum.org/agenda/2018/07/fish-stocks-are-used-up-fisheries-subsidies-must-stop.
6. FAO. FAO Yearbook. Fishery and Aquaculture Statistics 2018. 2020: FAO: Rome.
7. Gjedrem, T. The first family-based breeding program in aquaculture. *Reviews in Aquaculture*. 2010. 2(1): 2–15.
8. The Global Salmon Initiative. Sustainability report. 2020. https://globalsalmoninitiative.org/en/sustainability-report/.
9. Winther, U. et al. Greenhouse gas emissions of Norwegian seafood products in 2017. 2020. SINTEF Ocean. https://www.sintef.no/contentassets/25338e561f1a4270a59ce25bcbc926a2/report-carbon-footprint-norwegian-seafood-products-2017_final_040620.pdf/.
10. barentswatch.no. Employment. What impact does Norwegian aquaculture have on the environment, the economy and society? 2020. https://www.barentswatch.no/en/havbruk/employment.
11. International Salmon Farmers Association. Salmon farming: sustaining communities and feeding the word. http://www.salmonfarming.org/cms/wp-content/uploads/2018/06/ISFA-JUNE-2018-REVISED-FOR-FOR-WEB1.pdf.

CHAPTER 2

The world salmon industry

2.1 Introduction

The salmon industry consists of wild-caught salmon, from traditional fishing and recreational fishing, and farmed salmon. Wild salmon can be caught in the territorial waters of several countries but the only significant commercial fishing is that of Pacific salmon (genus *Oncorhynchus*). The main fishing countries are the USA, Canada, Russia and Japan. Farming of salmon is carried out in selected countries in both the northern and the southern hemispheres that possess suitable natural conditions. A number of terms and units are used to quantify the size of the industry. One unit used to measure the quantity of salmon harvested is whole fish equivalent (WFE). This is a standardized method for weight of round bled fish, which means fish that are starved and bled out. It is based on a conversion factor of typically a 6–8% reduction of live, unstarved weight. Another term is head on gutted (HOG), also named gutted weight equivalent (GWT) [4], which refers to fish that are starved, bled and gutted, and the conversion factor is in the area of a 16–20% reduction of live weight.

2.2 Farmed and wild-caught salmon

In 2019, the total supply of farmed salmon was around 2.5 million tonnes GWT while wild salmon catches were around 0.9 million tonnes GWT [1], so the tonnage of farmed salmon was almost three times that of the wild catch. Atlantic salmon is by far the predominant farmed species with an annual production volume of 2.3 million tonnes GWT, compared to farmed Pacific salmon with a production volume around 0.2 million tonnes GWT. Large farmed rainbow trout, with a weight exceeding 2–3 kg, are also produced for the same market with a production volume slightly below 0.3 million tonnes GWT.

Of the wild-caught salmon, all is Pacific salmon; pink salmon (*O. gorbuscha*) has the largest tonnage (almost 0.5 million tonnes GWT) followed by chum (*O. keta*) and sockeye (*O. nerka*). The main catches of pink salmon are taken off the coast of the USA (Alaska) and Russia. Pink salmon has a smaller harvesting size and a lower price than the other salmonids. The smallest reported catches were of Chinook salmon (*O. tshawytscha*), also called king salmon,

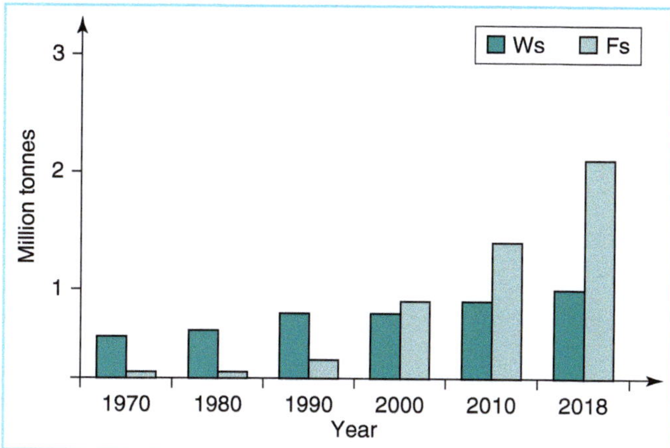

Figure 2.1 Development of the global salmon production of wild-caught and farmed salmon (data from [1]).

but this is a high-priced salmon species and a more direct competitor to Atlantic salmon. The volume of wild-caught salmon has been quite stable over the recent decades, fluctuating around 0.8 million tonnes GWT (Figure 2.1). An insignificant amount of Atlantic salmon is caught in the sea, mostly based on historical and cultural tradition, but due to the decline in wild salmon stocks this has gradually been more and more restricted and today only represents a very limited volume compared to the peak value of commercial Atlantic salmon fishing in 1973 of 3.5 million individual salmon [2]. In most places, fishing of wild Atlantic salmon ranges from commercial fishing to a cultural tradition, and catch and release is also becoming more popular. The catch and release fishing of Atlantic salmon is also growing more common in recreational fishing in rivers, and is a tool for protecting the declining wild salmon stocks. Recreational fishing of Atlantic salmon is very popular worldwide and as a rough estimate there are around 220,000 salmon anglers, which in total account for an equivalent of more than 1.6 million days of salmon fishing in a year [2]. In 2017 the estimated overall catch was 380,000 salmon, and the overall expenditure of the anglers was between €300 million and €500 million [2]. This means that each wild-caught salmon contributes a value in the region of around €1000. This value indicates that salmon angling can have a large impact on the local communities that host good salmon fishing spots, but also demonstrates the size of the overall catchment of wild Atlantic salmon compared to the farmed salmon harvest. The overall number of wild-caught salmon is in the same range as that contained in just a few salmon sea cages in aquaculture. If compared with the total harvested tonnage for farmed salmon of 2.44 million tonnes GWT, estimating an average individual weight of 4.5 kg gives a number of almost 550 million individuals. This shows that wild salmon catches are less than 0.5% of the farmed salmon harvest.

2.3 Farmed salmon production

As already discussed, Atlantic salmon is the species that accounts for by far the largest proportion of farmed salmon by quantity, with harvesting of 2.44 million tonnes GWT in 2018 [3] (Figure 2.2). Some Pacific salmon is also farmed, mainly coho salmon (*O. kisutch*) produced in Chile for the Japanese market, at around 0.2 million tonnes GWT. Small amounts of Chinook

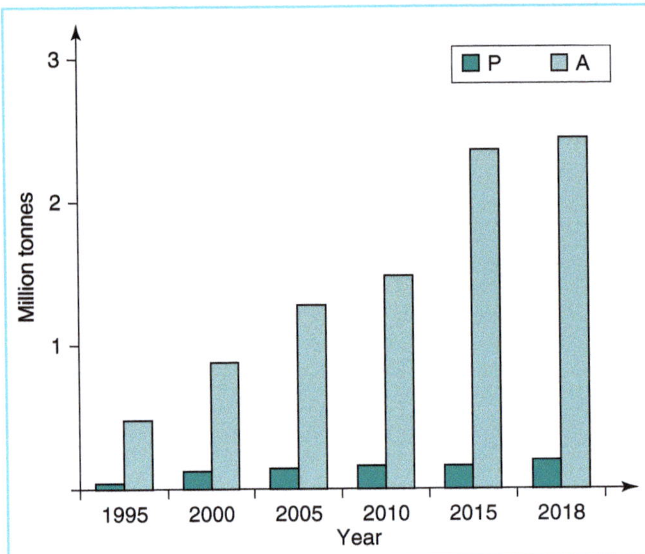

Figure 2.2 Development of farmed Atlantic salmon and farmed Pacific salmon production (data from [1]).

salmon are farmed in New Zealand and Canada. The largest producer of farmed salmon in the word is Norway, followed by Chile and Scotland [1,3].

The production of farmed Atlantic salmon has, in the main, increased continuously since the inception of the industry in the 1970s but there were notable reductions in 2009, 2010 and 2016 (Figure 2.2) (see Chapter 3). The reason for the drop in 2009 and 2010 was a collapse in Chilean salmon production due to an outbreak of the disease infectious salmon anaemia (ISA), resulting in high mortality. The drop in 2016 was a result of large mortalities caused by algal blooms. From 1995 until 2019 the yearly average growth in production was 8%, but from 2010 to 2019 it reduced to 7% and expectations for the period 2019 to 2023 are for a yearly growth of 3%. Therefore, a continuous but reduced rate of growth is expected [4].

2.4 Atlantic salmon production

Only a limited number of countries have the natural conditions suitable for production of Atlantic salmon in sea cage culture, which until today has been the predominant method used for farming of large salmon for the consumer market. The few tonnes that have been produced in land-based farms and in closed cages is not notable compared to the tonnage produced from sea cage farming.

Countries that have suitable natural conditions for salmon farming in sea cages have areas with acceptable seawater temperature and a sheltered coastline with islands, fjords and bays, and are found within a latitude belt on the northern hemisphere (40–70°) and southern hemisphere (40–50°) [5]. The following countries, ordered by tonnage (see Figure 2.3), have an appreciable production of farmed Atlantic salmon: Norway, Chile, the UK, Canada, the Faroe Islands, Australia, Russia, the USA, Ireland, and Iceland. However, with the introduction of land-based farms for ongrowing Atlantic salmon and the use of RAS technology, more countries will have

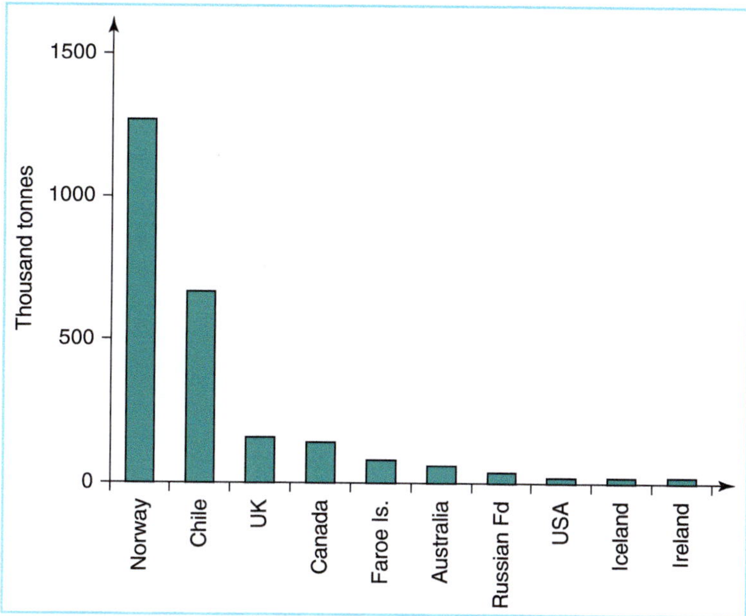

Figure 2.3 Major producers of Atlantic salmon (values from [1]).

the opportunity to produce market size Atlantic salmon in the future. The introduction of off-shore cages that can tolerate higher waves will also increase the number of countries that can efficiently produce salmon, as the need for a sheltered coastline will be removed.

Atlantic salmon are well adapted for farming and have a number of advantages compared to other fish species including ease of domestication, a good growth rate, a reasonably good resistance to disease, a taste that appeals to consumers, the possibility for use as a base for a number of consumer food products and a fairly high price.

Atlantic salmon is, however, also a highly appreciated species for recreational fishery, as in freshwater it bites actively on flies and lures. This has resulted in trials for the introduction of Atlantic salmon in a number of countries where the species is non-native in order to improve the appeal of and opportunities for recreational fishing. Juvenile farms producing Atlantic salmon fry or smolt from imported eggs for local restocking have been established in many countries worldwide solely to serve the needs of recreational fishing. The decline in wild salmon stocks has also resulted in an increased focus on local restocking in areas where Atlantic salmon is native, and farms producing salmon specifically for local restocking have also been built in these native areas. Special farms with the sole purpose of storing wild salmon genes to safeguard the genetic pool have also been established due to the observed decline in the wild salmon stocks.

2.5 Structure of the Atlantic salmon farming industry

The structure of the Atlantic salmon industry varies slightly between the salmon producing countries, but overall there has been a trend towards consolidation with a reduction in the number, but an increase in the size, of individual companies. When the salmon industry began, in many of the salmon producing countries it consisted of small companies, typically family-owned. In Norway, for example, a requirement for a farming licence was that the farm should be mainly locally owned. In the beginning, the most important driver in a number of the salmon producing countries was to strengthen rural development, especially in small rural communities that were based on traditional fishing and agriculture. As an example, there were almost 1000 producers of ongrowing salmon in Norway in 1990 but by 2015 the number had decreased to 160; however, in this period production increased severalfold [6]. Scotland and Chile, where salmon farming started slightly later than in Norway, did not have imposed requirements for local ownership which made it possible for the industry to grow faster because the financing was easier. However, today most of the smaller family-owned companies have disappeared in the main salmon producing countries. The industry is now controlled by large multinational companies, several of which are listed on the stock exchange. The five largest salmon farming companies produce almost 50% of the total Atlantic salmon production in the world [4] (Figure 2.4). In Norway the ten largest companies produce nearly 70% of the country's total volume while in Chile the ten largest produce almost 90% [4]. The large salmon producing companies of today are also vertically integrated and encompass the handling of the entire production cycle from egg to slaughter and also processing, branding and marketing with export-oriented activities.

2.6 The salmon producing countries

2.6.1 Norway

Norway is the largest producer of Atlantic salmon in the world, supplying 1,357,304 tonnes GWT in 2019 with a first-hand value of NOK 68.1 billion for unprocessed salmon [7]. Ongrowing salmon production takes place along almost the entire coastline, with the exception of south-east Norway. Norway has roughly 25,000 km of coastline [8] that is sheltered, with

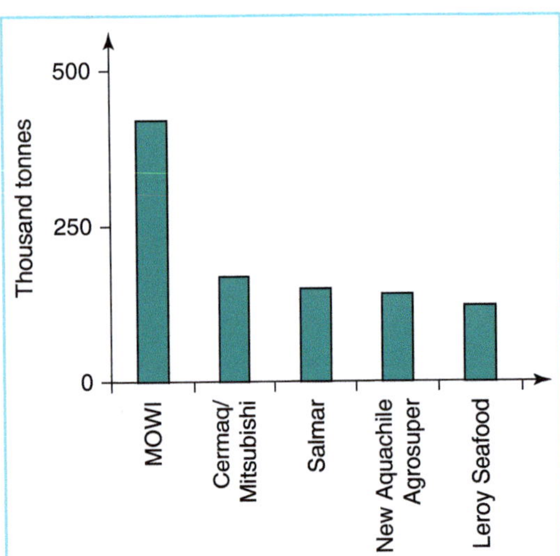

Figure 2.4 Production volumes for some of the largest salmon companies in the world (data from [4]).

islands and a number of fjords and bays suitable for sea cage farming. The seawater temperature is fit for salmon production thanks to the Gulf Stream that passes the coast, bringing temperate water. The availability of freshwater with rivers and lakes for smolt production is also good and the topography ensures access to high quantities of elevated water, illustrated by the large number of hydroelectric power plants established in Norway. This has made it possible to establish water flow through smolt farms at low cost as there is no need to pump water. The eutrophication status along the coast is also good, meaning it is possible to release outlet water from the smolt farms with limited restrictions. Atlantic salmon is a native species in Norway and is a very popular species for recreational fishing, which is of great economic importance for the tourist industry in rural communities. In addition to a large number of Norwegian anglers, 10,000 foreign anglers also visit to fish in Norwegian rivers [9]. The large rivers in Norway also represent the largest spawning grounds in the world for wild Atlantic salmon [9].

However, during recent years Norway has been struggling significantly with sea lice problems which have limited the growth of the industry. Salmon prices have increased in recent years, showing that world supply is below demand. Owing to the sea lice problems, Norway has introduced a traffic light system, and farms in some parts of the country must reduce production (see Chapter 3).

2.6.2 Chile

Chile is the second largest salmon producing country in the world with a production of over 661,000 tonnes GWT in 2018 [3], and the industry represents a significant part of overall Chilean exports. Salmon farming has grown quickly in the country and has been the fastest growing agro-industry [10]. Chile, like Norway, has a long, sheltered coastline with islands, fjords and bays. The southern part of the coastline (Puerto Month to Punto Arenas) has water temperatures suitable for Atlantic salmon farming. Of the salmon producing countries, Chile has the most favourable seawater temperatures in parts of the coast, higher winter temperatures and, overall, more stable seawater temperatures, varying between 10°C and 14°C [4]. Stable, suitable water temperatures can shorten the ongrowing production time in seawater by a couple of months. Chile also has plenty of available freshwater, of good quality and with temperatures fit for smolt production, in both lakes and rivers fed by Andean watersheds. The temperature in the lakes even makes smolt production in cages possible all year round (Figure 2.5). In 2017 14% of the smolt produced in Chile was farmed in cages in lakes [11], although this production has decreased during recent years due to environmental concerns. In the Pacific Ocean bordering south America, the world's largest fishery of fish species suited for production of fish meal and fish oil, oil-rich species such as anchoveta, takes place. This means that the salmon feed factories have access to large quantities of locally produced fish meal and fish oil. Plant production is also large in south America, so there is easy access to a high volume of locally produced raw vegetable material for feed production.

In the southern regions of Chile, where the salmon industry is currently growing, the infrastructure is not very well developed. Here the salmon farming industry has been an important contributor to the improvement of infrastructure, as well as an increase in local employment in the area.

There are no native salmonid species in Chile, so all salmonids occurring in the wild have been introduced. However, the release of salmonids in Chilean freshwater has not been a complete

Figure 2.5 In Chile, part of the smolt production takes place in cages in lakes.

success. Rainbow trout have successfully been introduced in the wild for natural stocking and coho salmon have been introduced with partial success, but for Atlantic salmon there have only been negative results. Farmed salmon production in Chile is dominated by Atlantic salmon but also includes some coho salmon, which was the original species used when the salmon aquaculture industry in Chile started in the 1980s.

In the early part of this century, Chile had almost the same production volume as Norway but the ISA crisis, which started in 2008, resulted in a dramatic drop in production. In recent years it was expected that Chilean production would have increased, but problems have occurred almost continuously, with diseases (e.g. salmon rickettsial septicaemia [SRS]) and sea lice being important factors. The result has been the use of antibiotics and pesticides for treatment. The most recent widespread problem has been harmful algal blooms (HAB), which have killed millions of salmon in the sea cages. As an example, there was a major event in 2016 with algae (*Pseudochattonella cf. verruculosa*) killing almost 40,000 tonnes GWT of salmon [11]. Global warming is also affecting the industry negatively because high summer temperatures reduce oxygen levels in the cages in the northern part of the salmon growing area. A number of sea cages in Chile are therefore now adding additional oxygen during summertime.

2.6.3 Scotland

Scotland is the third largest salmon producing country after Norway and Chile. Production in 2018 was 166 thousand tonnes GWT [3]. Production in sea cages is located off the north and west coasts of Scotland and around the large islands, such as the Orkney, Hebrides and Shetland archipelagos [12]. The water temperature in these areas is favourable for salmon production. Smolt production is spread around the Scottish Highlands. Five large companies account for over 90% of the total salmon produced in Scotland. Both sea cages and tanks/race ways are used for salmon smolt production but there has been an increase in land-based smolt farms [13].

Atlantic salmon is a native species in Scotland, and travels up Scottish rivers to spawn. As in Norway, recreational fishing of wild Atlantic salmon in Scottish rivers is an important and well-known attraction.

The industry employs a large number of people both directly and indirectly, with over 10,000 employees reported in 2017, and is an important industry supporting local settlements in the Scottish Highlands and islands.

2.6.4 Canada

Canada is the fourth largest salmon producer in the world today, with a production volume of 123,000 tonnes GWT in 2018 [3]. Production takes place both on the east coast (Atlantic Ocean) and on the west coast (Pacific Ocean). On the east coast it extends north from the US border (coastlines of New Brunswick, Newfoundland and Nova Scotia) until the temperature falls too low, and on the west coast it also extends north from the US border (the coastline of British Columbia) where water temperatures are favourable. On both the east and west coasts are some sheltered coastlines favouring sea cage farming. The main farmed species here is also Atlantic salmon, but some Chinook and low quantities of coho are also farmed on the west coast.

Atlantic salmon is a native species on the east coast but not on the west coast. On the west coast there has been a lot of public discussion about the sustainability of sea cage farming of Atlantic salmon due to the possibilities of escape and possible unwanted stocking in the rivers of what is a non-native species, and also regarding environmental issues. Today it is even proposed that salmon farming in traditional sea cages on the west coast should be prohibited.

As in other salmon producing countries the industry is an important creator of jobs in rural communities. In Canada this includes Indigenous communities where the aquaculture industry has created important full-time jobs. On the west coast the First Nations communities are an important part of the development of the aquaculture industry, helping to ensure sustainability and local involvement. Farming in First Nations areas has drawn negative public discussion. The wild salmon catches of Pacific salmon are of great importance for the First Nations and everything that affects this will be disputed.

2.6.5 USA

Atlantic salmon are grown in sea cages both on the east and west coasts of the USA in states near the Canadian border, such as Maine and Washington. Here the seawater temperatures are within an acceptable range.

Atlantic salmon is a native species on the east coast but not on the west, where it is imported. The growth in the salmon farming industry has been inhibited by strict environmental regulations and also by a limitation in the number of good sites for traditional sea cage farming.

The overall salmon aquaculture industry on the west coast is quite large, with more than 100 smolt farms. However, most of them are established with the purpose of producing Pacific salmon fry and smolt for wild stock enhancement to improve both commercial fishing and recreational fishing of salmon, as each is of great importance for this region. In Alaska traditional Atlantic salmon farming is prohibited due to the possible environmental consequences, but here the fishing of Pacific salmon is of great economic importance.

The reported farmed Atlantic salmon production volume was 16,000 tonnes GWT in 2018 [3].

2.6.6 Faroe Islands

The Faroe Islands have good natural conditions for sea cage farming of Atlantic salmon; the temperatures are favourable and Atlantic salmon is a native species. For the Faroe Islands, the aquaculture industry is of vital importance as it accounts for over 30% of the overall export income [14]. This is also of major importance for the creation of jobs in rural areas. In total, slightly below 80,000 tonnes GWT of Atlantic salmon was produced in 2018 [3].

Today there are three companies farming salmon, and the industry is recognized with well-developed vertical integration, meaning that the companies control all processes from roe production through all farming operations and on to processing and export.

Sea cage farming is performed in the archipelago where there are suitable sheltered areas between the islands. However, the number of sites suitable for traditional sea cage technology is therefore limited by the small size of the archipelago and there is little scope for growth. The introduction of cages tolerating rougher conditions will be an important development and would drive further growth.

2.6.7 Tasmania

The farming of Atlantic salmon in Australia takes place in the south east of the Australian continent, around the island of Tasmania. Here the production is of great economic importance for the area and also for rural development. Atlantic salmon farming is larger than all other animal production in Tasmania, and salmon is the most valuable product in Australia within the entire seafood sector.

The water temperature has historically been favourable and there are also protected sites for salmonid production in sea cages. However, rising seawater temperatures are starting to cause problems. Farms are moved further south and to more exposed water to avoid excessively hot water during summertime.

Today there are three main companies involved in the farming of Atlantic salmon in Tasmania, and the total production of Atlantic salmon in 2018 was reported to be slightly over 60,000 tonnes GWT [3].

2.6.8 Ireland

Ireland has a water temperature and coastline partly suitable for sea cage salmon farming, even if some of the western fjords are quite exposed. Production has been fairly stable during recent decades, while in Scotland, with quite similar conditions, the salmon industry has grown considerably during the same period. Reasons for the stagnation in growth is related to environmental concerns: environmental organizations challenging development and a lack of governmental support have been important hurdles. In general, salmon farming has a poor image in Ireland and during recent years almost no new licences have been granted for sea cage farming. Since 2005 the average time to obtain approval for a new licence for farming of Atlantic salmon has exceeded eight years [15].

Production in Ireland in 2018 was reported to be around 12,000 tonnes GWT [3].

2.6.9 Iceland

Aquaculture in Iceland actually started as sea ranching: farming of Atlantic salmon, a native species, for release in the sea for later recatching when it returned. However, the recatch rate was so low that as a private business it was not economically sustainable. The result of this attempt at sea ranching was that a number of hatcheries and smolt farms were established around Iceland. In the 1990s large land-based ongrowing farms were established, as was the case in Norway at this time, but such land-based farms were not economically sustainable in either country at that time (Figure 2.6). However, since then some sea cage farms for Atlantic salmon have been established around Iceland. However, even in the fjords the shelter is poor and this limits the number of suitable sites fit for traditional sea cage technology. Recent improvement of sea cage technology,

Figure 2.6 A number of land-based ongrowing farms were established in Iceland at the start of the 1990s.

allowing tolerance of rougher weather and higher waves, brings promise. This has been noted by the large international salmon producing companies, and growth looks possible in the future. Iceland has good access to freshwater, so smolt production is not an impediment to industrial growth. Iceland has also the advantage of the possible utilization of hot geothermal water for heating, which will reduce smolt production costs.

Atlantic salmon is, as mentioned, a native species in Iceland and protection of the wild salmon stocks has been important. The sea ranching trials with natural stocks has also been an important part of this.

In 2018 Atlantic salmon production was reported to be 13,448 tonnes GWT [3].

2.6.10 Other salmon producing countries

The Russian Federation is quite a new addition to the countries producing a significant amount of Atlantic salmon; in 2018 the reported production was 20,556 tonnes GWT. There has been a high percentage increase during recent years and several farms have been established. There is also some sea cage production of salmon and rainbow trout in Sweden and Finland. In New Zealand there is some production of king salmon to a size of 3–6 kg in sea cages: by volume 16,291 tonnes GWT in 2018 [3]. Japan also has some production of coho salmon, 18,000 tonnes GWT in 2018 [3], which adds to the quite sizeable farmed coho production in Chile of 148,521 tonnes in 2018 GWT [3].

References

1. FAO. Online query panels. http://www.fao.org/fishery/topic/16140/en.
2. NASCO. *State of the North Atlantic Salmon*. 2019. North Atlantic Salmon Conservation Organization (NASCO): Edinburgh.
3. FAO. *FAO Yearbook. Fishery and Aquaculture Statistics 2018*. 2020. FAO: Geneva.
4. MOWI. *Salmon Industry Handbook*. 2020. https://corpsite.azureedge.net/corpsite/wp-content/uploads/2021/05/Salmon-Industry-Handbook-2021.pdf.
5. FAO. *Salmo salar* cultured aquatic species information programme. 2020. https://www.fao.org/fishery/en/culturedspecies/salmo_salar/en.
6. Steinset, T.A. Fra attåtnæring til milliardindustri in Samfunnspeilet. 2017. Samfunnsspeilet/Statistisk sentralbyrå.
7. SSB. Aquaculture statistics of Norway. 2020. https://www.ssb.no/en/fiskeoppdrett.
8. CIA. *The World Factbook, Coastlines*. 2020. CIA: Washington, DC.
9. Liu, Y., Olaussen, J.O. and Skonhoft, A. Wild and farmed salmon in Norway – a review. *Marine Policy*. 2011. 35(3): 413–418.
10. Aguayo, B.E.C. and Barriga, J. Behind certification and regulatory processes: contributions to a political history of the Chilean salmon farming. *Global Environmental Change*. 2016. 39: 81–90.
11. Quinones, R.A. et al. Environmental issues in Chilean salmon farming: a review. *Reviews in Aquaculture*. 2019. 11(2): 375–402.
12. Kenyon, W. and Davies, D. Salmon farming in Scotland. Scottish Parliament Information Centre (SPICe) Briefings. 2018.
13. Munro, L.A. Scottish fish farm producion survey 2019. 2020. Marine Scotland Science.

14. International Salmon Farmers Association. Salmon farming: sustaining communities and feeding the word. http://www.salmonfarming.org/cms/wp-content/uploads/2018/06/ISFA-JUNE-2018-REVISED-FOR-FOR-WEB1.pdf.
15. Colins, S. What ever happened to the Irish salmon farming industry? *The Irish Times*. Dec 26, 2019.

CHAPTER 3

The history of salmon farming

3.1 Introduction

Salmonid farming has a long history (for a review see, for instance [1,2]). The life history of salmon was first described in the 15th century, and artificial fertilization and hatching were first carried out in Germany in the late 17th century. The first salmonid hatchery was established in the mid-18th century, with the purpose of producing salmon for wild stock enhancement rather than farming. In the 1950s many hatcheries were established worldwide for the same purpose. The hatcheries produced fry for release in rivers and lakes, both in order to strengthen a declining wild salmon stock and to support increasingly popular recreational salmon fishing. Atlantic salmon hatcheries outside the species' native areas were also established during this period to produce fry with the aim of establishing local populations, for instance, in Tasmania, due to the popularity of the species for recreational fishery. These attempts failed, as have a number of similar trials for the establishment of Atlantic salmon stocks outside its native areas. The knowledge established through these experiments was, however, of great use to those later developing commercial Atlantic salmon farming. However, the aim of farming for natural stock enhancement differs a lot from the farming of large salmon for the consumer market. When farming for enhancement purposes, the performance of individual salmon in terms of growth rate and feed utilization is of lesser importance. The breeding stock is also very different; in commercial aquaculture the fish used have been bred for optimal performance under aquaculture conditions for generations, and have been selected for attributes such as optimal growth rate and feed utilization. When breeding fish for wild stock enhancement, the most important requirement is that the fish are adapted to the wild and can survive in the site conditions where they are to be released. As a result, for wild stock enhancement typically a locally adapted strain of salmon, which has been bred for generations to survive in a particular site, is used for breeding. This means that the broodstock salmon is collected from the same river where the young are to be released and no selective breeding is carried out.

3.2 Beginnings in Norway

Atlantic salmon farming as recognized today, developed to produce salmon for the consumer market, actually sprung from rainbow trout (*O. mykiss*) farming, which had started much earlier. The trout farming knowledge base was of major importance when commercial Atlantic salmon farming started in Norway at the end of the 1960s (history reviewed in [3,4]). Farming began with pond farming of rainbow trout in freshwater in southern Norway. These initial operations were based on the experiences of farming portion-size rainbow trout taken from Europe and Denmark in particular. However, the water temperature in winter in Norway was lower, and therefore so was the growth of the trout. As the seawater temperature was higher than that of freshwater during winter, seawater was pumped up into the ponds where the rainbow trout were grown and mixed with the freshwater with the aim of increasing the water temperature, and consequently the growth rate of the fish. The results of this were promising. The Vik brothers in Sykkulven [5] also observed that rainbow trout could be gradually adapted to live in seawater when they reached a certain size, and in addition documented that the trout grew faster in seawater than they did in the pure freshwater ponds. The trout could also reach a size of several kilograms before slaughter. They built concrete ponds and also started rearing fish in simple floating wooden boxes, even applying for a US patent for some of their findings [6]. In the mid-1960s areas in the sea close to the land were fenced in by poles and nets due to a lack of a regular freshwater supplies. At this time, activity was limited to rainbow trout [7].

Atlantic salmon achieved better market prices than trout and so it was natural that aquaculturalists began to wonder if it was possible to replace the farming of trout with Atlantic salmon. However, there was a limited amount of salmon smolt available; that being produced was intended for restocking purposes. At the same time researchers (such as Skjervold and Gjedrem) investigating animal breeding were highly interested in fish (rainbow trout and salmon) as laboratory animals due to their large number of offspring compared to traditional livestock. They sought to study the factors related to the development of breeding programmes and the effects of breeding on animals and replaced the mice or the other animals that traditionally had been used as laboratory animals with fish. The results, however, were still intended to be transferred and used in breeding programmes for traditional animal livestock. They also made the interesting observation that there was a large genetic variation in the national salmon populations, which could potentially provide a high genetic gain from generation to generation. Their research activity had generated some smolt, which was used to start the breeding of high performance rainbow trout and Atlantic salmon specially adapted to farming conditions in the 1970s in Norway [8,9]. The base for the national breeding programme was the collection of broodstock salmon from a number of Norwegian salmon strains in 40 different rivers to find the best candidates [10]. This also included crossbreeding of different strains. This early start of breeding has been one the major factors in the success of the Atlantic salmon farming industry. Today almost 100% of the farmed salmon in the world is based on genetically improved stocks from selective breeding programmes.

The development of production methods continued and in the late 1960s tests were carried out with Atlantic salmon in closed straits in two areas of the sea off the west coast of Norway [11]. The straits were closed with grids made of steel rebars to keep the salmon contained but to keep water flowing. The first salmon were harvested from this production in 1971 and a large proportion of the salmon slaughtered during the first part of the 1970s came from those test straits. Production in the test straits ended in 1987 due to lack of proper fish control and the challenges of keeping a satisfactory oxygen concentration.

Almost at the same time, in 1970, Atlantic salmon smolt were transferred to simple octagonal homemade sea cages (Grøntvedt at Hitra), later named the Grøntvedt cages. This marked the first commercial sea cage farming of Atlantic salmon. Compared to ponds, it was a low-cost method and achieved much better control of fish and water quality in comparison to closed straits (Figure 3.1). The basic principles of the sea cage used today have remained the same 50 years later, which shows the importance of this development.

The near simultaneous establishment of breeding programmes experimenting with Atlantic salmon and the economic interest farming of Atlantic salmon in the sea were major reasons for the fast development of the Atlantic salmon farming industry in Norway.

There was a distinct pioneer phase with salmon and rainbow trout farming until the start of the 1970s, when salmon farming started to grow. It grew slowly at first, and started as a small-scale rural industry with a requirement for local ownership. Interest in the establishment of farms increased and in 1973 farming licences were introduced, which became a requirement for new start-up salmon sea cage farms. Those who had already established farms were granted licences automatically. During the rest of the 1970s and the 1980s growth increased, with up

Figure 3.1 Simple cages used at the start of the salmon farming industry.

to 50% annual growth in some periods. Growth was so fast that a limit was set on new licences for ongrowing sea cage farming, and there were also problems obtaining enough smolt. The economics of production were good, but in 1990 problems selling the end product arose. Farmed salmon production had grown faster than the development and opening of new markets. The result was the freezing of large volumes of salmon because of a lack of demand for the product. During this period, the salmon industry experienced a price drop and several companies went bankrupt. These problems drove a restructuring of the Norwegian salmon industry and industrialization accelerated. The industry developed from small family-owned businesses to large enterprises, a trend that has continued until today. Several of the companies that grew in this period are now large salmon production companies listed on the stock exchange.

When the salmon farming industry in Norway began, it was quite easy to obtain licences for ongrowing, but this is no longer the case. There was a great interest in starting salmon farming in the 1970s and 1980s and the licences became an important governmental tool to control industry growth and avoid overproduction and market collapse. At the end of the 1970s and the start of the 1980s there was a moratorium on new ongrowing licences due to a lack of smolt and market evaluations, which suggested restraint. In response Norwegian farmers and investors started to invest in salmon farming in other countries. In 2005 the Aquaculture Act was introduced in Norway with the aim 'to promote the profitability and competitiveness of the aquaculture industry within the framework of a sustainable development and contribute to the creation of value on the coast'.

In recent years only a very limited number of new sea cage farming licences have been granted, and they have typically been issued with special conditions, such as the green licences, research licences and development licences. Licensing was and remains an important tool for controlling the development of an environmentally sustainable industry and for promoting improvement and innovation [12].

3.3 Other salmon producing countries

The development of Atlantic salmon sea cage farming in the other salmon producing countries started almost simultaneously with Norway, and then developed partly with the technical experience and capital originating from Norway, particularly the use of sea cage technology. In Pacific water there was, however, another problem: Atlantic salmon was not a native species. There had been a number of trials involving the import of Atlantic salmon for stocking of rivers and lakes for recreational fishing, but these were generally unsuccessful when it came to the local establishment of a reproductive stock. The species was therefore not unknown, and hatcheries had been established for the growing of salmonids for attempts to establish local stocks. However, the salmon farming industry in several countries started with the growing of Pacific salmon (coho), not Atlantic salmon.

3.3.1 Canada

The history of salmonid aquaculture in Canada is old; studies on incubation and hatching were carried out in the 18th century, when the first hatcheries were also established. During the

1950s, federal and provincial hatcheries were established throughout Canada with the purpose of enhancing wild stocks. Production was high, and approximately 750 million trout and salmon were released [13]. In Canada, commercial salmon farming activities started in the 1970s on both the east and west coasts [13]. The geography and water temperatures were quite similar to conditions in Norway, which had been proven to be suitable. On the west coast Pacific salmon, coho, was grown with eggs from cultivation farms in British Columbia. The coho mature early and at a smaller size than Atlantic salmon, and competition with wild salmon fishing resulted in rather low prices. The infrastructure was not as good as in Norway in remote areas, and it was not until the 1980s that there was a noticeable growth; then Chinook salmon were also included as a farmed species. Neither of the Pacific salmon species were as easy to adapt to farming conditions as Atlantic salmon, and neither showed the same performance; when the real industry started to grow later it was due to the introduction of Atlantic salmon, which until 1985 had not been permitted [14]. In 1993 the production of Atlantic salmon surpassed the total farmed production of the Pacific salmon species. The same pattern of industrial development was observed in Canada as in Norway, with the industry starting at a small scale with local ownership but gradually consolidating; the large international Atlantic salmon producing companies have since taken over many farms.

On the east coast of Canada, New Brunswick Atlantic salmon is a native species and the industry started with this species at the end of the 1960s. However, winter temperatures were quite low and the amount of sheltered coast was also limited; this, among other factors, restricted the growth of the industry.

The growth and development in salmon aquaculture in Canada during recent years has been restricted by authority regulations. The escape of Atlantic salmon from sea cages on the west coast has been an important factor here. Conflicts with the commercial wild salmon fishing industry, recreational fishing and conflicts with the Indigenous communities have also prevented growth [15–17].

3.3.2 USA

In the USA, as in Canada, hatcheries for producing juvenile salmon for national enhancement programmes to strengthen wild stock and improve recreational fishing were established at the end of the 18th century. In 1900, as an example, Washington state released 50 million fry [18]. During the 1950s there was, again as in Canada, a new boom in the establishment of hatcheries for wild stock enhancement, and through this, sound knowledge and practice of cultivation was established (Figure 3.2).

The large-scale fishing of Pacific salmon also represented a valuable knowledge base at the start of the aquaculture industry in the USA and Canada. In the early 1970s private salmon ranching on the west coast of the USA was attempted, financed by several large companies. The aim was to make a profit from harvesting salmon when they return from their migration to the sea [1]. This therefore represented a direct alternative to sea cage farming. Until then, salmon ranching in the USA and Canada had not been a commercial activity but a purely government-controlled, public activity. The recatch quantity was, however, low and there were difficulties in controlling who caught the fish. There was therefore a lack of private economical sustainability, as has also been the case for other projects dealing with salmon ranching, for instance, in Chile and

Figure 3.2 A number of hatcheries have been established in the USA for wild stock enhancement.

Japan with coho salmon and in Iceland with Atlantic salmon. However, public salmon ranching expanded both in the USA and Canada; new hatcheries were built and those already in production increased in size, and until recently this has been the dominant form of aquaculture in the USA and Canada. The idea behind public salmon ranching in the USA and Canada was, however, to improve not only recreational fishery but also the traditional fishing of Pacific salmon in the Pacific Ocean. To illustrate the size of the industry, more than 2 billion juveniles are released from hatcheries in Alaska and Canada (BC), in addition to releases from facilities in Russia [19]. In 2010, for example, almost 50% of the commercial catches in Alaska were based on ranched salmon: salmon from controlled hatching in public hatcheries [19]. However, today there are concerns about the capacity of the north Pacific Ocean to hosting salmon ranching: how many salmon can be released to feed on natural food sources without affecting the natural ecosystem?

When commercial salmon farming was introduced the background and understanding of salmon biology was already good and there was much experience with juvenile salmon production. The development of sea cage farming started in the 1970s, both on the northwest coast (Washington) and on the east coast (Maine), close to the Canadian border. As in Canada, it started with Pacific salmon, coho and Chinook on the west coast but in the mid-1980s activities changed to Atlantic salmon, again similar to the situation in Canada.

A number of government-supported trials on the release of Atlantic salmon fry on the west coast were carried out in the 1950s and early 1980s. The aim was not only to strengthen recreational and commercial fishing, but also to see if it was possible to establish local stocks of Atlantic salmon [20], as this would strengthen recreational fishing in the long run. However, as in Canada these trials were unsuccessful and there was no return of Atlantic salmon to the rivers where the fry had been released. This is a common result with all trials worldwide in which Atlantic salmon have been introduced in areas outside their native Atlantic Ocean [20].

During the late 1990s there were some escapes from sea cages in Washington, which attracted negative attention due to the possible detrimental environmental effects. Also, at this time, the first evidence of natural reproduction of escaped Atlantic salmon in British Columbia was documented [21]. Despite this proof that it is possible for Atlantic salmon to reproduce outside their native area, to date no sustainable Atlantic salmon stocks have been established in the wild outside the species' native areas, and there is no absolute consensus of the consequences of this for Pacific salmon. Still, this has had important consequences for the growth of the salmon farming industry, which has been inhibited by environmental concerns to an even higher degree than in Canada [22]. Escape of salmon threatening the local salmon strains, disease transfer to local stocks, eutrophication, algal blooming, decline in fishing and conflict with marine mammals are all factors cited by environmentalists trying to inhibit the development of the salmon farming industry. The result has been ever-increasing costs due to regulatory compliance, and subsequently lower profitability [1]. In Alaska, fish farming for profit is prohibited by law due to the possible negative environmental impacts, but here, of course, the wild native salmon is of great economic importance due to large-scale commercial fishing.

3.3.3 Chile

The Chilean salmon farming industry started with the coho salmon, beginning in the late 1970s, and expanding in the 1980s. Japanese cooperation was important in this start up period for coho salmon farming [23]. Atlantic salmon eggs were then difficult to obtain and the price was higher. Coho was not a native species in Chile either, but was imported for the first time in 1921 for cultivation purposes, and growth and survival had been reasonably good [24].

The environmental conditions for salmon farming in Chile were also very good with rivers and lakes supplying freshwater of good quality and at suitable temperatures for fry production. The seawater temperatures were also favourable for ongrowing in the southern part of Chile, even better than in Canada and Norway. At first small fry were grown in local land-based farms, but as they grew, they were transferred into cages in lakes where they continued growth until smolt size (Figure 3.3). The lakes in the growing regions in Chile have suitable and stable water temperatures with no freezing during wintertime (unlike the rivers which have more unstable water temperatures), and good growth from fry to smolt was achieved. With this two-step smolt production, start-up and production costs were reduced. When smolt were released into the sea, the local sea temperatures in the area were also favourable for growing salmon, especially higher winter temperatures than the salmon farming regions in the northern hemisphere, such as Norway and Canada. The availability of sheltered sites for sea cage farming was also good during this period. Therefore, it was possible to produce salmon for a lower price than in the northern hemisphere, especially Norway, which was the major salmon producer at that time. Even though the production was just coho salmon at first, it was possible to produce this species with a higher growth rate than that which was achieved with Atlantic salmon in the northern hemisphere. The time from when eggs were taken into the coho salmon hatchery until slaughter at 2–3 kg was only 2 years. Mortality was also quite low compared to Atlantic salmon at that time: 15–20%

Figure 3.3 Smolt farm established in lakes in Chile after the first period on land.

in the seawater phase. During the 1980s and 1990s, the industry grew and in 1990 the first broodstock farm producing coho salmon in Chile was established. During the 1990s Atlantic salmon eggs were imported and production was gradually changed over to Atlantic salmon. By the end of the decade production of Atlantic salmon exceeded that of coho salmon. The market for the salmon was the USA and Asia and the price was quite good, but in 1998 an economic crisis started in Asia (Japan), the major coho salmon market, resulting in falling prices and difficult times for the Chilean salmon farming industry. Nonetheless, the industry survived and growth increased, and at the start of the 20th century the industry experienced rapid growth. Based on the better conditions for growing, many believed that Chile would surpass Norway as the major salmon producer. In 2004 salmon production in Chile was almost at the same level as in Norway. During the early 2000s more and more production shifted to Atlantic salmon and by 2006 63% of production was Atlantic salmon, while only 17% was coho salmon; the rest was rainbow trout. However, in July 2007 a real crisis in the Chilean salmon industry occurred when the first case of the disease ISA was identified in sea cage farms [25–27]. Many fish at this time were grown in sea cages in a very limited geographic area. Salmon farming was concentrated on a coastline of 300 km, while in Norway the equivalent area at that time was 1700 km of coastline [28]. Therefore, it was easy for the disease to spread to neighbouring farms. During a 3 year period, salmon production in Chile dropped considerably. In 2008 the production was 403,000 tonnes GWT, while in 2010 the slaughter tonnage had dropped to 130,000 GWT: a reduction of almost 75% due to the disease. It was a total collapse of the salmon industry. Based on this experience, a reconstituted salmon industry that was rebuilt in Chile focused on health issues, with regular health control, area coordinated disease treatment and limitation in fish densities in cages, sites and areas. The original industry was located in mid-Chile, around the city of Puerto Month and especially Chiloé Island. Now the area farmed has been extended and salmon ongrowing is now performed across a larger part of the Chilean coastline, including more southern regions, but here the infrastructure is still limiting the growth rate.

3.3.4 Scotland

Salmon farming has a long history in Scotland. The first attempts to hatch incubated salmon eggs were in 1838 and before the start of the 19th century over 18 hatcheries were producing Atlantic salmon in Scotland [29]. However, as in other salmon producing countries the fry and smolt produced were intended for enhancement of natural salmon stocks.

Sea cage farming in Scotland was also based on trout farming experience, and marine rainbow trout farming was introduced in the mid-1960s. As salmon farming was developing in Norway, a switch towards Atlantic salmon also occurred in Scotland and salmon farming started almost simultaneously with Norway. In 1971 the first tonnage of farmed Scottish Atlantic salmon was harvested. From the start, the farms in Scotland were larger in size than in Norway because regulations for farm size and local ownership were not as strict. During the first part of the 1970s total Atlantic salmon production in Scotland was higher than in Norway [30]. Development was, however, quite slow during the 1970s and Norway overtook Scotland as a larger producer of Atlantic salmon. During the 1980s salmon production in Scotland increased, partly due to European funding programmes [14].

Since the 1980s there has been a gradual increase in Scottish Atlantic salmon production. For a long time, Scotland was the second largest salmon producer in the world after Norway, but in 2000 it was surpassed by Chile, which has a much greater length of coastline suitable for production of Atlantic salmon in sea cages.

3.3.5 Tasmania

The history of farmed salmonids in Tasmania started in the 18th century when salmon and trout were imported and hatched with the purpose of establishing local stocks fit for recreational fishing, but without success. In the 1960s rainbow trout eggs were imported with the purpose of establishing commercial aquaculture, and at the start of the 1980s the first sea cage farms with rainbow trout were established [31]. In 1984 the first Atlantic salmon eggs were fertilized based on earlier imported salmon from Canada for the purpose of aquaculture, and this represented the start of the Atlantic salmon farming industry in Tasmania [32]. At first it was partly a joint venture with Norwegian investors. Since then there has been a slow growth in the Atlantic salmon farming industry.

3.3.6 Faroe Islands

The development of salmon farming in the Faroe Islands started with trials, partly government initiated, during the 1970s when there was a decline in the traditional fishing industry:

an industry that was of major importance for the economy of the Faroe Islands. It was until the 1980s that commercial farming started, based on knowledge from Norway. Salmon eggs from Norwegian breeding stocks were an important contribution. After this there was a steady growth throughout the first years, after some ups and downs due to lack of smolt, diseases and some periods of low prices. During recent years, however, growth has been reasonable but there are limitations for further expansion due to the length of the coastline that limits available sea cage sites.

3.3.7 Ireland

Atlantic salmon farming in Ireland, as in Scotland, started in the mid-1970s with trials. Industrial development started in the 1980s with financial support from different sources, including from the European community. The industry grew quite fast initially, but growth has gradually decreased and over the last few years it has been very limited. There are limited good sheltered sites for sea cage ongrowing, with several areas exposed to severe weather, but the major reason for the slow growth has been the public opinion against sea cage salmon farming.

3.3.8 Iceland

Farming of Atlantic salmon in Iceland started in the mid-1980s but due to the limited number of good sites for sea cage farming with the existing cage technology there was low growth. At the start of the 1990s, a number of land-based ongrowing farms were established, as was also the case in Norway. The economics of this production method was unsatisfactory in Iceland, as in Norway, and resulted in bankruptcy or farm shutdown. Iceland has, however, a long tradition of wild stock enhancement and between the 1970s and 1980s a large programme for sea ranching of Atlantic salmon was developed. However, the economics of sea ranching of salmon was not sustainable, as has also been seen in other places [33,34]. Smolt farms had been established for this purpose, and there was the great benefit of the possibilities for utilizing geothermal water to speed up salmon growth and thereby production. During recent years a new optimism has been observed for salmon farming in Iceland, partly due to the development of cage technology suitable for more exposed sites.

3.4 Development of the Atlantic salmon farming industry

3.4.1 An industrial adventure

The development of Atlantic salmon farming over the last 50 years has, as the figures show, been an industrial adventure, and no other animal production has shown the same growth. It has been a stepwise increase, with some important findings that have created the industry of today. The history of Atlantic salmon farming can be told from two points of view, from the land or from the sea, or smolt farming and sea cage farming. In general, several contributing factors have made growth possible but some major factors may be highlighted. These include developments in the areas of genetics and breeding programmes, vaccines and disease control, feed and feeding routines, and new technology. The development process has been relatively similar in all Atlantic salmon producing countries. Below is an overview of the important steps in the development of smolt and ongrowing sea cage farms. The focus has been on production and technological development, which has been of major importance for the development and location of the farms. The description is based on experiences from the Norwegian salmon industry, but similar trends have been observed in all salmon producing countries as the salmon industry today is an international industry, where the major actors have activities in several salmon producing countries.

Similar development stories hold true for the other main contributors to the growth in the salmon industry: feed development, development of vaccines and disease control efforts, and finally, breeding. On the subject of feed, the development of a tasty and well-composed start feed was very important at the beginning of the salmon farming industry. A major advance in salmon feed resulted from the introduction of high-energy feed, including more lipids that increase growth in the ongrowing stage (see also Chapter 10). The inclusion of more carbohydrates and a reduced amount of fish meal and fish oil in the feed were also factors that should be highlighted today, because this increases the amount of available raw material for salmon feed production. The continuously improved FCR is also a major progression for the economics and environmental impact.

From the start, diseases and parasites have been a great challenge to the salmon farming industry: a challenge that will certainly continue in the future with new diseases continuously emerging (see also Chapter 11). The development of vaccines has been one major factor aiding growth in the salmon industry. At first, vaccines were designed for protection from one disease, while today's multicomponent vaccines protect against a range of diseases. The development and use of DNA vaccines is and will continue to be an important tool in the future of the Atlantic salmon farming industry.

Development of breeding programmes and the selection of traits fit for farming have been very important for the development and economics of salmon production (see also Chapter 12). Today, farmed Atlantic salmon is in the 13th generation, meaning it has been possible to select the best traits and fish over 13 generations. In the beginning the breeding goals were growth rate, FCR and late maturation, but later many other traits were included in the breeding goals, including disease resistance and, recently, sea lice resistance. Biotechnological expertise has also been an important contributor to breeding during recent years and genomic selection has become an important addition to traditional family selection. The recent sequencing of the salmon genome will also be an important tool for further breeding progress. To illustrate the importance of breeding, two examples are further described. In one experiment, farmed salmon that had been through a breeding programme were compared to wild salmon that were the basis for the first generation of farmed salmon; 94 days after first feeding, the weight of the bred salmon was almost double that of wild salmon [35]. In another experiment, females from the 11th generation were crossed with males from the first and 11th generations. After 14 months in sea, the average weight of the crossing with the first generation was 2.1 kg and the 11th generation was 3.7 kg [36].

3.4.2 Developments in smolt farming and farms

The development of smolt farms has, as mentioned, been a stepwise process, in which the most important factor during recent years has been access to enough freshwater at the production sites and efforts to find ways of producing more smolt from the same volume of incoming freshwater.

This is due to an increased demand for smolt to raise in the sea and for smolt with an ever-increasing size; both of which require more freshwater. As a result, some of the sites used for smolt farming at the start of the salmon farming industry have been abandoned.

3.4.2.1 Moving away from 2+ production

In the late 1970s and 1980s the majority of today's smolt farms were established in Norway. A typical smolt farm then had a production of around 500,000 smolt per year. Production was slow and mainly produced what are termed 2^+ smolt. These were released into the sea in springtime, just over 2 years after hatching. The natural water temperatures of the freshwater sources were used, without any form of heating, which resulted in very low water temperatures during the winter. The smolt that were transferred into the sea cages were quite small, with a typical size of 20–40 g.

This temperature regime resulted in a long production time, which in turn caused production to be rather ineffective. As a result, the major objective at the time was to reduce the production time: to produce fish in a shorter time in order to make the production more efficient and to get more smolt on the market. This was done by heating the water supplied to the eggs, the start-feeding fry and the early fry, which are all periods where a limited amount of water is needed. In addition to increasing growth, this made start feeding simpler because the start-feeding fry were more active. By heating up the water during winter when the natural temperature was low, the biological process was speeded up. The result was the production of what are termed 1^+ smolt, fish that need only slightly over 1 year in the smolt farm before transfer to the sea cages in springtime. Heating systems therefore became an important part of the smolt farms. This originally took the form of electric or gasoline boilers, but later combinations of a number of heat exchangers and heat pumps became the norm, which were actually large effective energy plants (high coefficient of performance, COP (Figure 3.4).

Figure 3.4 Energy plants were introduced on smolt farms to heat up the water to increase production speed and fish size.

3.4.2.2 0+ and light manipulation

The next question that arose was how production of smolt could be increased even more without using more water. There was a lack of smolt on the market, and in addition a larger smolt size was desired. At many sites there was now a lack of water in late winter, which limited the increase in production. At this time of year, the standing biomass was at its highest on the smolt farm, and the water consumption was also the highest compared to that naturally available from the water sources on the sites. This was the start of the change from 1^+ smolt to 0^+ smolt or ½ year smolt. By producing such fish, it was possible to send out fish as 0^+ in the autumn to the ongrowing sea cage farms and thereby decrease the biomass of 1^+ fish in the smolt farm in the winter. This resulted in the smolt farm having a higher production of individuals in total, delivering both 0^+ and 1^+ smolt. However, to achieve this the smolt had to be produced in the autumn, out of season for traditional smolt production, therefore the term 'season independent smolt production' or 'season smolt production' was introduced. To achieve this the salmon had to have reached a minimum size and the smoltification process had to be completed sooner. The smoltification process now had to take place during summer and early autumn, rather than the natural time for Atlantic salmon of winter and spring. The solution to producing smolt in the autumn was to start with light manipulation, meaning the creation of an artificial spring in the autumn (Figure 3.5). This was followed up with increased water heating to get the salmon to the suitable size for smoltification. In addition to increased water heating, the fish were size graded in late spring and only the largest were set on a light control programme. A number of different light manipulation programmes were developed and optimized. One light programme used with good success was 6 weeks with 12 hours of light followed by 4–6 weeks with 24 hours of light, depending on water temperature, before transferring as smolt. By following this production regime, the winter biomass was reduced because a proportion of the fish were already smolt and so could be sent out from the farm to sea cages in the autumn. The smolt farm could now almost double their production on the same water flow. Around 50% of the fish produced on the site were transferred in the autumn.

3.4.2.3 Liquid oxygen and carbon dioxide removal

Smolt demand was still increasing and the smolt farms were already delivering maximum output from the natural water sources available on the site. In order to increase the production even further, the next step was to break the limitations of nature. This involved introducing pure oxygen. By adding around 3–4 ppm pure oxygen gas to the incoming water supplying the ongrowing tanks, smolt production could be approximately doubled on the same waterflow, depending on water temperature. By continuing to add more oxygen, the production could theoretically be increased even more. However, as is shown later, this brought new problems. Plant water oxygenation was therefore installed. At first the farms invested in plants for the self-production of oxygen gas based on the sieving of air. Later, the infrastructure for delivery of liquid oxygen from centrally located suppliers was developed. The farms then needed a storage tank for liquid oxygen and an evaporator for converting liquid oxygen to the oxygen gas that was used to superoxygenate the water (Figure 3.6). Today, purchasing liquid oxygen is the most common method but some farms also produce oxygen themselves. The advantage of self-production is that the price of oxygen can be lower, but the farm needs to be competent in the running and maintenance of the oxygen production plant. Some plants used really high supersaturation of oxygen gas in the inlet water, far more than today's recommendations, but still with reasonable growth.

Figure 3.5 Light was introduced on smolt farms to produce smolt out of season.

With the supply of water required for each kilogram of fish in the farm reduced through to the addition of pure oxygen the next hurdle was identified. There was an accumulation of metabolic waste products (carbon dioxide [CO_2] gas) in the salmon production tanks due to the reduced amount of new water added, and the concentrations of dissolved carbon dioxide gas in the water reached levels that affected the performance of the salmon. Reduced growth and reduced animal welfare became a problem. A rough estimate says that for each unit of oxygen added, one unit of carbon dioxide will be released. The next step was therefore to add carbon dioxide strippers that remove carbon dioxide gas from the water into the surrounding air. This lowered the concentration of carbon dioxide in the water to a satisfactory level for salmonids. This was typically done by pumping part of the water in the tank over the carbon dioxide stripper and returning it to the tank: a type of internal recycling of the tank water. By the removal of carbon dioxide from the water, the production of smolt (both in number and size) could be increased even more on the same amount of new water. The baseline for acceptable carbon dioxide levels to ensure the

Figure 3.6 Use of liquid oxygen was introduced to increase production; the tanks with liquid oxygen towers became a characteristic of smolt farms.

welfare of Atlantic salmon was also set during this development phase, but it started with values above today's recommendations.

3.4.2.4 Introduction of recycling aquaculture systems

Once again, with more fish being produced on the same amount of new incoming water the next problem arose. The industry not only wanted more smolt produced using the same amount of new water, but also wanted a much larger smolt size for transfer to the sea. One reason for this was that a larger smolt is more robust against sea lice, which was now a major problem. The time required in the sea, when fish are exposed to sea lice and diseases before reaching slaughter weight, was also reduced by transfer of a larger-sized smolt. Because the quantity of new water added per kilogram of fish in the tanks had decreased, another metabolic waste product became a problem. There was not enough new water supplied to dilute the concentration of ammonia (NH_4^+/NH_3) to acceptable levels for optimal salmon growth and welfare. Removal of carbon dioxide did not help this situation, because the removal of carbon dioxide increases the pH of the water, changing the equilibrium between ammonium and ammonia which results in greater levels of ammonia, a more toxic substance to salmon. Solutions to the ammonia problem now had to be found.

The solution to the problem of high concentrations of ammonia was to install biological filters. The biological filter used hosted a special type of aerobic bacteria culture that removes ammonia from the water, or more accurately transfers ammonia to the far less toxic form, nitrate. This meant that the nitrogen was still in the water but was less toxic. However, to utilize this type of biological filter hosting the correct bacterial culture in a beneficial way, the organic particles (faeces and waste feed) had to be removed prior to the filtering, so a particle filter had to be installed. By removing organic particles carbon would be removed from the flow but most nitrogen will still be in the water, and by this would a correct carbon–nitrogen ratio be achieved. The biological filter also lowered the pH of the water and therefore a regulation step to handle this had to be included, as did a unit to control the water temperature (heating in winter, cooling in summer) to achieve optimal growth. The farm design was therefore transformed from a flow through system to a system that recycled or reused the water, termed a recycling aquaculture system (RAS).

This is today's state-of-the-art design for smolt farms; a large number of RAS smolt farms have been built or are in the process of being rebuilt in all countries producing Atlantic salmon (Figure 3.7).

By utilizing RAS instead of flow through systems, the design of smolt farms was changed but the number and size of smolt delivered from the same water source could be greatly increased. The supply of new water was reduced to 1–5% of what was used originally in a flow through system, so the degree of water reuse is 95–99%. However, current fish density demands and the degree of reuse are continually stressing the system, and toxic levels of nitrate may therefore be reached in the near future. A denitrification step, transferring nitrate to nitrogen gas that can be released into the atmosphere, may need to be introduced. What has been observed through the development process putting stress on the amount of new water supplied is that it will always be in the best interests to have as much freshwater as possible available on the production site, in order to have a backup if something fails.

The introduction of RAS bring the advantages of large economies of scale, meaning that the cost to produce smolt is reduced in larger farms. The result is that the old smolt farms are

Figure 3.7 RAS is installed in a number of farms to increase production without using more new water.

being rebuilt and increased in size. Yearly production capacity above 10 million is not unusual and farms producing over 15–20 million smolt are being built today. The enormous development from the start of the industry when a large smolt farm produced 0.5 million smolt per year is readily evident.

3.4.3 Developments in sea cage farming

As for smolt farms, some distinct steps in development can be identified from the start of sea cage farming until today. The steps are similar in all salmon producing countries, partly because the large salmon producers have farms in several countries but also because the large technology suppliers are international companies.

3.4.3.1 Small cages in sheltered sites

When salmon farming began, the sea cage farms for ongrowing of salmon were located inside very sheltered fjords. The waves were small and so the water currents were weak. Very simple cage constructions were used: wooden sticks or steel rails and simple air-filled plastic buoys or unprotected Styrofoam cubes were used for buoyancy. Sizes were small and net bags with a typical size of 5 m × 5 m × 5 m were used. The fish density in the cages was low and so was the total biomass on the sites.

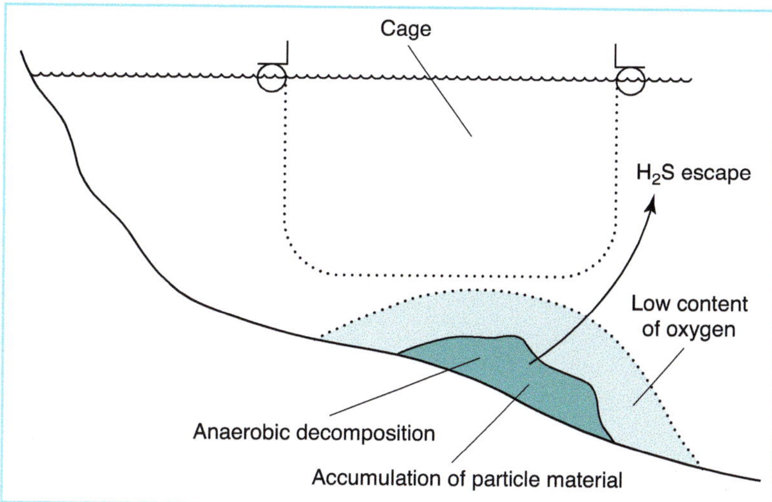

Figure 3.8 Very sheltered sites were used and accumulation of waste below the cages was a possibility with reduced water quality as a result.

As the industry started to grow and more salmon were produced, the cages increased in size and so did the amount of fish on each site, but the same sites were still used. The result was overloading of the production capacity of the sites. The low self-cleaning capacity of the sheltered sites became a problem and waste accumulated on the sea floor below the cages (Figure 3.8). This resulted in lack of oxygen and anaerobic decomposition with production of hydrogen sulphide (H_2S) gas. This became a problem at several sites, and mortality due to toxic H_2S gas released from sediments occurred. Large propellers were tested to provide a current to ensure that enough new water was flowing through the net bag. However, this represented additional production costs and so was not an attractive option.

3.4.3.2 Exposed sites and certification of equipment

The solution to this problem was to move the cages to more weather-exposed sites, and therefore only a limited number of the sites used at the start of salmon sea cage farming are still in use today (Figure 3.9). The sites that are now used have better water currents to handle and dilute the metabolic products released by the salmon, especially the particulate matter, faeces and feed waste. However, the sites that were now being used are also less sheltered and more weather exposed. The simple cage construction technology was not suitable for these more exposed sites, and physical breakdown of sea cages occurred. At the same time attention to the problem of escaped salmon increased. Therefore, laws and regulations for the construction and management of sea cages were introduced. Requirements were set for cage construction and strength based on the environmental conditions of the site to ensure the cages and mooring system used were adapted to the wave heights and water currents on the site (included in regulations such as Norway's NS 9415), so it became a requirement to perform a site classification. This included collection of data on the highest environmental loads including water current and climate (wind and calculated waves) over a 50 year period. In addition, bottom mapping of the site became a requirement. The data collected was then used for classification of the site and was used to calculate the requirements for moorings, cage collars, net bags and eventually feeding barges. This has almost eliminated problems with construction failure.

Figure 3.9 More weather-exposed sites were taken into use for sea cage farming.

3.4.3.3 Fallowing of sites and introduction of MOM

Another problem in the early days of the industry was the limited number of sites in use for production, so the fish were raised on the same sites each year. As problems with disease increased, the result was an easy transfer of diseases between the groups/generations. The solution to this was to have only one group or generation per site. In this way, direct disease transfer from one group to another was eliminated. This was also a way to reduce the impact of any negative incidents, because with a number of sites over a relatively large area, negative incidents would not normally occur on all sites at the same time, thereby distributing the risk.

The use of multiple sites led to a new possibility: the resting or fallowing of sites (Figure 3.10). Owing to the high biomass on the sites, accumulation of waste and changes in the bottom fauna below the sea cages would occur. By using the strategy of resting/fallowing the sites between stocking, the site had the opportunity to return to its natural conditions, so therefore three sites were typically used: two with salmon from two different years and one site that was resting. Later this was modified so today the requirement for fallowing is at least 2–3 months between stocking, dependent on the conditions of the site. The purpose of resting is to allow the sites to return to their original natural conditions, and by this both maintain nature and improve the production results as the site will be in a better condition for further salmon production.

This period saw the introduction of the MOM system: modelling–ongrowing fish farms–monitoring. The purpose of this was to inhibit the overload of the farming sites with particles and nutrients (NS 9410). However, the system was also used to adjust the discharge from the sea cage farms to the carrying capacity of the site. This ensured that the farms had no negative environmental impact on the bottom fauna below the cage areas. Large penalties were imposed when this regulation was not followed. The system included a number of modules from the

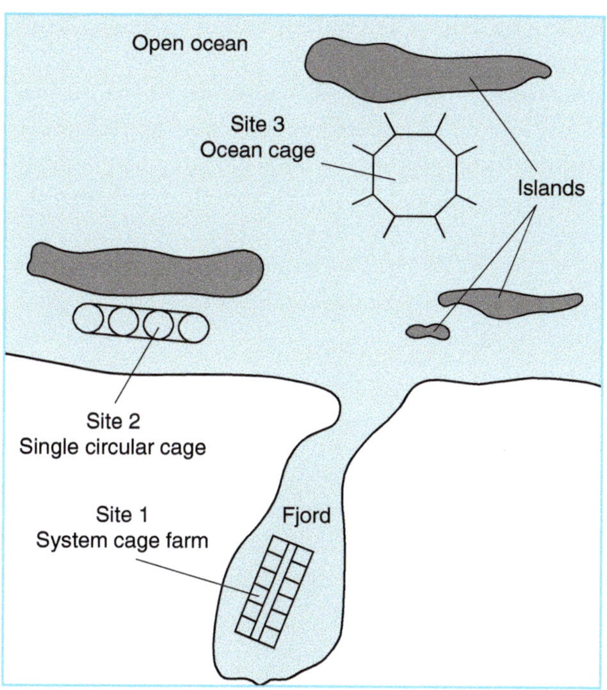

Figure 3.10 Several sites were used which allows fallowing of sites for periods so the bottom fauna below the cages keeps its natural condition.

start, including MOM A (voluntarily performed by farmer), MOM B (compulsory) and MOM C (to be performed by an external accredited company if high levels were measured on MOM B or on sites with a high biomass). This methodology has been further developed up to today's NS 9410.

3.4.3.4 Converting licences to maximum allowed biomass

In Norway the first licences for sea cage farming were set according to a fixed production volume. The volume was calculated in a standardized way, independent of the real net bag volume. The following equation was fixed and used to calculate the farming volume: 0.85 × surface area of the net bag × 5 m deep. This was appropriate when the industry began, but then deeper nets were introduced so it was possible to have more fish in each individual cage. As the market for salmon increased, supporting increased production, deeper and deeper nets were used. This was not included in the calculated volume, which was fixed to 5 m regardless of what the depth actually was. There were also examples of quite creative net bag design in this period. The net bags were made with a bottle-like form, and in this way the used surface area was quite low while the real production volume was very large. From 1 January 2005 there was a change in the licence description and farming volume was no longer used. Instead the limitation was for allowable biomass, and the term MAB or maximum allowable biomass was introduced to define new requirements for a licence. The earlier volume was limited and had increased gradually from 3000 m^3 at the start of the industry in steps up to 12,000 m^3 per licence. This was now changed and 1 m^3 volume was set to 65 kg in Norway, except in the north of Norway (Troms and Finnmark) where it was set to 75 kg due to lower water temperatures and subsequently lower expected growth. Each licence was therefore converted to a MAB of 780 tonnes or 900 tonnes, respectively. This was also applied to the production site and today 4 MAB can be kept on one production site with the requirement that MOM B is fulfilled. If more fish could be supported at the site, which is possible for good sites, MOM C is then required.

With the advent of a fixed value for the quantity of fish each company could have in total based on its number of licences came higher requirements for management and optimization. Improved coordination between freshwater and marine operations became of interest to increase the turnover per MAB/licence. This included several differentiated (in size) smolt transfers to sea cages, and ideally with these transfers spread over as large a part of the year as possible. Again, this required more 0^+, larger smolt, and it also increased the benefits of size grading during marine ongrowing (especially in relation to slaughter). The permitted production of salmon from one license/one MAB is from 1000 tonnes to slightly above.

3.4.3.5 Artificial light in the sea cages

Another important developmental step in sea cage farming was the introduction of artificial light in the sea cages during the winter season (Figure 3.11). This resulted in increased performance, including a delayed and reduced degree of early maturation, better and more steady growth, and improved slaughter quality. At first, lights were positioned above the water surface of the cages and directed down into the sea, but later this was changed to submerged lights; today light emitting diodes (LEDs) are also used to reduce energy consumption.

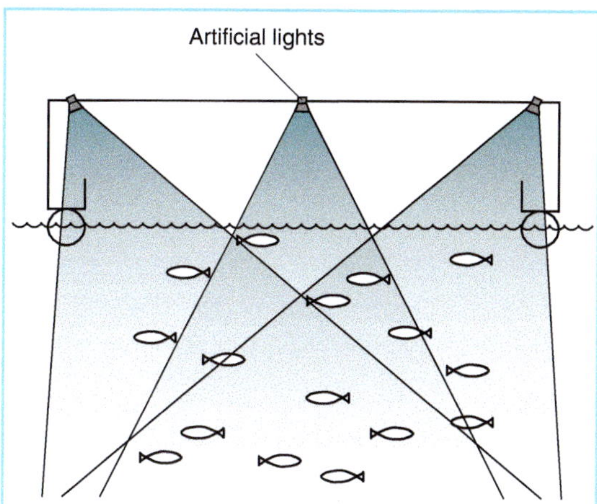

Figure 3.11 Artificial light was introduced to increase the growth in sea cage farms and to avoid early maturation.

3.4.3.6 Feeding systems

A tremendous development has also taken place regarding feeding. Initially hand feeding was used, and feed was supplied to the farms in small bags of 25–50 kg; the cages were small and therefore so was the amount of feed. Over the years, the amount of feed that has to be delivered to each sea cage farm has increased severalfold due to increased cage size and fish biomass. Hand feeding was therefore soon replaced with automatic feeders. At first these were simple, only filling the feed hoppers but still using small feeding bags. The feeders were controlled by simple controllers, controlling run and stop time (Figure 3.12). Gradually, automatic feeding systems were introduced, which was a great advancement (Figure 3.13). The feed was taken directly from feed silos on land, which later were also installed on large feed barges close to the cages. From the silos the feed was transported and distributed into the individual sea cages, and the correct amount of feed for each sea cage was automatically weighed out. Water was used as a medium for transporting the feed pellets at first, but later air was used, which is the predominant method today. By installing automatic feeding systems, all human feed handling on sea cage farms was eliminated. Lately the hatches on feed silos on the barges have also been automated with provision for remote control. In this way they can be filled automatically from vessels transporting feed.

There have also been developments in feed transport from the feed factories to the ongrowing farms; today, large boats specially designed to only transport fish feed are commonly used. They

Figure 3.12 Simple automatic feeders from the early days of salmon farming.

are also equipped with a dynamic position system and combined with remotely operated hatches on the feed silos allow for automatic non-contact delivery of feed to the feed barges on the sea cage farm.

3.4.3.7 Feeding barges

The sites used at first were located close to the land, and many had direct access via walkways. Others were only small distances from a land base and small boats were used for handling the sea cages. Small feed sacks of 25–30 kg were used and transported by boat from land. As the size of the cages increased, more feed was needed, and the sites were more weather exposed so required the use of larger boats that could tolerate rougher seas. The distance from the land base also increased. It became impossible to transport all feed to the sea cages with the traditional boats. It was also impossible to use feeding systems with silos on land because the tubes required for the transport of feed from land would become too long. Fallowing of sites was also introduced, resulting in the feeding systems remaining unused for the fallow period. These were all factors that were important in the introduction and use of large feeding barges. The automatic feeding system and the silos were now installed on a barge that was anchored up close to the sea cages (Figure 3.14).

The barges were also equipped with a number of other facilities for personnel working the farms, such as personal rooms, toilets, a control room, small workshops and equipment (for instance, dead fish handling), and many of the functions that until then had been handled by the

Figure 3.13 Feeding systems were introduced due to the large amount of feed required on the sites where biomass was increasing.

land base were now transferred to the barge. Some of the barges also included sleeping rooms so the operators could live on the barge for short time periods. The size of the barges has increased since their introduction, and today they are large steel or concrete constructions tolerating rough weather conditions and with increased capacity for storing of feed.

3.4.3.8 A growing service industry

At the start of the salmon farming industry the sea cage farmers did most of the work on the farm themselves. This included tasks such as net cleaning, net service and setting out, and controlling mooring systems. Today these services and others are increasingly outsourced to specialized service companies, which are either independent or a division of the parent company (Figure 3.15). This, of course, is due to the ever-increasing size of the cages, feeding barges and mooring systems that require larger and more specialized equipment for handling. Owing to the increasing size and cost of the equipment, the need for more continuous use also became a requirement. Today there are a number of specialist companies dealing only with services to ongrowing sea cage farms. Examples of these are diving services, net handling, handling of moorings, and environmental monitoring. An industry has also been developed based on well boat services. Today well boats are used for transport of smolt, slaughtering fish, sea lice treatment, and grading and counting. Shipping companies that operate only well boats have been established and the size of the well boats continues to increase. Today well boats with an overall length above 100 m and with a well volume over 7000 m^3 are built.

Figure 3.14 Feeding barges became an important part of the sea cage farms.

Figure 3.15 A service industry supplying sea cage farms, including diving services, has developed.

3.4.3.9 More monitoring and centralized control

A trend that has been seen from the start of the salmon industry and continues today is the use of increasing numbers, and the sophistication of, sensors, instruments and monitoring systems. Typically, this is seen with the development of the feeding equipment in sea cage farming described above, but it can also be seen with monitoring equipment. Sensors providing environmental data, such as temperature, oxygen and salinity, became more commonly included on the individual cages. Video cameras were installed in cages for control of feeding and also to monitor behaviour. Operators watching the video feeds were traditionally in the control room, typically on the feed barges, but eventually moved to land when feed was supplied from there. In this way the feed operators moved away from the cage to indoor control rooms, so instead of standing on the cage watching the feeding they could sit inside and monitor video pictures from cameras lowered inside the individual cages. They could then stop feeding based on what they saw on the monitors. This trend has continued and with improved internet access, which today is a must on all sea cage farms, new possibilities for remote control of operations on the farm have opened up. One important operation is the control of feeding and the feeding equipment. The result is that the feeding operators can now be located totally independently of where the sea cage farm is located. More recently large centrally placed feeding centres have been established, which look like large industry control rooms, to control feeding on a number of sea cage farms across regions (Figure 3.16). In the future, highly skilled feeding operators located in cities may be responsible for feeding salmon in sea cages across large geographical areas, either of one entire country or possibly even across several countries. The cages can also be equipped with all type of sensors and machine learning can be used in addition to operators' skills to optimize feeding.

3.4.3.10 The traffic light system

In recent years, the sea lice situation has generally not been sustainable in almost all salmon producing countries. Precautions have been taken. An example is the traffic light system instigated

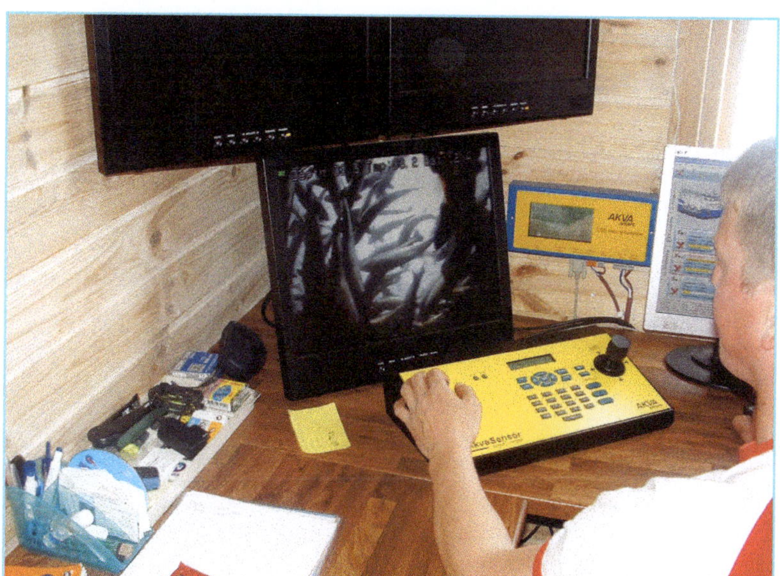

Figure 3.16 Feeding centres serving several cage farms became a natural part of the sea cage farming infrastructure.

in Norwegian sea cage farming. The Norwegian coast has been divided into 13 zones based on the effects of the salmon industry on the environment, and in particular the sea lice situation. If the traffic light is green, there is a possibility for an increase in production on the sites in the region. A yellow light indicates no increased capacity and if the light is red, salmon production has to be decreased. The light is updated every second year based on the situation in each zone. In 2017, for example, eight zones had a green light and could increase production, three had a yellow light and two had red lights. The red-light zones were on the southwest coast and were a result of the sea lice situation. However, individual farms in a red-light zone who want to produce in closed containment systems (CCS) into which sea lice cannot enter may still have the opportunity to increase production.

References

1. Knapp, G., Roheim, C.A. and Anderson, J.L. *The Great Salmon Run: Competition Between Wild and Farmed Salmon*. 2007. TRAFFIC North America.
2. Stead, S.M. and Laird, L. *The Handbook of Salmon Farming*. 2002. Springer: Dordrecht.
3. Hovland, E. et al., *Over den leiken ville han rå*. 2014. Fagbokforlaget: Havbruksnæringens historie. Trondheim.
4. Hovland, E. and Møller, D. *Åkeren kan òg være blå. Et riss av havbruksnæringens utvikling i Norge*. 2010. ABM-utvikling; Kystverket; Riksantikvaren; Fiskeridirektoratet.
5. Berge, A. The salmon farming pioneers: the Vik brothers. *Salmon Business*. 2017. salmonbusiness.com.
6. Oddmund, V.K. and Birger, V.O. Method for breeding fish, such as salmon or sea trout. 1968. Google Patents.
7. Berge, A. *Pionérene: Erling Osland*. 2014. ilaks.no.
8. Gjedrem, T. The first family-based breeding program in aquaculture. *Reviews in Aquaculture*. 2010. 2(1): 2–15.
9. Gjedrem, T. and Baranski, M. *Selective Breeding in Aquaculture: An Introduction*. Vol. 10. 2010. Springer: Dordrecht.
10. Bakke, H. and Melingen, G.O. Avlsarbeid – viktig bidrag til fremgang i norsk fiskeoppdrett. *Genialt*. 2010. Bioteknologirådet.
11. Berge, A. The breakthrough for farmed salmon. *Salmon Business*. 2017. salmonbusiness.com.
12. Lekang, O.-I., Salas-Bringas, C. and Bostock, J. Challenges and emerging technical solutions in ongrowing salmon farming. *Aquaculture International*. 2016. 24(3): 757–766.
13. *Farming the seas – A timeline*. https://www.dfo-mpo.gc.ca/aquaculture/sector-secteur/frm-tml-eng.htm.
14. Willoughby, S. *Manual of Salmonid Farming*. 1999. Wiley: Chichester.
15. Page, J. Salmon farming in First Nations' territories: a case of environmental injustice on Canada's west coast. *Local Environment*. 2007. 12(6): 613–626.
16. Gerwing, K. and McDaniels, T. Listening to the salmon people: coastal First Nations' objectives regarding salmon aquaculture in British Columbia. *Society & Natural Resources*. 2006. 19(3): 259–273.
17. Schreiber, D. First Nations, consultation, and the rule of law: salmon farming and colonialism in British Columbia. *American Indian Culture and Research Journal*. 2006. 30(4): 19–40.

18. Northwest Power and Conservation Council. Hatcheries. https://www.nwcouncil.org/reports/columbia-river-history/hatcheries.
19. Grydeland, O. Salmon ranching explained. *Fishfarming Expert*. 2012. fishfarmingexpert.com.
20. Amos, K.H. and Appleby, A. *Atlantic salmon in Washington State: A fish management perspective*. 1999. Department of Fish and Wildlife: Washington, DC.
21. Volpe, J.P. et al. Evidence of natural reproduction of aquaculture-escaped Atlantic salmon in a coastal British Columbia river. *Conservation Biology*. 2000. 14(3): 899–903.
22. Knapp, G. and Rubino, M.C. The political economics of marine aquaculture in the United States. *Reviews in Fisheries Science & Aquaculture*. 2016. 24(3): 213–229.
23. Hosono, A., Iizuka, M. and Katz, J. *Chile's Salmon Industry*, Vol. 10. 2016. Springer: Japan.
24. A case study of the salmon industry in Chile. Report. 2006. United Nations Conference on Trade and Development. https://digitallibrary.un.org/record/574403?ln=enhttps://iseralaska.org/static/legacy_publication_links/greatsalmonrun/SalmonReport_Ch_5.pdf
25. Figueroa, J. et al. Addressing viral and bacterial threats to salmon farming in Chile: historical contexts and perspectives for management and control. *Reviews in Aquaculture*. 2019. 11(2): 299–324.
26. Asche, F. et al. The salmon disease crisis in Chile. *Marine Resource Economics*. 2009. 24(4): 405–411.
27. Ibieta, P. et al. Chilean salmon farming on the horizon of sustainability: review of the development of a highly intensive production, the ISA crisis and implemented actions to reconstruct a more sustainable aquaculture industry. *Aquaculture and the Environment – A Shared Destiny*. 2011. 215–246.
28. Aguayo, B.E.C. and Barriga, J. Behind certification and regulatory processes: Contributions to a political history of the Chilean salmon farming. *Global Environmental Change*. 2016. 39: 81–90.
29. Scottish Salmon. The history of Scottish salmon farming. https://www.scottishsalmon.co.uk/facts/innovation/the-history-of-scottish-salmon-farming.
30. Ellis, T. et al. Trends during development of Scottish salmon farming: An example of sustainable intensification? *Aquaculture*. 2016. 458: 82–99.
31. https://www.utas.edu.au/tasmanian-companion/biogs/E000050b.htm.
32. http://www.salmonfarming.org/cms/wp-content/uploads/2018/06/ISFA-JUNE-2018-REVISED-FOR-FOR-WEB1.pdf.
33. Arnason, R. *The Economics of Ocean Ranching: Experiences, Outlook and Theory*. 2001. FAO: Geneva.
34. Isaksson, A. Salmon ranching: a world review. *Aquaculture*. 1988. 75(1–2): 1–33.
35. Jin, Y. et al. Comparative transcriptomics reveals domestication-associated features of Atlantic salmon lipid metabolism. *Molecular Ecology*. 2020. 29(10): 1860–1872.
36. AquaGen. *Stor økning i laksens veksthastighet etter 11 generasjoner med avl* in *Aquagen kunnskapsbrev*. 2020. AquaGen.

CHAPTER 4

Salmon biology: morphology, anatomy and physiology

4.1 Introduction

The aim of this chapter is to give a very basic overview of salmon biology (the science of the living organism) including morphology (form), anatomy (structure and relationship between body parts) and physiology (function of the body parts). Some of the topics related to physiology require a good knowledge of biochemistry to fully understand them and are therefore not described in depth. The aim here is not to go into great detail but to provide to better understand salmon production and the possibilities for production optimization. For more in-depth knowledge there is plenty of literature and textbooks on the biology, anatomy and physiology of fish in general, and also specifically of salmonids (for example [1–7]). A number of web pages also provide ample information about this topic (for example [8–11]).

4.2 Salmonids

Salmon are finfish and belong to the salmonid family (Salmonidae). The family also includes trout (*Salmo trutta)*, chars (*Salvelinus*), freshwater whitefish and graylings (*Thymallus*). Atlantic salmon (*Salmo salar*), the focus of this textbook, and Pacific salmon are included in the salmonid family. The Pacific salmon (*Oncorhynchus*) can be divided into the following six species: Chinook salmon, also named king salmon (*O. tshawytscha*), chum salmon (*O. keta*), coho salmon (*O. kisutch*), masu salmon, also named Japanese salmon (*O. masou*), pink salmon (*O. gorbuscha)* and sockeye salmon (*O. nerca)*. The salmonids are teleost fish (Teleostei).

4.3 Ectothermic and carnivorous

The salmon is, as are other finfish species, cold blooded, also known as ectothermic. This plays an important role in the metabolism of the salmon. In nature salmon are adapted to live in temperatures between 0°C and 20°C. The temperature also dictates the metabolic rate: a higher rate at higher temperatures. When the temperature drops below 8°C the metabolic rate becomes low, as does the growth. In the upper part of the temperature range the metabolic rate increases but the oxygen content in the water decreases. Therefore, it is difficult for salmon to obtain enough oxygen for a maximal metabolic rate at the high end of the temperature range. Oxygenation levels also affect the growth rate, which drops at high temperatures. The optimal temperature for growth lies in the range 14–16°C, but the optimal temperature under farming conditions with optimized environmental conditions is to date not fully documented. Additional adverse consequences of being in the upper part of the temperature range include increased risk of disease and mortality (temperatures above 20–22°C represent the temperature limitation of the species and failure of body functions may start to occur). The velocity of any rise or drop in temperature is also important since it affects the ability of the salmon to regulate its body functions. There are examples of Atlantic salmon being kept at temperatures higher than the recommended level after acclimatization, and this topic is very interesting with regard to the wild Atlantic salmon stock's ability to handle global warming [12,13]. The available amount of dissolved oxygen in the water is also a contributing factor [14]. Optimal temperature is also dependent on fish size and stage of development.

Atlantic salmon is a carnivorous fish species, meaning that its digestive system is adapted to mainly eat meat (fish); this dictates its food requirements.

4.4 Lifecycle

Salmon are an anadromous species, meaning that they live partly in freshwater and partly in seawater during their life cycle (Figure 4.1). The eggs develop and hatch in freshwater, and the fry grow up to smolt in freshwater; they then move to seawater for further ongrowing, and before spawning return to freshwater again. For wild salmon this process occurs as the mature salmon travel up the rivers for spawning in the autumn season. Here they make spawning nests in the river gravel, and the eggs are fertilized by milt from the male before being dropped into the nest and covered with gravel, where they remain until the next spring. The hatching fry find their way up from the gravel and start to search for food in the river, where they will then stay for some years. The size to which they grow is dependent on food availability and the conditions in the river. After some years, they become smolt; this occurs in the spring. A smolt is a salmon that can tolerate seawater, so is therefore ready to live in the sea. By this stage, special cells have developed on the gills that enable the salmon to control the salt/ionic balance (described in section 4.13). The behaviour of the salmon when reaching the smolt stage changes from swimming against the water flow to swimming with the water flow: from countercurrent to co-current. This results in the smolt following the river out to the ocean. Here they start their migration: searching for food and growing in size for one or more years before reaching maturity and returning to the same river that they left in order to spawn.

Wild anadromous fish are characterized by relatively slow growth in the freshwater phase because of the limitation in natural feed and the low water temperatures, especially in winter. In rivers the number of predators is quite low and it is easy for the fry to find shelter, therefore many salmon survive to reach the sea. Here the number of predators is high and there is no shelter and so the salmon travel in schools. However, the supply of natural food is very good. The salmon in the sea therefore have a large potential for fast growth and do not need to remain in the sea, where they are exposed to predators, for long before returning to freshwater for spawning.

The schooling behaviour is also a way to reduce predation in the sea [15]. This life cycle is a major advantage for anadromous species in general: they grow up in a nutrient-poor environment but with a reduced number of natural predators, then enter the sea where a nutrient-rich environment enables fast growth, but with a higher predation risk and limited shelter. The fast growth reduces the period that the fish need to be in this predator-rich area. Migration to the sea may also be a step in ensuring further development of the species because the nutrients available ensure good growth, resulting in large fish that have the potential to produce viable eggs and sperm to ensure survival of the species.

A limited number of males do not become smolt but undergo a very early direct maturation. These are termed dwarf males. This is a natural mechanism to ensure survival of the species. If fewer males return from migration, the dwarf males will ensure that several males are available to continue the species. As they are small, they can also go between the full-size males and females during the spawning process, releasing their milt. Owing to this they are also named sneakers. The number of dwarf males varies greatly from river to river [16] and is dependent on the availability of natural food in the rivers, among other factors.

4.5 Morphology

The morphology of salmon varies a lot between different life stages (Figure 4.1). On hatching, they start as yolk sac fry. The yolk sac is a package of nutrients that the fry use for growth, so the newly hatched fry do not need to eat during the first period of life. This is a great advantage that salmonids have over many fresh- or seawater fish species. When the salmon fry start to eat, they have already become quite large and robust with a well-developed digestive system, in comparison to other fresh- and seawater species that are very little developed at the time of hatching. This is a great advantage in aquaculture, as the species that need to start eating at once need to be supplied with live feed. Formulated dry start feed, as used for salmon, is not yet developed

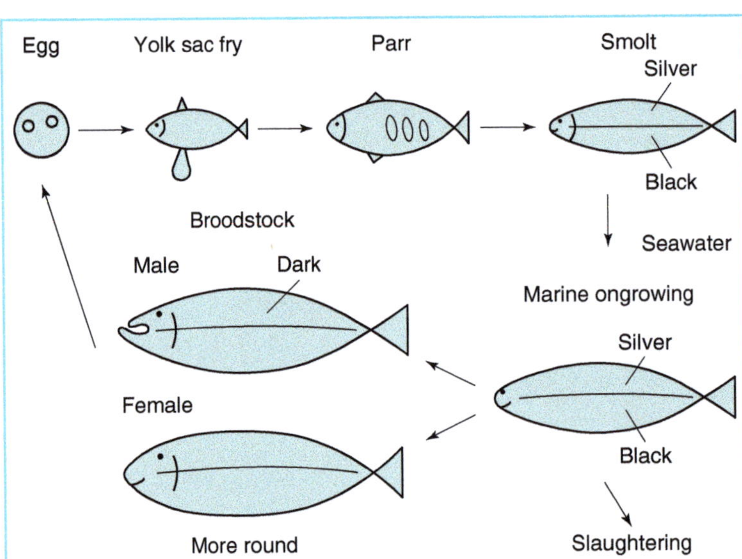

Figure 4.1 Steps in the life cycle of salmon with morphological signs.

for these species. When the yolk sac of the salmon is consumed and feeding starts, the fry are termed start-feeding fry. Alevin is another name used for these fry. After a while, the fry develop what are called 'finger marks' on their sides and are now termed parr. For wild salmon that remain in the river at this stage, the parr marks offers camouflage against predators. The fry now become territorial, remaining in the river in sheltered areas and feeding on food within the passing current. The breast fins are therefore large at this stage so the fry are able to stay still in quite strong currents [16]. When reaching smolt stage, the next stage of development, the size of the breast fins will be relatively reduced. At this stage the finger marks disappear, and the fish get a black back and a silver belly. The fish are now preparing for life in the sea, and the colour change is part of this preparation. In the sea they become pelagic, and change behaviour from that of a territorial individual to living in schools with other salmon, searching for prey. If looking at the school of salmon from below, against the sky, the silver undersides make it difficult for predators to see the fish against the light. Predators looking from above or from the air will have the same problem seeing the fish, because from above the sea will look black, as will the salmon. The size of the fish within the school is quite similar.

4.6 Condition factor

The shape of the salmon is like a 'torpedo', a shape well adapted for energy-effective swimming with low friction in water. Salmon can swim long distances and swim fast. However, the shape of the salmon is dependent on the life stage of the fish. In the freshwater phase, the salmon is shorter and broader than it is in the seawater phase. Considering the behaviour of the wild salmon in the river, it is not moving long distances as is the case during migration when it enters the sea. The shape of the salmon can be expressed with what is called Fulton's condition factor (CF). This is a specific ratio between the length of the fish (l) given in centimetres and the weight of the fish (w) given in grams. The length is defined as fork length (from nose tip to middle of the caudal fin) and the weight is the total body weight ([17]).

$$CF = \frac{w^{*}100}{l^{3}}$$

In general, a high CF value indicates a good shape. A fish that has not obtained enough feed will be skinny, long and thin and have a low CF value, while a fatter fish will have a higher CF [18–21]. If the CF is extremely high, indicating a short length and a high weight as seen in the humpback

Example

A smolt is 110 g and the length is 22 cm. Calculating the condition factor:

$$CF = \frac{w^{*}100}{l^{3}}$$

$$CF = \frac{110g^{*}100}{(22\ cm)^{3}} = 1.03$$

salmon (pink salmon), it may be an indication of vertebral deformation in an Atlantic salmon. During the different life stages the salmon will change shape, and typically has a higher CF when small, which decreases during smoltification and then increases again during the growing phase in the sea: for example, CF 1.1 at smolt stage and 1.6 at slaughter stage. The CF may also vary with season, being higher when the water temperature is higher due to increased growth [22,23].

4.7 Fins

The external anatomy of the salmon comprises a body with eight fins, each of which has a different purpose (Figure 4.2). The dorsal fin, ventral fins, pectoral fins, pelvic fin and anal fin provide stability for the salmon when swimming. The caudal fin is the main propelling fin and moves from side to side, ensuring momentum. The two pectoral fins, one on each side, allow side movement and some locomotion. Salmon also have an adipose fin, the function of which is not well documented; it may have a function related to stability. In the past this fin was cut to recognize individuals in fish trials, and salmon can live 'well' without this fin. Nonetheless, the cutting of the adipose fin can result in problems with wounds and fungus, and is questionable from an animal welfare point of view. All fins have spines and they are linked directly to the muscles, not to the skeleton of the salmon.

4.8 Sight, hearing and taste

The lateral line or sideline organ of the salmon is the fish 'ear', one of the places where the fish hears sound [24,25]. This organ is very sensitive to vibration, movement in the water and sound (all of which are actually vibration waves). There are a number of sensory cells along the lateral line that are all connected. Owing to the fact that the sideline organ detects vibration or pressure waves it can also be visualized as a sonar or echo sounding organ, as it sends out pulses and receives feedback. In this way, the fish can detect its surroundings and avoid hitting things, even those it cannot see. The otoliths or the inner ear of the fish is also an organ connected with hearing. The two otoliths are located in cavities on either side of the head, at the start of the backbone [26–28]. The otoliths also have an important function in balance control. They can be used to age fish as they have year rings, like in trees.

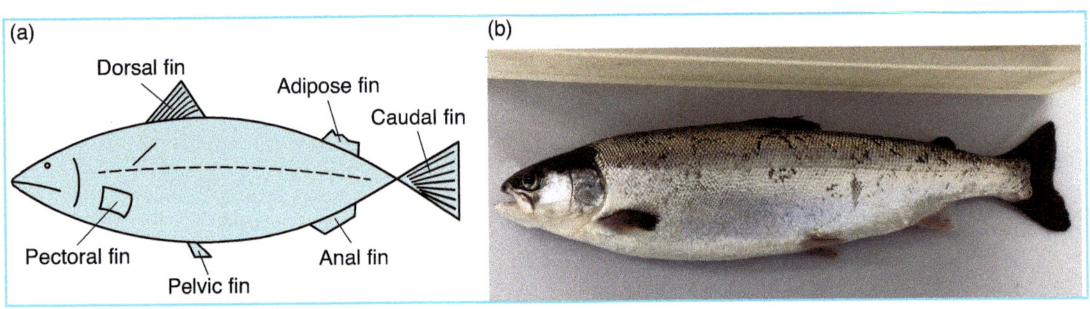

Figure 4.2 Names of the fins on a salmon, illustrated with an ongrowing salmon.

The head of the salmon also contains the mouth, eye, nostrils and gill cover. Salmon have sharp teeth, but they are not used for chewing. The salmon is a meat eater (carnivorous), and in the wild the teeth are used to catch and hold the prey. The nostrils are one of the places where the fish can detect odour. These are located in the nasal cavity of the salmon, and detect odour as the water flows through the cavity. The sense of smell is highly developed and is of major importance for the fish when searching for food. The salmon's vision is best directly forward within a distance of 1 m, but it can also see to the sides and above. The eye can detect colour and is specially developed to see with low light because light gradually reduces further down the water column. The fish can also see above the water surface, but not very clearly. The eye of the salmon lies outside the body, and focus is achieved by moving the lens in and out. However, sight is not the major sense.

The fish can detect taste. Taste cells are located both in the mouth and also in cells outside the mouth, in the skin of the head (nostrils). The salmon is therefore able to taste food before it enters the mouth.

4.9 Skin and mucus

Salmon have skin, including scales that protect the fish, covering the body. The skin can be divided into several layers (Figure 4.3). The outer layer is the epidermis, which contains epithelial cells and mucus cells that are continuously produced. This is also called the slime layer or mucus layer, but actually consists of living cells. The mucus protects the fish from infection, disease agents, parasites, and harmful substances in the water, and therefore the function and thickness of the mucus layer is important. It also reduces the drag resistance when salmon swim through the water. The next layer is the dermis. This contains the pigment cells (colour) and the base for the scales. The scales cover the surface of the fish like roof tiles, and are separated plates with a high calcium carbonate content. They have a degree of flexibility, and can bend. Fish can be aged by looking at the scales, which show year rings, like in trees, just as for the otoliths. The scales may also be used for classification of wild versus farmed salmon. In farmed salmon the distance between the year rings will be larger compared to a wild salmon if the growth has been better, which most probably is the case, but the distance between the individual's rings will certainly be more equal due to the constant supply of adequate food under farming conditions, which results in an equal growth rate all year around. Closest to the muscle is the skin layer named the hypodermis. Overall, the skin and scales have an important function in regulating the ionic balance of

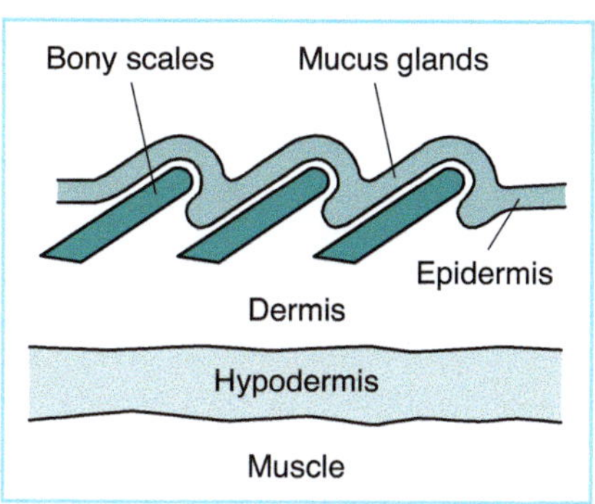

Figure 4.3 The different layers of the skin of the salmon.

the fish and the mucus cells are, as mentioned, important for protecting the fish against disease. Damage to the skin and scales compromises this protective layer, resulting in problems with ionic balance and susceptibility for fungi to attack. Rough handling may therefore lead to the destruction of the mucus layer and will result in an increased probability of disease. In the worst case, the scales may also be lost. When salmon undergo smoltification, the scales become loose. Therefore, gentle handling in this period is critical as rough handling may result in loss of scales and problems with maintenance of ionic balance when the smolt enter the sea.

4.10 Muscle

Below the skin, the fish consists of muscle (fish meat), internal organs and the skeleton. The main component is the muscle (Figure 4.4). In salmon this accounts for approximately 70% of the body. A single muscle cell is named a myofibril and looks like a cylinder with a diameter of 1–2 μm. A number of myofibrils stacked together make up a muscle fibre and a number of muscle fibres form the muscle. A stack of muscle fibres is also called a myomere or myotome. What differentiates the myofibrils/muscle fibre in salmon from those in humans and other mammals is the length. In salmon the myofibrils are short, for example, 20 mm in salmon ready for slaughter, while in humans they can be several hundred millimetres. In salmon, like in other fish, the myomeres or myotomes are separated by a thin layer of connective tissue named myosepta or myocommata. The thickness of this is about 2 mm.

Looking deeper into a separate myofibril, it can be divided into a number of what are called sarcomeres. The sarcomeres lie side by side in the direction of the length of the myofibril. They are composed of the proteins actin and myosin, and a single sarcomere extends between two areas called z-lines. The actin and myosin are responsible for muscle contraction. A simplified explanation of muscle contraction is that a nerve signal is sent to the sarcoplasmic reticulum surrounding the myofibril, and from here calcium ions are released into the muscle cell. This results in the opening of sites on the actin thread where the myosin head can attach and drag on the actin thread, reducing the size. In this way the distance between two z-lines is shortened. The sarcomeres will then contract, as will the myofibril and therefore the entire muscle. For the movement of the myosin head, energy is needed. For this, ATP (adenosine triphosphate, a chemical energy carrier in the cells) is used; when it degrades to APD (adenosine diphosphate) energy is released (30.6 kJ/mole)[29](described further in Chapter 10). A salmon muscle can only contract, not elongate. However, when the fish are swimming they are able to move their bodies from side to side, including the caudal fin. This is possible because the muscle on one side will contract; however, on the other side the muscles do not elongate as this is impossible. It is here where the connective tissue between the muscle fibres, the myocommata, plays an important role. This tissue because of its high collagen content has some flexibility, so it is able to stretch like a rubber band. In this way it is possible for the salmon's body to move from side to side by contracting on one side and stretching on the other. The connective tissue also helps the fish to straighten out when the muscle contraction is released.

The muscles in the salmon's body can be divided into what is called light and dark muscle, or white and red muscle (Figure 4.5). The main body mass of the salmon is light muscle, but there are areas of dark muscle along the lateral line and at the base of the fins. The difference between

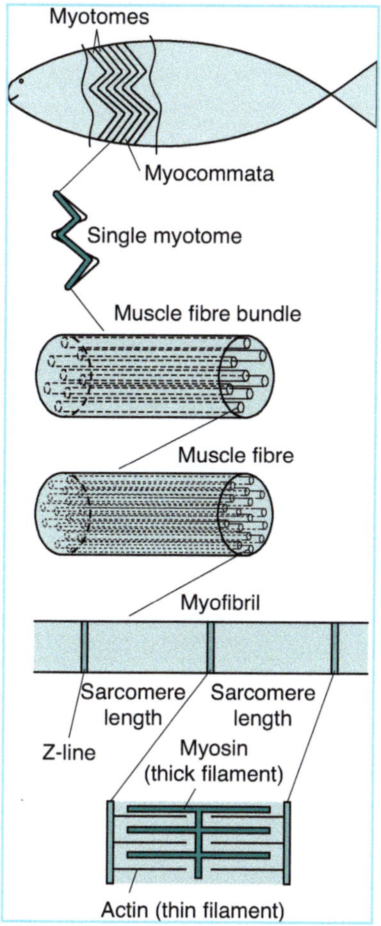

Figure 4.4 The muscle of a salmon.

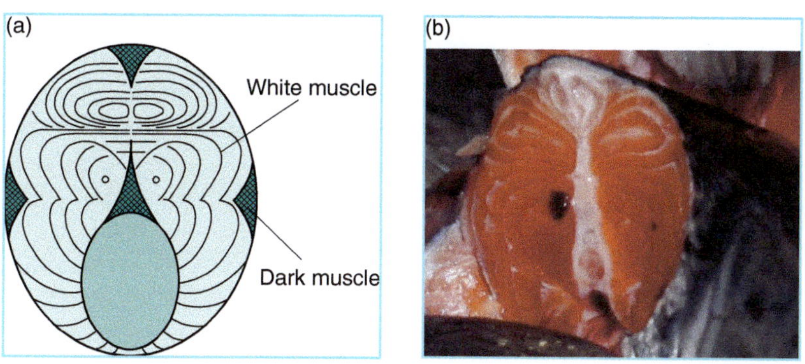

Figure 4.5 Location of white and dark muscle.

the light and dark muscle is in its structure and function. Dark muscle is used for continuous swimming, and is also called slow muscle. There is a high flow through of oxygen, meaning more blood vessels and a high content of haemoglobin (transporting oxygen), therefore it looks darker. The energy used for the regular muscle contraction is taken from the breakdown of lipids and therefore this muscle type has a higher fat content (described further in feeding chapter). White muscle can be termed quick muscle. It can be used for fast, short periods of swimming but cannot be used for continuous swimming over a longer time like the dark muscle. There are fewer blood vessels in light muscle compared to dark muscle, and it uses glycogen/glucose as the energy source for contraction. This form of stored energy is easily released and depleted but glucose is rapidly "transformed" to energy (ATP) (see also Chapter 10).

4.11 Buoyancy

Salmon have a swim bladder located below the spine. This is a sac containing gas that can be filled or drained in order to adjust the buoyancy of the fish. Taking gas into the swim bladder increases buoyancy so the fish will move upwards, and vice versa. This reduces energy consumption when swimming, and gives salmon the ability to stay still in the water. The swim bladder takes in gas from the surface and therefore the salmon need to regularly surface. The swim bladder is filled through the pneumatic duct, a connection linking the swim bladder with the digestive tract. Species that have this function are named physostomous fish. The fish can also release gas through this duct to reduce buoyancy. The first time the swim bladder is in function is when the fry come up to the surface searching for their first feed. Fish species that do not have swim bladders, such as tuna, must swim all the time. The swim bladder also functions as a resonating chamber that can produce and receive sound.

4.12 Blood system

The salmon has a simple circulatory system, compared to mammals that have a double system (Figure 4.6). One circulatory system goes from the heart to the lungs where mammals take up oxygen and the other system goes from the heart to the body. The pressure of the blood in the circulatory system is also higher in mammals than in fish. In fish, the blood is pumped from the heart through the ventral aorta, over the gills, through the dorsal aorta and out to the individual cells and then back again to the heart: a single circuit system. The blood provides the cells with oxygen and takes up metabolic waste products such as carbon dioxide and ammonia from the cells (section 4.12). The heart is located slightly behind the gills. Oxygen is taken from the water and transferred to the blood through the gills, and carbon dioxide and ammonia go in the opposite direction from the blood through the gills' surface and out into the water.

4.12 Gills

The gill cover or operculum covers the gills of the fish. The gills are the respiratory organ, through which oxygen is transferred from the water into the fish and respiratory products from the fish are released in the form of the gases carbon dioxide and ammonia.

The gills consist of four gill arches on each side (Figure 4.7). On each gill arch there are a number of gill filaments, and on each filament there are a number of lamellae. In this way, an

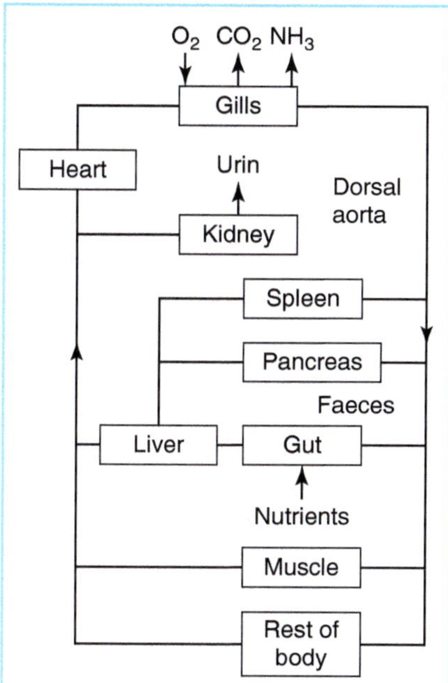

Figure 4.6 Simplified salmon blood system.

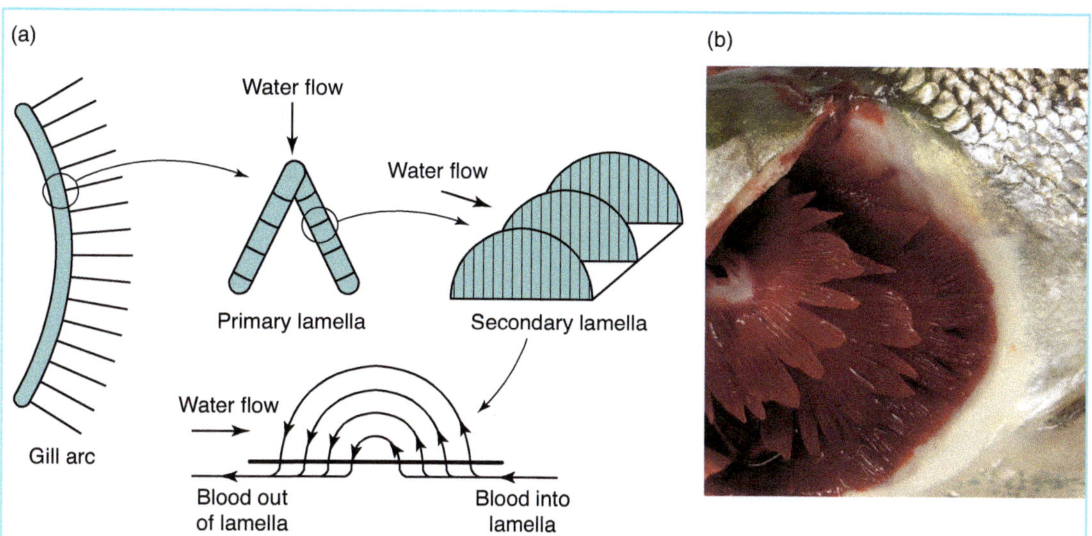

Figure 4.7 Gills of a salmon.

extremely large surface area is created for contact between the water and the gill surface. It is through this surface the gas transfer takes place. When blood enters the gills from the heart it goes out in the gill arcs, and from here to the filaments and then to the individual lamellae. Gas transfer takes place when the blood is travelling through the lamellae. The thickness of the cell layer is very thin here, which allows gas to be transported across by diffusion. The mechanism for the oxygen gas transfer is based on the partial pressure gradients over the gills; the higher oxygen pressure outside in the water results in oxygen diffusing into the blood. The mechanism has a countercurrent flow pattern, meaning that the oxygen-poor blood enters at the rear of the lamella and the blood flows countercurrent to the water flow direction.

The water flows over and past the gills because the fish swim with open mouths and the water is then pressed over the gills and out of the gill cavity. In this way oxygen-rich water is taken in and water rich in metabolic products exits the gills. However, salmon do not need to swim to create the flow of water over the gills. The fish may 'stand still' and create an active flow of water over the gills by the opening and closing of the mouth combined with a size reduction of the gill cavity. The fish opens the mouth and expands the gill cavity, then closes the mouth and reduces the size of the gill cavity. In this way the water is pushed out and over the gills like a piston pump and therefore the fish do not need to swim to achieve gas exchange.

4.13 Controlling ionic balance

Salmon are special because as mentioned above they are an anadromous species, meaning they can live both in freshwater and in seawater. The ionic concentration is different in these habitats. The salmon has a salt content of around 1%. Freshwater has a very low salt content, less than 0.2%, while seawater has an ionic concentration of around 3.4 %. In freshwater, therefore, ions leak out of the fish and into the water due to the ionic balance, and in seawater the opposite occurs (Figure 4.8). There will always be a process to equalize the concentration as no ionic concentration gradient can remain over time unless actively inhibited, because there will always be areas where ions can leak. At the same time, water will try to compensate for the concentration difference and will move in the opposite direction, going from the water and into the fish in freshwater. This is called osmosis. In seawater, however, the opposite process will occur; the fish will be exposed to a high ionic pressure from outside, and water will move out of the fish in order to equalize the concentrations. Special mechanisms in the salmon, described later, are responsible for handling the ionic balance and this also represents a major difference between salmon parr and salmon smolt, or the freshwater and seawater phases of the salmon's life.

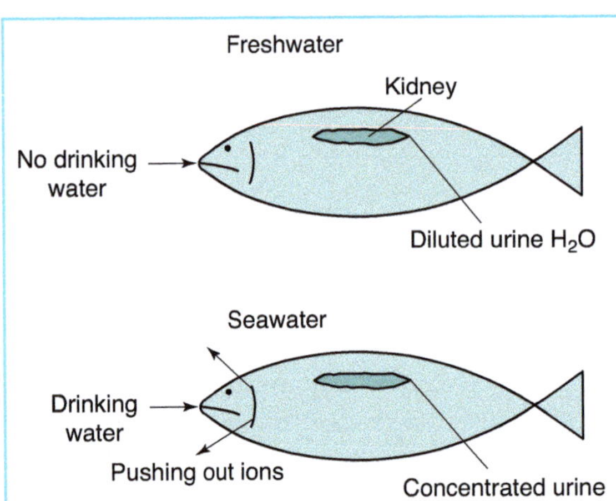

Figure 4.8 Control of ionic balance in freshwater and seawater.

In the freshwater phase, the salmon needs to take ions into the body, because ions are leaking out through all surfaces and at the same time freshwater is being moved into the body by osmosis due to the difference in the ionic concentration. The skin is quite dense, but water enters through other surfaces not covered by skin such as the gills, inside the mouth and the stomach. The salmon is therefore not drinking much water when in freshwater, because water is taken in by these means. The incoming water is removed in the kidney and excreted to avoid too low an ionic concentration. The salmon will try to take up as many ions as possible from the water before it is excreted. The result is that salmon in freshwater urinate a lot and the ionic concentration of the urine will be very low. Even in freshwater, there are some salts or ions in the water which enable the salmon to survive, as its metabolic processes require ions and it is not able to survive in deionized water. The other gateway through which ions (salt) can enter the body of the salmon is through food, and most of the ions that the salmon needs are actually obtained this way.

In seawater, however, ions move into the fish due to the higher external ionic pressure. There will be water loss through all surfaces (apart from skin) because water will move by osmosis to compensate for the higher external ionic pressure. Therefore, the fish needs to drink a lot of water to compensate for the continuous loss of water, so drinking velocity is increased for salmon in seawater compared to freshwater [30]. The result of this is that even more ions enter into the body fluids. Therefore, in seawater the fish needs to actively move the ions out of its body against a positive gradient, which is an energy-requiring process. There are special cells in the fish's gills that are responsible for this active movement of ions out of the fish, which are called chloride cells or mitochondria-rich cells (MR cells), because the number of mitochondria is higher due to the increased energy requirement. Small-sized salmon, fry and parr have not developed or activated these cells and they are not fully functioning until the salmon reaches the smolt stage and has adapted to life in seawater. If salmon are placed into seawater before these cells are functioning, they will die due to dehydration and accumulation of ions. Activation of these cells is therefore an important part of the smoltification process. The urine of salmon when in seawater is rich in ions because they are also excreted in the urine.

In salmonids, the volume of blood is estimated to be in the range of 5–6% of total body volume, and 30–40% of this consists of blood cells [16,31]. The body liquid, where blood is the major component, can be divided into intracellular (within the cells; 40% of the body liquid) and extracellular (around the cells). The ionic concentration of the intracellular and extracellular liquids differs significantly. The concentration of sodium and chloride ions (Na^+ and Cl^-) is high outside the cells while potassium (K^+) has a high concentration within the cells. To maintain this ionic difference, the salmon needs to pump ions against an ionic gradient. The enzyme Na^+/K^+ ATP-ase located in the cell membrane plays an important role in the mechanism (described below). This enzyme, a transport protein, functions as a pump and uses a large part of the energy that the fish produces, whether the fish is in freshwater or seawater.

From this point of view, the most energy-effective environment for the salmon to remain in is brackish water. The natural Na^+ concentration in the blood is 11% and the Cl^- concentration is 8–9 % [16]. The overall osmotic pressure in the body liquid, the number of ions, is quite constant (see osmolality below). It is typically a bit lower in freshwater than in seawater, but the difference is limited. Normal values for the major ions Na^+ and Cl^- for salmonids in freshwater are within the ranges 135–155 mM and 115–135 mM, respectively, while the corresponding values in seawater are within the ranges 145–165 mM and 130–150 mM [16]. Values that are too high or too low will present problems for the salmon, >20–30% concentration [16].

Osmolality is a term used to show the ionic balance of the salmon. It is defined as the number of osmotic active particles (ions) per kilogram of pure water. Seawater has a high osmolality, around 1000 mOsm/kg (milliosmole per kilogram of water), while freshwater has a much lower osmolality, typically up to 5. In salmon, the osmolality in the blood remains in the same range whether it is swimming in seawater or freshwater, and is typically from 300 to 400 mOsm/kg (highest in seawater and lowest in freshwater). A smolt that is not ready to move to seawater will not have the capability to regulate this concentration [32].

The functioning of the Na^+/K^+ – ATP-ase (sodium potassium adenosine triphosphatase) pump, also referred to as the sodium potassium pump, is of major importance in maintaining the ionic balance in the fish (Figure 4.9). The K^+ in the cells is essential for many body functions, and has to be pumped back into the cells again following leakage. This is combined with pumping Na^+ out of the cells (pump 1). The biochemistry behind these pumps or enzymes (proteins) is quite advanced, and only a simplified version is described below (for more information see [33–36]). The enzyme Na^+/K^+-ATPase (NKA) is located in the cell membranes. In the chloride cells, or MR cells, on the gills the concentration of NKA enzymes is very high compared to other cells [37]. These cells are activated during the smoltification process to enable the salmon to live in seawater with its higher ionic concentration. In addition to NKA, two other proteins are important for the maintenance of the salt balance in cells: the Na-K-2Cl cotransporter (NKCC) and the cystic fibrosis transmembrane conductance regulator (CFTR) [35,38–40]. The NKCC actively transports ions (Na^+, K^+ and 2 Cl^-) into the cell (pump 2) (Figure 4.9). This sets up a potential with the higher positive potential outside, due to the sodium, and the higher negative

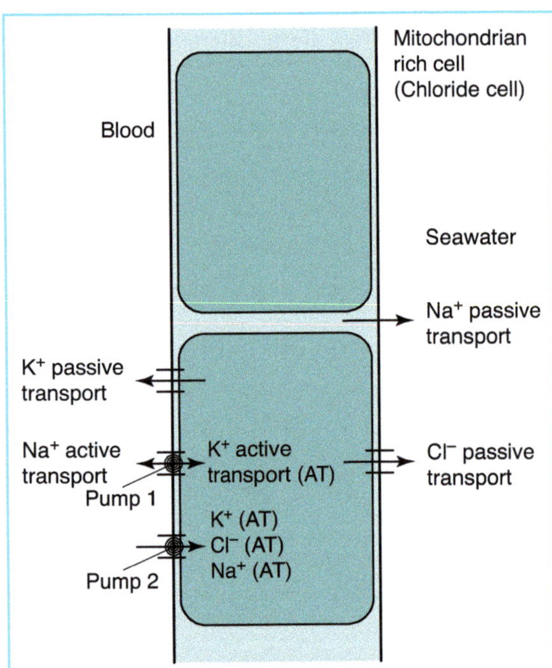

Figure 4.9 Mechanisms that maintain the ionic balance in the mitochondria-rich cells in the gill (based on [34,41]).

potential inside, due to the chloride. The chloride then leaves the cells through the CFTR channel and is released into the surrounding seawater. In this way, chloride ions are removed from the blood. Due to the electric potential that is set up, sodium is released through paracellular transport. This is through the space between cells and is a passive transport mechanism.

When the salmon is drinking seawater to compensate for water loss due to osmosis, it will take in the major ions Na^+ and Cl^- along with other ions which must also be excreted, for example Mg^{2+} and SO_4^{2-} which are excreted in the faeces [42].

4.14 Kidney, liver and spleen

The kidney of the salmon is located just below the spine. The kidney in the fish can be divided in two. The head or first part is responsible for production of red blood cells, which differs from humans where these are produced in the bone marrow. The main purpose of the second part of the salmon kidney is to filter the blood to remove toxic substances, and to ensure the correct ionic balance of the fish. As discussed above, in seawater ions are removed from the blood to ensure acceptable levels for the fish, and the salmon excrete ion-rich urine. In freshwater the kidney removes water rapidly and dilute urine is produced.

The kidneys also have an important role removing toxic ammonia (NH_3) produced from the metabolism of protein in the cells. However, fish excrete most of the nitrogenous waste as ammonia (gas) through the gills rather than as urea (N_2H_4CO) as in mammals. Around 80–90% of the nitrogen is released as NH_3 through the gills. It is only fry that have the typical urea cycle and release ammonia as urea. There is an equilibrium between NH_3 and NH_4^+ (ammonium) in the blood. In the salmon, blood pH is slightly above 7, and most of the ammonia will therefore be in the ionic form, NH_4^+. This means that the released NH_3 is transported as NH_4^+ in the blood. The problem when it is in the form of NH_4^+, a cation, is that it may act as a substitute for K^+ which is of major importance for cell function; therefore, high concentrations of NH_4^+ are undesirable. When NH_3 is converted into NH_4^+ it also requires H^+ so the pH will increase as H^+ is removed from the blood. It also inhibits production of some important enzymes. The major proportion of the total ammonia therefore occurs as NH_4^+ in the blood being transported to the gills, and is not excreted in urine like in mammals. When it reaches the gills, due to the higher gas pressure of NH_3 inside the gills than in the water it diffuses through the gills (a simplified explanation; for more information see, for example [43, 44]).

The liver purifies the blood. All blood coming from the stomach and digestive system passes through the liver. The blood enters with many nutrients and substances, some toxic. In the liver the nutrients are broken down to simpler molecules that are more easily absorbed in the rest of the body. Toxic substances and other unwanted substances, such as the ions in seawater, are sent to the spleen and kidneys. The liver has a number of functions important for the survival of the salmon. These include production of glucose that is stored in the liver as glycogen (an important energy store). Bile is also produced in the liver. This is transported to the gall bladder and further to the intestine where it is used for digesting food. Vitamins and minerals are also stored in the liver.

The spleen is a non-vital organ, the main purpose of which is removal of waste products and damaged blood cells from the blood. If the spleen is not functioning the salmon is more likely to succumb to disease; the spleen is therefore an important part of the immune system.

4.15 Stomach and intestine

The stomach and the intestine are part of the digestive system. Carnivorous fish, such as salmon, have a relatively simple and short digestive system. The food is transported quite fast through the system compared to herbivorous fish, such as grass carp, which have a much longer digestive system and can eat and digest plants. This is, of course, of major importance for the choice of diet for the fish. Salmon cannot digest substances that are difficult to break down, such as plants; if too much plant matter is consumed, problems may occur in the intestines. When food enters the digestive system, it first enters the stomach, where the pH is low. The pyloric caeca is an organ located between the stomach and intestine and consists of a number of small tube-like organs that create a large surface area where enzymes are released for digestion of the food. The major uptake of nutrients takes place in the intestine, a long tube-like organ going from the stomach to the vent. In salmon this is, as mentioned, quite short compared to omnivorous and herbivorous fish species (further described in Chapter 10).

4.16 Brain

Fish have a brain located in the upper part of the head (Figure 4.10). The brain of a salmon can be divided into three parts: forebrain, midbrain and hindbrain. The forebrain controls the sense of smell, the midbrain controls vision, learning and responses to stimuli, and the hindbrain controls movement, muscles and balance. The brain of the fish differs a lot from the human brain. Fish lack a neocortex, which is responsible for higher brain functions. Fish have nociceptors on the head and other areas of the body which function as signal cells. From here, signals are sent to the brain for interpretation and subsequent response. There is, however, some discussion as to whether the brain of the fish is developed in a way that enables it to actually feel pain [45–49]. It is known that signals from negative stimuli reach the brain which result in avoidance behaviour by the fish, but it is still under discussion whether this amounts to the fish experiencing pain.

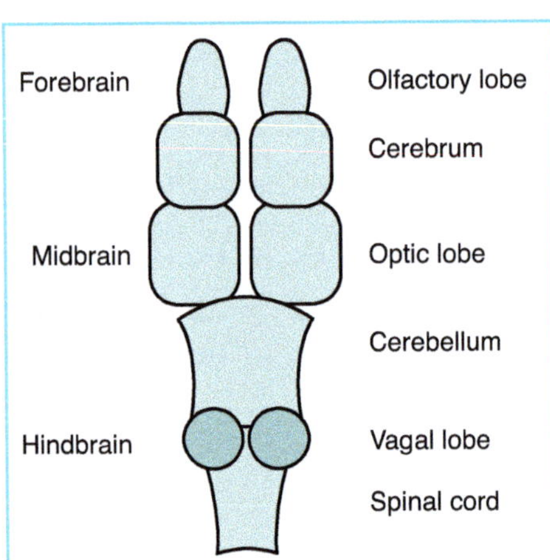

Figure 4.10 Brain of the salmon (partly based on [50, 51]).

4.17 Skeleton

The skeleton starts with a cranium that is linked to the backbone (spine) going from the head to the tail. Inside the backbone is the spinal cord, which is directly connected to the brain in the cranium. From the backbone, the ribs extend to enclose the abdominal cavity. Salmon also have pin bones, which can be considered a type of rib (fake ribs) that are present in the middle of the muscle.

4.18 Gonads

The gonads of the salmon, the reproductive organs responsible for producing gametes, are the eggs or roe for the females, and milt, milk or sperm for the male (see Chapter 5 on broodstock farming for more detail).

4.19 Proportion by weight of different commercially valuable parts

At slaughter and further processing, different parts of the salmon are utilized. The weight of the different commercially valuable parts that can be utilized is therefore of interest [52]. The main part is the two fillets, representing up to 60% of the live weight, but this is highly dependent not only on the fillet type and how much it is trimmed but also on the salmon size and the CF. The head accounts for approximately 10% of the live weight, the viscera around 10.5% and the blood around 2%. After filleting, the frames represent around 10%, the trimmings 2% (dependent of fillet type), the belly flaps 1.5% and the skin 3.5% [52].

4.20 Organ sizing

A number of indexes can be used to show the size of different organs compared to the weight of the salmon. The gonad somatic index (GSI) shows the weight of the gonads in relation to the weight of the fish. This is used to assess the development of the maturation process. It increases through the maturation period and differs greatly between the sexes, being much higher for females due to the production of eggs compared to milt.

$$\text{GSI (\%)} = 100 * (\text{gonad weight/fish weight})$$

The heart index (HI) gives the relative weight of the heart compared to the body weight, while the liver somatic index gives the relative weight of the liver compared to the body weight. Both indexes can be used to see if the size is in the correct ratio or if they are over- or underdeveloped compared to normal, for instance during disease diagnostics.

References

1. Hart, P.J.B. and Reynolds, J.D. *Handbook of Fish Biology and Fisheries: Fish Biology*. 2008. Wiley: Chichester.
2. Bone, Q. and Moore, R. *Biology of Fishes*. 2008. Taylor & Francis: Abingdon.
3. Lacopo, P. and Riemma, M. *Salmon: Biology, Nutrition, and Consumption*. 2010. Nova Science Publishers: Hauppauge, NY.
4. Burton, D. and Burton, M. *Essential Fish Biology: Diversity, Structure, and Function*. 2017. Oxford Universtiy Press: Oxford.
5. Evans, D.H., Claiborne, J.B. and Currie, S. *The Physiology of Fishes, Fourth Edition*. 2013. Taylor & Francis: Abingdon.
6. Woo, P.T.K. and Noakes, D.J. *Salmon: Biology, Ecological Impacts and Economic Importance*. 2014. Nova Science Publishers: Hauppauge, NY.
7. Amin, A.B., Mortensen, L. and Poppe, T.T. *Histology Atlas, Normal Structure of Salmonids: A Colour Atlas, English, German, French and Spanish Legends*. 1992. Akvapatologisk Laboratorium AS: Bodø.
8. CABI. Invasive species compendium. https://www.cabi.org/isc/search/index?q=Salmo%20salar.
9. Froese, R. and Pauly, D. (eds). FishBase – electronic publication. 2019. https://www.fishbase.se/summary/Salmo-salar.html.
10. Noguera, P.A. et al. Fish necropsy manual. 2015. https://www.necropsymanual.net/en/.
11. PSKF. Salmonids in the classroom, salmon dissection. https://www.pskf.ca/sd/.
12. Kriseweb. Water temperature and Gulf of Maine Atlantic salmon. http://www.krisweb.com/krissheepscot/krisdb/html/krisweb/stream/temperature_sheepscot.htm.
13. Fishbio. Atlantic salmon can take the heat. *The Fish Report*. 2014. https://fishbio.com/field-notes/the-fish-report/atlantic-salmon-can-take-heat.
14. Stehfest, K.M. et al., Response of Atlantic salmon *Salmo salar* to temperature and dissolved oxygen extremes established using animal-borne environmental sensors. *Scientific Reports*. 2017. 7(1): 1–10.
15. Handeland, S.O. et al., Osmotic stress, antipredatory behaviour, and mortality of Atlantic salmon (*Salmo salar*) smolts. *Canadian Journal of Fisheries and Aquatic Sciences*. 1996. 53(12): 2673–2680.
16. Heggberget, T.G. et al. Smoltifisering hos laksefisk. *NINA forskningsrapport*. 1992. 31: 1–42.
17. Nash, R.D.M., Valencia, A.H. and Geffen, A.J. The origin of Fulton's condition factor—setting the record straight. *Fisheries*. 2006. 31(5): 236–238.
18. Einen, O. et al. Feed ration prior to slaughter—a potential tool for managing product quality of Atlantic salmon (*Salmo salar*). *Aquaculture*. 1999. 178(1): 149–169.
19. Einen, O., Waagan, B. and Thomassen, M.S. Starvation prior to slaughter in Atlantic salmon (*Salmo salar*): I. Effects on weight loss, body shape, slaughter- and fillet-yield, proximate and fatty acid composition. *Aquaculture*. 1998. 166(1): 85–104.
20. Hamre, K. et al. Antioxidant vitamins, minerals and lipid levels in diets for Atlantic salmon (*Salmo salar*, L.): effects on growth performance and fillet quality. *Aquaculture Nutrition*. 2004. 10(2): 113–123.
21. Herbinger, C.M. and Friars, G.W. Correlation between condition factor and total lipid content in Atlantic salmon, *Salmo salar* L., parr. *Aquaculture Research*. 1991. 22(4): 527–529.
22. Sutton, S.G., Bult, T.P. and Haedrich, R.L. Relationships among fat weight, body weight, water weight, and condition factors in wild Atlantic salmon parr. *Transactions of the American Fisheries Society*. 2000. 129(2): 527–538.
23. Nordgarden, U. et al. Seasonally changing metabolism in Atlantic salmon (*Salmo salar* L.) I–Growth and feed conversion ratio. *Aquaculture Nutrition*. 2003. 9(5): 287–293.

24. Bleckmann, H. and Zelick, R. Lateral line system of fish. *Integrative Zoology*. 2009. 4(1): 13–25.
25. Mogdans, J. Sensory ecology of the fish lateral-line system: morphological and physiological adaptations for the perception of hydrodynamic stimuli. *Journal of Fish Biology*. 2019. 95(1): 53–72.
26. Rodríguez Mendoza, R.P. Otoliths and their applications in fishery science. *Croatian Journal of Fisheries: Ribarstvo*. 2006. 64(3): 89–102.
27. Thomas, O.R.B. and Swearer, S.E. Otolith biochemistry – a review. *Reviews in Fisheries Science & Aquaculture*. 2019. 27(4): 458–489.
28. Morales-Nin, B., Review of the growth regulation processes of otolith daily increment formation. *Fisheries Research*. 2000. 46(1–3): 53–67.
29. Biological Dictionary. Adenosine Triphosphate (ATP). https://biologydictionary.net/atp/.
30. Usher, M.L., Talbot, C. and Eddy, F.B. Drinking in Atlantic salmon smolts transferred to seawater and the relationship between drinking and feeding. *Aquaculture*. 1988. 73(1–4): 237–246.
31. Smith, L.S. Blood volumes of three salmonids. *Journal of the Fisheries Board of Canada*. 1966. 23(9): 1439–1446.
32. McCormick, S.D. et al. Differential regulation of sodium potassium pump isoforms during smolt development and seawater exposure of Atlantic salmon. *Journal of Experimental Biology*. 2013. 216(7): 1142–1151.
33. Zadunaisky, J.A. The chloride cell: the active transport of chloride and the paracellular pathways, in *Fish Physiology*, W.S. Hoar and D.J. Randall (eds). 1984. Academic Press: Cambridge, MA, pp. 129–176.
34. Evans, D. A brief history of fish osmoregulation: the central role of the Mt. Desert Island Biological Laboratory. *Frontiers in Physiology*. 2010. 1(13).
35. McCormick, S.D., Regish, A.M. and Christensen, A.K. Distinct freshwater and seawater isoforms of Na+/K+-ATPase in gill chloride cells of Atlantic salmon. *Journal of Experimental Biology*. 2009. 212(24): 3994–4001.
36. McCormick, S.D. Smolt physiology and endocrinology, in *Fish Physiology*. 2012. Elsevier: Amsterdam. p. 199–251.
37. Karnaky Jr, K.J. Structure and function of the chloride cell of *Fundulus heteroclitus* and other teleosts. *American Zoologist*. 1986. 26(1): 209–224.
38. Marshall, W.S., Lynch, E.M. and Cozzi, R.R.F. Redistribution of immunofluorescence of CFTR anion channel and NKCC cotransporter in chloride cells during adaptation of the killifish *Fundulus heteroclitus* to sea water. *Journal of Experimental Biology*. 2002. 205(9): 1265–1273.
39. Marshall, W.S. Na+, Cl–, Ca2+ and Zn2+ transport by fish gills: retrospective review and prospective synthesis. *Journal of Experimental Zoology*. 2002. 293(3): 264–283.
40. McCormick, S.D. et al. Influence of salinity on the localization of Na+/K+-ATPase, Na+/K+/2Cl- cotransporter (NKCC) and CFTR anion channel in chloride cells of the Hawaiian goby (*Stenogobius hawaiiensis*). *Journal of Experimental Biology*. 2003. 206(24): 4575–4583.
41. Evans, D.H., Piermarini, P.M. and Choe, K.P. The Multifunctional Fish Gill: Dominant Site of Gas Exchange, Osmoregulation, Acid-Base Regulation, and Excretion of Nitrogenous Waste. *Physiological Reviews*. 2005. 85(1): 97–177.
42. Di Giulio, R.T. and Hinton, D.E. *The Toxicology of Fishes*. 2008. CRC Press: Abingdon.
43. Wilkie, M.P. Ammonia excretion and urea handling by fish gills: present understanding and future research challenges. *Journal of Experimental Zoology*. 2002. 293(3): 284–301.
44. Wilkie, M.P. Mechanisms of ammonia excretion across fish gills. *Comparative Biochemistry and Physiology Part A: Physiology*. 1997. 118(1): 39–50.
45. Braithwaite, V. *Do Fish Feel Pain?* 2010. Oxford University Press: Oxford.
46. Rose, J.D. et al. Can fish really feel pain? *Fish and Fisheries*. 2014. 15(1): 97–133.

47. Key, B. Why fish do not feel pain. *Animal Sentience*, 2016. 1(3): 1.
48. Vettese, T., Franks, B. and Jacquet, J. The great fish pain debate. *Issues in Science and Technology*. 2020. 36(4): 49–53.
49. Chatigny, F. The controversy on fish pain: a veterinarian's perspective. *Journal of Applied Animal Welfare Science*. 2019. 22(4): 400–410.
50. Näslund, J. et al. Early enrichment effects on brain development in hatchery-reared Atlantic salmon (*Salmo salar*): no evidence for a critical period. *Canadian Journal of Fisheries and Aquatic Sciences*. 2012. 69(9): 1481–1490.
51. Monks., N. Fish brain. *Wikimedia*. https://commons.wikimedia.org/wiki/File:Fish_brain.png.
52. Stevens, J.R. et al. The rise of aquaculture by-products: increasing food production, value, and sustainability through strategic utilization. *Marine Policy*. 2018. 90: 115–124.

CHAPTER 5

Broodstock and the broodstock farm

5.1 Introduction

Broodstock farms are specialized farms that deliver eyed eggs to smolt farms. Eyed eggs are eggs where the embryo have reached a development step where the eyes of fish are visible as two black spots. Selective breeding in some form or another is always involved in their activity, and they are either owned by breeding companies or are part of integrated companies that handle the entire production line in house, from egg to salmon ready for the consumer market. Breeding is involved to ensure that the broodstock farms are delivering eyed eggs from the best possible breeding material. Owing to this specialization only a few broodstock farms supply the entire salmon industry, and they are international companies.

Traditionally, eyed egg production was seasonal, based on the natural spawning time of salmon, and the eyed eggs were delivered in the winter. Today, however, seasonally independent production is also possible due to manipulation of spawning time to induce spring spawning. This is carried out in land-based farms with full control of light and water temperature to artificially create spawning conditions in the spring.

A successful hatchery operation is based on healthy, good-quality broodstock. It is therefore of major importance that the feed, water quality and other factors related to welfare conditions are optimized for the brood fish [1].

Atlantic salmon approach maturity in spring and early summer, normally from the age of 3 years under farming conditions with a good supply of food, but this varies to some degree and can also be manipulated. When a normal production cycle is followed, the fish approach maturity in the spring and will be ready for spawning in the autumn, typically in October in the northern hemisphere (dependent on water temperature and light conditions). When the maturation process is complete, the salmon are ready for spawning and produce eggs (females) and milt (males). In the wild, mature salmon in the sea will typically stop eating in July to August in the northern hemisphere before entering the rivers in the autumn for spawning.

Synchronization of the maturation of the different sexes is an issue, and too early a maturation of males presents a problem in the production of salmon for the consumer market. Therefore, manipulation of maturation is a topic of great interest.

5.2 The production sites

A salmon broodstock farm normally includes two sites: a sea cage farm site and a land-based farm site where the salmon are kept in tanks (Figure 5.1). The fish are usually kept in traditional cages in seawater during the gradual development of the maturation process. It is important to have a low fish density during this period, lower than in the commercial ongrowing farms. This is to ensure that the fish are not exposed to any stress during this period nor are kept at densities that result in the fish sustaining wounds. It may, however, also be a fully land-based process where the fish are kept in tanks on land through their entire life, including the seawater phase after smoltification, and then returned to freshwater (or brackish with limited salinity) when close to spawning. In a land-based farm, it is possible to control both water temperature and light conditions and, in this way, control the time that maturation starts. When the salmon are kept in sea cages, the traditional method until recent years, it is important to be especially careful when moving the fish from the sea cages to the land-based tanks when close to the spawning point to avoid any wounds. This can result in osmotic problems and also present an opportunity for fungal infection. The fish are normally moved from the cages to the land-based facilities with a freshwater supply 1 month or so before the stripping time. In this period, it is very important that the brood fish are provided with adequate feed of the correct composition. It is especially important to supply the essential vitamins and minerals that ensure optimal development of the gonads (the organs that produce the gametes: testes and ovaries). The individual salmon that make up the broodstock are specially chosen for good performance, and are

Figure 5.1 A broodstock farm where the mature salmon are in land-based tanks.

tested for disease through the selective breeding programme. This is to avoid using fish that are carrying diseases and so reduce the risk of vertical disease transfer (the transfer of disease from parents to offspring).

During the maturation process the brood fish are susceptible to disease due to reduced immune defences and their stress tolerance is also reduced. They are also susceptible to fungal infections at this time and therefore special care must be taken during all handling. Fungal infection will spread rapidly from one fish if other fish have wounds. Treatment can be carried out with, for example, formalin or hydrogen peroxide in dosages of 100 ppm and 500 ml/m^3, respectively [2].

Water quality for the broodstock is important, and as a minimum the same requirements as for ongrowing of salmon for the consumer market is required. As mentioned, the immune defence of the salmon may be reduced at maturation, and poor water quality can be the triggering factor for a disease outbreak. High water temperatures (18°C and above), even for short periods, can have a negative impact on the development of the eggs [3].

5.3 Feeding broodstock

A successful spawning is based on a healthy good-quality broodstock, and for this the composition of the supplied feed is of major importance. Changing the feed from traditional ongrowing salmon feed to special broodstock feed should be done at least 1 year before maturation to ensure good development. Broodstock feed has a higher protein and fat content (typically including more fish meal and fish oil) and a reduced content of vegetable material. The astaxanthin content is increased to ensure there are adequate levels for production of viable eggs. There are also some changes in the addition of vitamins (such as C and E) and minerals (such as selenium) to the feed.

The feed provided to the broodstock will directly affect the quality of the egg and sperm, which are of major importance for success in fertilization, hatching and fry development.

It is important to consider the effect that maturation has on the overall metabolic rate of the salmon. The fish get a somatic growth boost when entering the maturation period. This is due to the deposition of large lipid reserves to ensure enough energy stores for the entire maturation period. Later in maturation, most of the supplied nutrients and energy will be used for gonadal development, until the salmon stop feeding at the end of the maturation period. This can be observed in the reduction of growth and increased FCR in the late stages of maturation. Salmon in the maturation period are not suitable for the consumer market due to a high reduction in meat quality. In addition, the salmon develop distinct secondary sexual characteristics that also reduce the suitability for entering the market.

5.4 The maturation process

5.4.1 Changes in the salmon during maturation

Maturation is, as mentioned, the process that ends with the salmon releasing its gametes, with the female and male releasing eggs and milt, respectively. The maturation process involves a number of great changes for the salmon – morphological, histological and physiological. The first maturation represents the puberty of the fish, and since Atlantic salmon under farming conditions are only used once (one-time spawner), the two terms are equivalent. In nature, some female salmon may survive after the first spawning to spawn again, and for the male this is common, but under farming conditions all are killed following first spawning. The onset of puberty includes the start of oogenesis in the females and spermatogenesis in the males [4–6].

During the maturation period the male and female salmon develop distinct morphological differences (Figure 5.2). It is most easy to see this in the male, which develops a distinctly prominent lower jaw and a more pointed head, and the size of the males is typically a bit larger than the females. The colour also changes from silvery to a darker brownish colour. The female becomes wider, due to the development of the eggs, and the head looks smaller. Females also go through a colour change, getting darker.

The visual morphological signs develop some time into the maturation period. It is, however, possible to identify sex much earlier by looking at the gonad development. The male has a pair of testes that produce sperm while the females have a pair of ovaries in which the eggs are produced. Measuring the gonad development by gonad histology, endoscopy, sex hormone analysis, expression of genes related to maturation and ultrasound are all methods for early identification of sex and may also identify how far the maturation process has progressed [7–9]. A number of these methods, however, involve killing of the fish or significant handling (blood sampling, endoscopy). Ultrasound represents a non-invasive method for determining sex and can also be used for monitoring the progress of development of maturation in both male and female salmon [7,10].

In the later part of the maturation period the salmon need to be transferred to freshwater, or at least to brackish water. Under wild conditions this is when the salmon travel up rivers for spawning. They stop eating, and after this point lose weight. The further development of the gonads is now dependent on energy stores that the salmon established earlier in the maturation

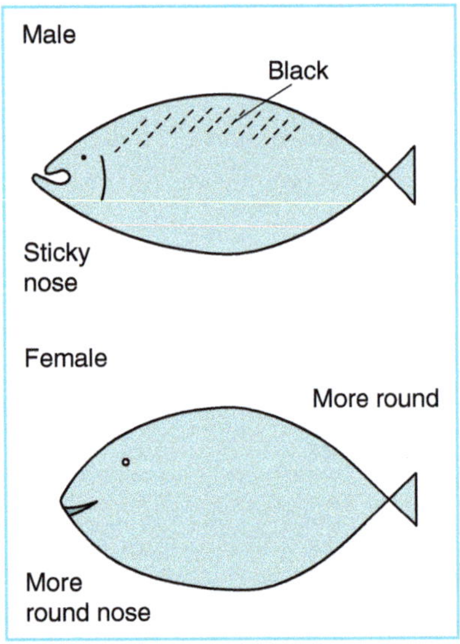

Figure 5.2 Male and female salmon develop distinct morphological signs when maturing.

period. There is a reduction in the activity of the chloride cells in the gills, so the salmon are no longer able to remain in seawater. They will suffer if not moved to freshwater, and welfare will be reduced. Salmon will spawn even if kept in full seawater, but the success of spawning and fertilization will be reduced and from a welfare perspective it is not recommended. This period is stressful for the salmon because of the large changes occurring in the maturation process; the immune response is reduced and the salmon are more susceptible to disease.

The colour of the muscle changes, becoming pinker, and the meat quality drops because the salmon need pigments and lipids for other purposes: pigments for colouration of eggs and external colour, and lipids for energy. If slaughtering mature fish, the meat quality will differ a lot from the traditional salmon delivered to the consumer market. As mentioned, the exterior colour will also change to the spawning colour, so will not be silver. This is undesirable in comparison to the usual silver-coloured salmon that enters the market. Chemical changes occurring in the flesh during the maturation also change the smell and taste of the flesh in a negative way for consumers [11]. Economic losses will therefore occur if mature salmon (with distinct markings) are sent to the consumer market. This is also the reason for the importance of avoiding early maturation in commercial ongrowing salmon production. Flesh from early maturing salmon is downgraded and of significantly lower value. Looking at the composition of the fish the dry matter (DM) content drops (from 34% to 26%), the fat content in the flesh is highly reduced, energy reserves drop (from 12% to 5%) and there is also a reduction in the protein content (from 22% to 19%) (values in parentheses are used as examples from [12]).

The growth rate and feed utilization will also be affected during the maturation period. At the start of the period the growth rate will surge [13], but later the salmon will reduce their feed intake and the feed utilization will drop. The salmon will now direct all energy towards developing eggs and milt.

The behaviour of the salmon also changes during this period as the fish are going from schooling behaviour to individual behaviour. The males become aggressive towards other males, and the females subordinate the males [11].

5.4.2 Timing of maturation

The maturation or puberty of salmon is controlled by the brain–pituitary–gonad (BPG) axis [14–18] (Figure 5.3). Signals to the BPG axis control the activation of maturation or puberty through release of hormones. The signals depend on a number of factors that can be divided into internal factors related to the fish, and external factors related to the environment [19,20]. Important internal factors are fish size, growth rate, fat stores (energy status) and genetics. The major external factors are light conditions/photoperiod, temperature conditions, food, food supply and hormones in the water. The sex of the salmon is also a factor, and some males will typically mature before the females. Therefore, early maturation of males presents a problem in traditional Atlantic salmon sea cage farming for the consumer market.

The combination of the external and internal signals stimulate the brain to produce gonadotropin-releasing hormone (GnRH), which activates the pituitary to produce and release the gonadotropins (GTH), follicle-stimulating hormone (FSH) and luteinizing hormone (LH); these hormones stimulate the production of the sex steroids (androgens, oestrogens and progestogens), which in turn stimulate gamete (eggs and sperm) production in the gonads [15–18,21].

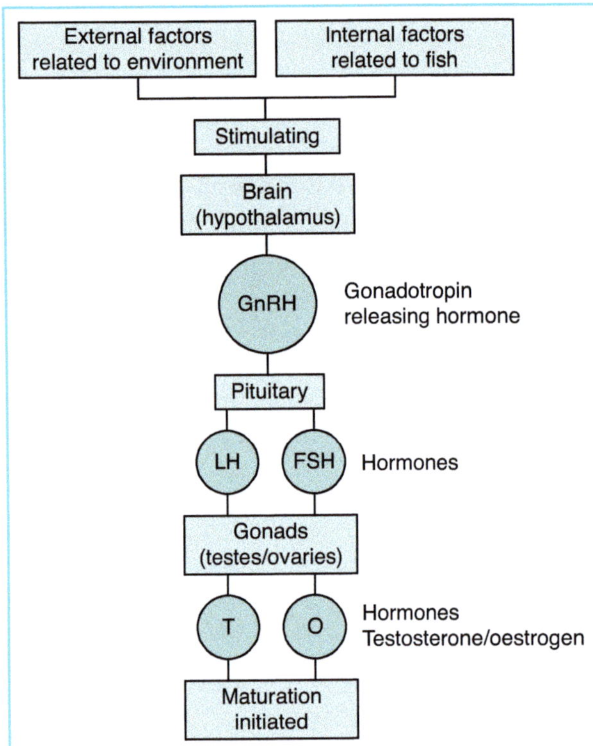

Figure 5.3 The brain–pituitary–gonad (BPG) axis.

Atlantic salmon naturally spawn once a year in the autumn when day length becomes shorter and water temperature drops. The trigger for maturation has already begun in the spring when there is a natural increase in water temperature and day length; this continues throughout the autumn with naturally reducing day length and water temperature. A very limited number of male salmon will mature early as parr in the freshwater stage [22,23]. These are also named sneakers (see Chapter 4). A small number of males also mature as post-smolt, either just before or just after transfer to seawater [23–25]. Both water temperature and fish size seem to be important triggers for post-smolt maturation in males [25]. Fast growth may therefore be a trigger for early male maturation. Further, some males mature during the first autumn in the sea, and some mature in the autumn after one winter in the sea. These males are named grilse. After two winters in the sea even more salmon will naturally mature in the autumn; the number is highly dependent on the time of year that transfer of the smolt to seawater occurs, the seawater temperature and the growth of the fish, but the number can be significant. In the autumn after three winters in the sea many males and females will mature, and all salmon intended for the consumer market must be slaughtered prior to this.

The production of the gametes (eggs and sperm) is named gametogenesis. The maturation process can be divided into three distinct phases: oogenesis/spermatogenesis; vitellogenesis (females); and ovulation and gamete release [6]. Oogenesis is the formation of the egg or ovum and starts with the oocyte being formed in the female. Spermatogenesis represents the same process in males, where the sperm are formed. Vitellogenesis involves the process of deposition of

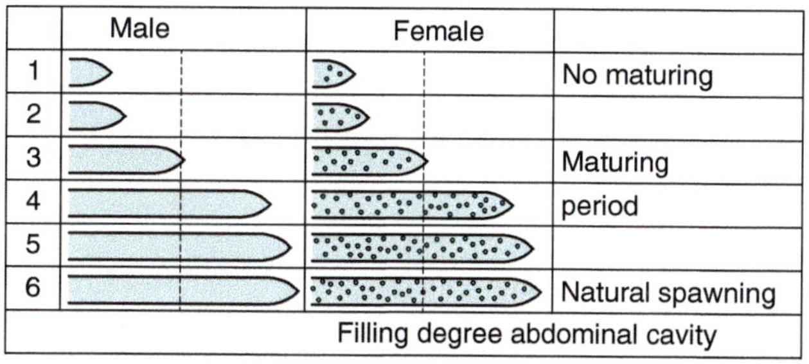

Figure 5.4 Stages in maturation of salmon (based on [28,29]).

nutrients in the egg. The yolk protein precursor, vitellogenin, is produced in the liver of the fish and transported to the egg in order to produce yolk proteins. This is a process where more and more nutrients are deposited, which become the yolk sac of the fry when the eggs hatch. Owing to this deposition the need for protein in the feed for broodfish is higher than normal. In the final part of maturation, ovulation occurs in the females. The tissue in the ovarian follicles surrounding the eggs will degrade and the eggs will be released into the abdominal cavity. This process is controlled by the hormones LH and FSH.

The maturation period can be further separated into different stages based on shape, colour and other developmental differences of the gonads [26,27]. One way to separate the stages is based on the proportion of the gonads in relation to the abdominal cavity, and here six stages can be identified [28] (Figure 5.4).

Another way to separate into stages includes five stages, starting with an immature phase, followed by preparation, maturing and spawning phases, and finishing with a depletion or spent phase [30].

The gonad development may also be classified according to the gonadosomatic index (GSI). This is the weight of the gonads in relation to the weight of the fish (excluding gut).

$$GSI = \frac{Weight\ of\ gonads}{Weight\ of\ fish\ (excluding\ gut)} * 100$$

Females typically spawn at a GSI above 20. The amount of milt is, of course, lower but typically a GSI for males is around 3–6%, while the value for mature male parr is around 9% [31].

5.4.3 Manipulating and controlling maturation

By controlling the time that maturation commences it will be possible to control the stripping point, when the eggs and milt can be released. This makes it possible to manipulate the time for delivery of eyed eggs from the broodstock farm and is also what makes seasonally independent egg production possible. Control of maturation is also a solution to the problem of early natural maturation. The reason to control the maturation process therefore differs between broodstock farms and ongrowing farms producing salmon for the consumer market, but the biological basis is the same. However, when producing for the consumer market, maturation is of no interest and therefore it may be beneficial not just to delay maturation, but to totally prevent it. This results in the production of sterile or non-reproductive salmon. This will also be of interest in preventing the negative consequences of escaped salmon. For the broodstock farm, maturation is required and the salmon need to be of a certain size. They should be above 5 kg, but much over 10 kg will result in heavy fish to handle. For progress in selective breeding a short generation interval is a positive outcome.

Manipulation or synchronization of maturation can be done by stimulating the BPG axis. This can be achieved either by manipulation of the external or internal factors affecting the release of hormones controlling maturation or by directly administering artificially made hormones to the salmon. Manipulation of light conditions (photomanipulation) and manipulation of water temperature are examples of two external factors that may stimulate the release of hormones in the BPG axis, and by controlling these factors the initiation of maturation can be controlled. This can be done in order to enable spawning to occur from earlier in the autumn to later in the autumn or the beginning of winter: in the northern hemisphere from late August to January. In addition, it is possible to regulate the water temperature of the incubated eggs, thereby controlling the delivery date of the eyed eggs. The eyed egg stage is the natural point in time for sending the eggs out from the broodstock farm. By lowering the water temperature of the eggs, the development will be slowed down and the delivery date of eyed eggs can be delayed. In this way it can be possible to deliver eyed eggs for a longer period, such as between November and May in the northern hemisphere. If there is a wish to produce them outside this period, it is possible to fully reverse the production cycle on the broodstock farm. This means that the salmon must spawn in the spring rather than the autumn. To be able to achieve this the entire production has to be controlled, so requires a land-based farm with the ability to fully manipulate water temperature and light. By having both production regimes on a land-based broodstock farm, it is possible to supply the market with eyed eggs all year round.

Photoperiod manipulation (alteration of day/night cycles and length) is an important step in the control of the maturation point. By exposing the fish to additional light during the nights in periods of low light intensity, the degree of early maturation has been reduced in sea cages [32–35]. One photoperiod manipulation regime that has been used with success is to have long days in the spring followed by short days from mid-summer. Prolonging the days can be carried out from mid-summer, but this may reduce egg quality.

Water temperature is also a factor affecting maturation. If the salmon have no other signs of maturation early in the summer, high water temperatures in the sea in the late summer may result in abrupt maturation of the males in the autumn [36].

To be able to conduct the maturation process, the energy reserves in the body need to be above a certain level. If this is not the case, maturation will not take place even if all other conditions are met. Restriction of food intake of the salmon in the winter months will therefore help to inhibit early maturation [37,38]. Even if this strategy is possible, it is not recommended because it will result in an overall reduction in growth and raises questions regarding fish welfare. Experiments have also been conducted to inhibit early maturation by adding feed ingredients that promote metabolism of the salmon's energy reserves (fat), thereby reducing these reserves, and this has also given positive results [39]. Increased phosphorus levels in feed is another factor that may affect early maturation [40], as is vaccination of the fish, due to a temporary loss in appetite following vaccine administration [40].

Selective breeding has been, and will continue to be, an important tool in the control of maturation. It is known that there are variations between strains regarding the age of maturation. Late or delayed maturation is included as one of the goals in an overall breeding programme selecting for different traits [41]. Selecting against early maturation has shown a genetic gain, and the problem has been reduced during the last decades due to breeding.

5.4.4 Hormone-induced spawning

Hormones may also be used to directly manipulate and synchronize ovulation and spermiation [17,42–44]. This may also be termed induced reproduction [45]. Administration of gonadotropin-releasing hormone has been used to induce maturation, sometimes with inclusion of an anti-dopamine [46,47]. It is, however, also possible to use hormones further along the brain–pituitary–gonad axis, such as luteinizing hormone-releasing hormone (LHRHa) [48]. When using hormones, they can be directly administered to the individual fish by injection directly in the muscle or in the body cavity once or several times; pellet implant therapy is also a possibility.

Hormones specially designed for this purpose are available on the market, for example, Ovaplant, Ovaprime and LHRHa (for overview see [49]). The first represents a synthetic analogue of salmon gonadotropin-releasing hormone (sGnRHa), the second a mix of sGnRHa and domperidone, a dopamine antagonist, while the third is a synthetic version of luteinizing hormone-releasing hormone (see also 5.4.3, sterile fish).

5.5 The stripping process

When maturation develops the salmon are transferred to tanks on land with a supply of fresh or brackish water. Here males and females are kept in separate tanks. In this period, the ripeness of the fish should be regularly monitored. The females will be the most challenging because females must be stripped at the correct ripening point. It is easier with the males because they release milt over a much longer time period and therefore it is easier to find a suitable stripping point. In addition, the male may be stripped several times, not killed in the stripping process like females. Since the males can be stripped over a longer period of time it is not so important to check the degree of ripening, but in general the milt should flow freely and be released when the belly is touched. When close to the stripping point the females must be checked regularly, at least once a week. The ripening degree of the females can be checked by feeling the belly of the salmon, which should be soft and loose when touched since the females have released the eggs from the ovaries (ovulated) and the eggs are lying free in the belly cavity. By squeezing slightly on the belly, the eggs will be released as single eggs, not lumped together. If the eggs are still lumped together, it is too early. The correct point will occur some days after ovulation, and acceptable fertilization is within a period from 4 to 10 days after ovulation for salmonids in general [2,50,51]. If the ripening process has gone too far (overripe), the quality of the eggs drops, and they cannot be used. However, to discern the correct ripening point by palpation of the female is highly dependent on the operator's skill and experience.

To ensure good fertilization it is important to synchronize spawning from both sexes. Typically, males will be ready before the females, but are also fertile over a longer period of several weeks. It is also possible to store milt and eggs for a period of time in cold storage before fertilization, preferably the milt, even though this is not a recommended procedure. Milt may, however, also be stored for years by cryopreservation, and will retain fertility although there will be some degradation in motility. When cryopreserving, the milt are stored surrounded by liquid nitrogen at –196°C. Hormone injection prior to spawning may be used to synchronize spawning between sexes (described earlier). By having both sexes in the same water, a more synchronized spawning may also be achieved due to the hormones the fish release in the water.

The broodfish are stripped individually: eggs or roe from the females, and milt, milk or semen from the males. Before stripping, the broodfish are anaesthetized and culled (bled out). During the stripping process gentle handling is important to keep the eggs and milt in as good a shape as possible. Evaluation of the health status of the individual culled fish is performed to identify any infections. This is especially important due to the possibility of vertical disease transfer. Following anaesthesia, the fish need to be cleaned of blood and liquids and dried. It is especially important that the male is dried around the gut to avoid water mixing with the milt, because this can result in the sperm being activated too early. Often it is preferred to remove the ovarian liquid around the eggs to have better control. However, ovarian liquid stimulates increased activity of the sperm in the milt and can therefore be an advantage.

From the male, the milt is pressed out by hand; a typical volume is around 20 ml milt/kg body weight [52] (Figure 5.5). The milt should be checked by microscopy, to ensure both the virility of the sperm and the density of the sperm. Virility can be checked by adding water to the sperm to lower the potassium concentration. The sperm will then be activated and start swimming activity, and by looking at the activity, numbers and swimming speed, the virility can be evaluated. In practice this can be done by taking out a small dose of milt and placing it under a microscope together with some water. The activity of the sperm when stimulated by the freshwater lasts for 1–2 minutes. In nature, when sperm is released into the water they become very active for a short period of time before activity drops and they die. Because of this, water must not be added to the sperm prior to fertilization unless using wet fertilization. In commercial

Figure 5.5 Stripping of milt from male fish and eggs from female fish and addition of water to close the micropyle.

eyed egg production, dry fertilization is the only method used, where eggs and milt are removed individually and mixed under controlled conditions.

The eggs can by removed by gutting the fish manually and taking out the eggs. The eggs may also be pressed out of the abdominal cavity by gently stroking on the belly by hand several times. Pressurized air may also be used (Figure 5.5). By adding pressurized air with a needle into the abdominal cavity the eggs will be pushed out by the air pressure. Milt from several males can be used for each female to reduce the risk of poor or improper fertilization due to reduced milt quality.

The eggs of salmon are covered with a relatively hard membrane with some adhesivity, and are surrounded by ovarian fluid. The volume of the ovarian fluid typically represents 10–30% of the volume of the eggs for salmonids [53,54]. The viscosity of the fluid is high so that it encapsulates the spawned egg as a sticky layer [55]. The ovarian fluid is important for sperm motility, due to its chemical composition and reduction of the Ca^{+} concentration [55,56]. Ovarian fluid seems to be at its greatest concentration high on the surface of the eggs, especially close to the micropyle (the point at which the sperm enters the egg) [57,58].

5.6 Fecundity

Fecundity is defined as the number of offspring that an individual can produce. For Atlantic salmon this is represented by the number of eggs the female can produce. Since Atlantic salmon only spawn once under farming conditions this value equals the number of eggs from one spawning. The term 'total fertility rate' may also be used to describe this. As for other fish species, the number of offspring produced is large. The number varies with the size of the fish at maturation, but typical values are between 1200 and 2000 eggs per kilogram bodyweight. Compared to large brood fish, smaller brood fish produce more eggs per kilogram bodyweight, but the eggs are smaller [31,49]. The large number of offspring is a great advantage for performing selective breeding and for the possible breeding progress, but of course also depends on the individual variations in traits between the offspring.

The size of the eggs from Atlantic salmon is typically within the range 5–8 mm in diameter; they have an individual weight of around 0.1 gram and the ATP content has been measured at 0.0281 nmol/egg [59–61]. The egg size represents a great advantage for the salmon compared to, for instance, marine and freshwater species that have much smaller eggs and small, less developed fry. A volume of 1 litre of salmon eggs contains around 5000 eggs, but again this depends on the egg size. Typically, the egg size is dependent on the size of the salmon, increasing with size of the fish, even though some studies report no significant difference [59,62]. The reason for this might be due to the condition of the fish, as much of the available data is from wild salmon stocks. A larger egg size will also produce larger fry when hatching [31,59]. The egg size is also dependent on the strain of the fish [31]. During the incubation period of the egg, the size of the egg shrinks slightly because the embryo utilizes the nutrients in the yolk for development. Oxygen gas is transferred from the water to the egg during the incubation period.

The following equations were developed early on for calculation of the overall number of eggs and total volume of eggs spawned depending on salmon size (Table 5.1) [63].

Table 5.1 Estimation of volume of eggs after hardening and number of eggs per kilogram brood fish produced from female brood fish of different size (based on values from [63]).

	Fish weight (kg)									
	4.0–4.4	4.5–4.9	5.0–5.4	5.5–5.9	6.0–6.4	6.5–6.9	7.0–7.4	7.5–7.9	8.0–8.4	8.5–8.9
Volume of eggs per kg fish	0.288	0.282	0.279	0.275	0.274	0.271	0.269	0.268	0.266	0.264
Number of eggs per kg fish	1810	1734	1683	1632	1597	1560	1535	1506	1488	1466

Example

The weight of the salmon is 7 kg. What number and volume of eggs does it produce?

Number of eggs = 1150 * w + 2700
Volume of eggs (litre) = 0.24 * w + 0.3
w = Fish weight (kg)
Number of eggs: 1150 * 7 + 2700 = 10750
Volume of eggs (litre): 0.24 * 7 + 0.3 = 1,98 l

5.7 The fertilization process

Fertilization is defined as the process of merging of the two gametes whereby the somatic chromosome number is restored, and the development of a new individual is initiated. In order for this process to occur, the release of milt and eggs must be synchronized. In nature this occurs simultaneously in the rivers. For farmed salmon this is done by individual stripping of males and females and collection of gametes before the gametes are mixed.

The fertilization process can either be performed as wet fertilization or as dry fertilization. The difference between the two methods is whether water is used or not at the moment of mixing the gametes. Broodstock farms today only use dry fertilization, while the natural process occurring in rivers is wet fertilization.

There are several detailed approaches that can be used for the dry fertilization process of salmon and often the broodstock farms have individual procedures based on their experience. Here, one process is described (partly based on [64]). The eggs and milt are collected individually. Samples should be taken from the ovarian fluid for disease screening. Because the result of this screening is not immediately available, it may be necessary to destroy the fertilized eggs afterwards if disease is identified. The ovarian fluid can be removed before fertilization by collecting the eggs in a colander where the fluid drains through. The eggs are then rinsed twice in pure physiological seawater/isotonic water (90 g salt dissolved in 10 l of pure freshwater), which represents the same ionic concentration as the eggs. The purpose of this is to remove/dilute the potassium (K^+) released from any eggs crushed during stripping. Potassium, which also accompanies the milt when stripping, inhibits the sperm's motility [65–68]. Both water and ovarian fluid will have a positive effect regarding removal/dilution of the potassium concentration. Ovarian fluid may have additional positive effects [54]. Concentrations of K^+ above 6–80 mM have been shown to inhibit motility [69]. The addition of 1 mM Ca^{2+} solution has been shown to slightly prolong sperm motility, but may also inhibit motility in high concentrations, which can also be the case for magnesium [69,70]. Some farms, therefore, prefer to keep the ovarian fluid to ensure virility and afterwards use water to give the sperm a second boost [2]. The ionic strength of the isotonic water solution used to rinse the eggs must not be too weak because it can result in blocking of the micropyle and premature swelling of the eggs (described below).

It is possible to dilute the milt with special diluents containing various salts and buffers, called milt extenders. This can be required if the milt quality is suboptimal or if low amounts are available. The milt can be refrigerated for a short time (a few days) or cryopreserved for long periods of storage (years), both still giving good fertilization rates [69,71,72]. Unfertilized eggs may also be stored but only for a short period.

To achieve fertilization, the eggs and milt are mixed together thoroughly, and a new rinse with water may also be performed to ensure sperm motility. Recommended milt to egg ratio is around 2 ml milt per litre of eggs. One reported example is a milt to egg ratio of 1 ml milt to 1500–2000 eggs, or about 10 million sperm per egg [73]. The amount of sperm in 1 ml milt is reported to be in the range $1.9 * 10^{10} – 3 * 10^{10}$/ml [74].

The eggs are then rinsed again in isotonic water to clear any leftover milt remaining on the eggs, before they are disinfected, typically in an iodine solution (for example 90 g salt, 100 ml buffodine mixed in 10 l water for 10 minutes). Then the eggs can be transferred to the incubators and freshwater can be added. By disinfecting the eggs before transfer to the incubation unit, any microorganisms on the surface of the eggs will be reduced and, in this way, prevented from being transferred into the incubation units. Freshwater may also be added before disinfection but then the micropyle closes and the hardening process starts, and the eggs must be left for a period of time before being moved to the incubator.

The fertilization process now takes place inside the egg. The sperm enters the egg by passing through the micropyle. Each egg has one micropyle, which is a very small opening in the protective membrane surrounding the egg. When the sperm goes through the micropyle only the nucleus will enter; the tail, which may be considered as a propeller moving the sperm forward, will not pass. An egg where the sperm has entered is called a green egg or newly fertilized egg. When the sperm enters the egg and fertilization occurs, the cell starts to divide (mitosis), and this is the start of embryo development, or embryogenesis (Figure 5.6). Water is then added to the newly fertilized eggs. This will close the micropyle and no more sperm will be able to enter the egg. It is of major importance that the eggs are not exposed to freshwater before the sperm have entered the egg as the micropyle will close; the time window for this is about 1 minute [75]. This happens in nature (wet fertilization) but here the female releases the eggs in the water at the same time as the male releases the sperm; the sperm can therefore enter the eggs immediately, before the micropyle is closed. This is also the reason for the large number of sperm in the milt, which increases the chances of one sperm finding the micropyle within the limited time window before it is closed. Sperm motility and velocity is of major importance for

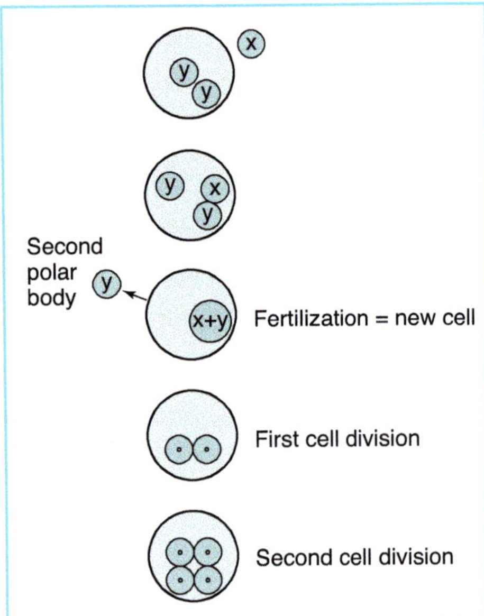

Figure 5.6 Fertilization process and start of the embryo development.

the chance of the sperm reaching the micropyle within this limited time window and for further fertilization success.

When water is added to the fertilized eggs they will start to swell as water enters an inner membrane in the egg. The outer layer of the egg becomes harder, and this is also important for the micropyle to close (Figure 5.7). It is of great importance to use good-quality water for this purpose, preferably disinfected. This process may also be called water hardening. Typically, the volume of the hardened eggs increases by around 25–30%. The water that is taken up by the

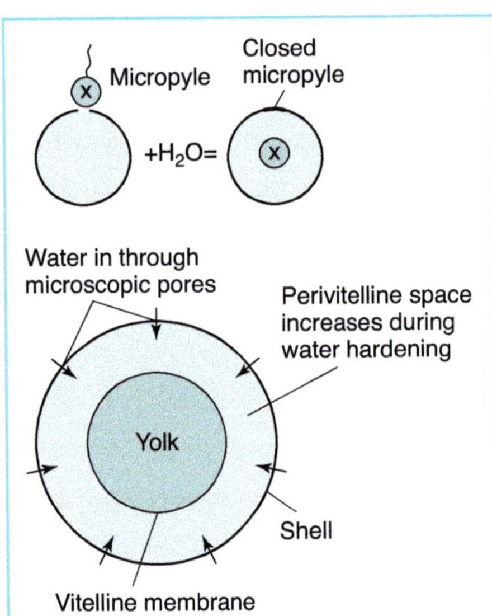

Figure 5.7 Closing of micropyle and swelling/hardening of eggs.

eggs during the swelling period surrounds the developing embryo and works like a protective cushioning layer against physical shock.

Within the swelling or hardening period, which takes 1–3 hours, the eggs should not be moved. They are very sensitive to vibration, which can cause the vitelline membrane (inner egg membrane) to burst, and water will then enter the yolk. This destroys the egg: the yolk within the egg will coagulate and become white (protein denaturation). The fertilized eggs may be transported prior to or after the swelling period but not during this period. If moving or transporting prior to swelling, the milt and unfertilized eggs must be kept separately and without access to water. If kept at a low temperature, there is a time window of several days. If moved after the swelling period, there is a limited time window of up to 48 hours where the eggs can be handled again. After this, they must not be handled, so the eggs must be transferred to the incubation units used in the hatchery within this time window. As described, surface disinfection of the fertilized eggs is performed to remove bacteria and other microorganisms that may be on the surface of the eggs. It is important to rinse the eggs after disinfection to dilute/remove the disinfection liquid from the surface of the eggs.

The fertilization degree is a value that indicates the proportion of the eggs that are fertilized; this value should be above 90%, meaning almost all eggs are fertilized. This can be checked by taking a sample of eggs from the incubation unit after a certain period and examining them under a microscope. For instance, by taking out eggs at approximately 100-hour degrees (temperature x hours) the eggs should have reached the two-cell stage and the germinal disc should be visible; eggs should be prepared in a solution of 7 g NaCl, 50 ml water-free (anhydrous) acetic acid and 1 l water [76].

5.8 Incubation of fertilized eggs

The fertilized eggs are, as already described, transferred to the incubation unit either pre- or post-hardening. The incubators are placed in a separate room with light control. The water in the incubation units is normally continuously replaced by letting new water flow through the units, typically in an upwelling flow (Figure 5.8). This to ensure that the eggs continuously have access to enough oxygen. It is recommended that the oxygen concentration in the outlet of the incubation unit is above 95%.

The fertilized eggs are left in darkness, because certain wavelengths of light can damage the eggs. This includes the lower part of the visual light spectrum, blue-violet light. In nature the eggs lie at the river bottom covered with gravel to protect from light, particularly ultraviolet light. The eggs should not be exposed to any movement to ensure that the developmental steps of the embryo can continue undisturbed. In the period from just after fertilization to the time of closing of the blastopore (later described) the embryo is especially sensitive to vibration. When the blastopore is closed it acts to provide a protective layer. This occurs slightly before reaching the eyed egg stage. At the start of the incubation period the eggs are also very sensitive to variation in environmental factors. They should therefore be kept at a stable, good water quality and at a water temperature that is as stable as possible. The water temperature during incubation should not be raised above 8°C, because this may result in malformations of the embryo [77,78]. The time taken for embryonic development, for instance the time to reach eyed egg stage or hatching

Figure 5.8 Incubation of eggs on broodstock farms is done in hatching cylinders.

stage, can to some degree be controlled by the water temperature. This can be done by lowering the water temperature below 8°C, for instance to 4°C. All changes in water temperature must be done gradually to avoid stressing the embryo.

After fertilization the egg develops through a series of well-defined stages. There is some difference in how the stages are described, but in total there are up to 30 stages from fertilization to hatching. This is the period where the embryo is formed and develops, or embryogenesis [79–81]. This can be summed up in three major stages: cleavage; gastrulation; and organogenesis (Figure 5.9) (for example [80]). Before fertilization, the sperm enters through the micropyle and then fertilization takes place and a diploid zygote, a fertilized egg, is formed. The egg is covered with an outer membrane, the chorion, that inhibits entry of substances into the egg, but is water permeable. It is within the chorion that the micropyle, a funnel shaped opening where the sperm enters, is located. An inner membrane, the vitelline membrane, surrounds the yolk. Between the membranes there is a water layer. This protects the egg from microbes and mechanical damage. Initially the yolk takes up a major part of the egg, but this is reduced as the embryo develops. The cleavage phase is where the first cell development starts, within an area named the animal pole located above the yolk. This is a yolk-free cytoplasm, which looks like a dome. The yolk, in this context, may be named the vegetable pole. The cell division starts by dividing in two, then again in two in successive divisions, so 2, 4, 16, 32, … The first steps are in a horizontal plane but thereafter also include the vertical plane. The first cell division occurs within the first 8–12 hours (temperature dependent). This process is only partial cell division, or meroblastic, and the cells form a disc with linked cells, the blastodisc, also named the germinal disc or blastoderm. The blastodisc therefore represents the part of the egg where the embryo starts to form. The next phase, gastrulation, includes the formation of the embryo including tissue formation. The tissue develops and the blastodisc expands, and when it surrounds the entire yolk, the blastopore closes. This stage is also known as epiboly. A new membrane will now surround the yolk. The last stage, organogenesis, includes formation of the organs in the embryo. Vascularization is also included here and is the process where the blood vessels are formed and cover the yolk. The different stages

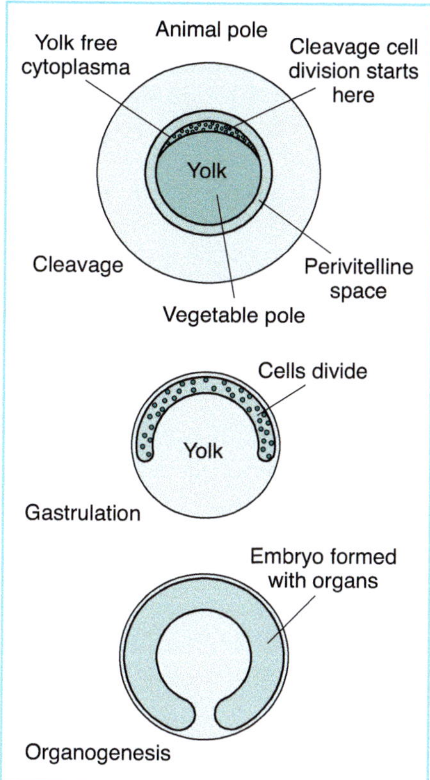

Figure 5.9 Development of the embryo within the egg.

can be observed by looking at the egg under a microscope. The first step that can be observed by the naked eye is the when the eyes are developed (phase 24). This stage is when the embryo begins to develop the pigmented retina in the eye, which becomes visible. The eggs have now reached a stage where they are more robust and can be transported. After the eyeing, further development occurs and the fish gradually gets more and more visible inside the egg. The heart starts to move and later the embryo also starts to move inside the egg. Fish eggs are soft, and in nature they are good food for predators, but by burying the eggs in the gravel, as is done in the wild, it is difficult for predators to find them.

5.9 Eyed egg production

The major purpose of the broodstock farm is, as mentioned above, production of eyed eggs. This means that when the eggs reach the eyed egg stage it is time for delivery to the hatcheries/smolt farms. The time required from fertilization to the eyed egg stage is dependent on the water temperature supplied to the incubated eggs. The sum or accumulation of the daily thermal units (TU) is a model used to estimate when certain stages occur. One TU is defined as 1°C over a period of 24 hours. If, for instance, the water temperature is 5°C for 1 day (24 hours) this will represent 5 TU. The sum or accumulation of the daily TU is described as accumulated temperature units (ATU), but the terms cumulative degree days, degree days or day degrees (DD, °D, D°) may also be used. As an example, if the water temperature is stable at 10°C each day for 20 days, the accumulated temperature units are 10°C * 20 day = 200 ATU. Normally the eyed egg stage will be reached after 225–230 degree days when the water temperature is within the range of 4–8°C. Having a stable water temperature of 8°C gives a time of slightly below 1 month from fertilization to eyed egg stage. Since the eggs are now more robust, they can be handled and transported out of the broodstock farm. They may, however, be kept longer at the broodstock farm, but once over 350–400 ATU they ought to be dispatched. The closer to hatching the eggs are when transported, the greater the risk of early hatching when in transit. Hatching occurs at around 500 ATU.

5.10 Water quality during incubation

It is important to maintain good water quality during the incubation stage. The water must not be oversaturated with any gases. At this stage this also includes oxygen; oxygenation should not be used. Oxygen bubbles may block the water supply or lift (disturb) the eggs. It is recommended to filter the incoming water for particles because they may settle on the egg surface, and to disinfect the water to prevent disease agents from entering the incubation units and to help prevent fungal problems.

It is common to use traditional flow through systems in this part of production, meaning that the water flows directly through the incubation unit, with upwelling flow, and to the outlet. However, it is also possible to use the water multiple times in RAS.

The incubators used in broodstock farms are designed for eyed egg production, so the eggs must be removed from the incubators before hatching. This makes it possible to store many eggs within a limited floor space by incubating the eggs in high cylinders where eggs are placed on top of each other. The incoming new water enters at the bottom of the cylinder, flows up through the egg layers and drains at the top of the cylinder [82].

5.11 Important maintenance during the incubation period

It is necessary to perform a number of working operations during the egg incubation period from fertilization to dispatch of eyed eggs from the broodstock farm. This includes fungal control, grading of eggs, counting eggs, shocking of eggs, control of eggs for survival, control of water quality and water velocity, and, at the end of production, packing and dispatching the eggs as eyed eggs to the customers: the smolt farmers.

5.11.1 Shocking of eggs

Both the few unfertilized eggs and the 'bad' fertilized eggs will go through some development, meaning they do not die straight after fertilization. This is of great importance because they will not be as good a substrate for fungal growth as dead eggs [76]. To detect if the eggs are satisfactorily fertilized or not, and whether the embryo has the potential to develop, an operation called shocking is performed. Shocking can only be done from the early eyed egg stage, as the blastula has to be closed and the eggs in a suitable condition to be handled.

Shocking of the eggs can be done by exposing the eggs to a mechanical force, like a rough vibration. A typical way to do this is to expose the eggs to a free gravity fall, for instance, by pouring the eggs into a water-filled bucket. In this way the eggs will fall a distance through the air, perhaps 50 cm, before they hit the water surface which applies a mechanical force to the eggs. Good eggs will tolerate this, but bad eggs will not and they will die, causing breakage of the vitelline membrane. This will cause the colour of the bad eggs to change from pink to white, due to coagulation based on protein denaturation. This colour difference between live and dead eggs can be utilized later to grade the eggs with the help of a photo selective diode. The eggs are sent through a grading machine where the grading principle is the colour of the eggs, and the machine will separate dead eggs from live eggs.

The process of shocking the eggs can be avoided by using a specially developed grading machine that uses image analysis technology to categorize the embryo based on real-time pictures of the embryo development: unfertilized eggs, dead eggs, and eggs with deformed embryos can then be graded out. In the grading machine the embryos are automatically graded one by one, with a capacity of over 100,000 eggs per hour.

5.11.2 Fungi control

The development of fungi (*Saprolegnia* spp.) in the incubation unit will present a problem if no precautions are taken [83–85]. Some eggs will always die in the incubation units and they provide an excellent nutrient substrate for fungal development. There are always a number of fungal spores naturally present in the water; these will find the dead eggs and start to grow on their surface. If nothing is done the fungal growth will continue and then spread to cover nearby eggs. Those eggs will then suffer from a lack of supply of new water and therefore oxygen, causing them to die; this process will continue as the fungi covers new eggs. The fungi is sometimes called cotton fungi, because when growing it looks like cotton.

There are several strategies to control the development of the fungi. The best, of course, is to prevent fungal spores from entering the incubation unit. This can be done by disinfection of the incoming water, which is always recommended at broodstock farms. Disinfection of water to kill fungal spores is, however, quite tricky, and large doses of disinfectant are necessary. The advantage is that the volume of water supply to the incubation units is quite low. It is also important to be aware that the number of fungal spores varies a lot between different freshwater sources, and in some sources there are few fungal problems. Since the eggs are typically incubated in large stacks in the broodstock farms, it is not possible to regularly remove dead eggs and by doing so remove the substrate for growth. Removal of eggs would, in any case, not be recommended as it would inevitably disturb the developing eggs during the stage when they must not be subjected to movement. Therefore, removal of dead eggs is not an option (see eyed egg to start feeding). The last option is the addition of chemicals to the incoming water that kill or restrict the development of any fungi inside the incubator. The treatment is performed by sending an upwelling flow of the chemical through the incubation unit.

In the past, malachite green was used for this purpose and this was a very effective chemical. However, it is carcinogenic, and traces could still be found in slaughter-size salmon; its use was therefore forbidden [86–88]. Chemicals used today include bronopol, formalin, hydrogen peroxide (H_2O_2) and salt [89]. The chemical is flushed through the egg layers at the concentration and for the time recommended by the supplier, for instance, a 1:600 dilution for 15 minutes for formalin (Parasite-S) [90]. It is important to remember that these are toxic chemicals, and must be handled according to the safety regulations provided by the supplier. Moving the eggs to place them in a salt solution may be carried out when the eggs have reached a developmental stage where they can tolerate handling, but this is typically not before they are sent out from the broodstock farm as eyed eggs. A salt solution with a salinity of 9.5% has been used for this purpose.

5.11.3 Counting eggs

The eggs are counted in and out of the incubation units and when sending the eyed eggs out from the broodstock farm. This is important both to know how many are sent to the customers but also for internal production control. Counting is typically done indirectly by measuring the volume of the eggs. A volume of 1 l of salmon eggs produced in aquaculture contains a relatively constant number of eggs, around 5000, so by measuring how many litres of eggs are placed into the incubation unit an estimate of the number of eggs in total can be found. Specially designed counting plates with a known number of bowl shaped depressions that capture a single egg may also be used (Figure 5.10). This plate is dragged through a quantity of eggs and the grooves in the plate are filled with eggs. Weighing the eggs is another method, as the weight of 1 egg is around 0.1 gram. Another method is use of a length scale: by knowing the number of eggs in a length and relating this to volume a more exact count of the number of eggs per litre can be calculated. One standardized length used for the scale is 25 cm, the Brofeldt length scale. By counting the number of eggs on this scale the number of eggs per litre can be calculated [91]. It is normally in the region of 40 eggs per 25 cm, so each egg has a diameter of 0.625 mm giving a total of around 5000 eggs per litre (Table 5.2).

Figure 5.10 Specially designed counting plates may be used for counting.

Table 5.2 Number of eggs per 25 cm and resulting number of eggs per litre (after [92]).

	Number of eggs per 25 cm							
	37	39	41	43	45	47	49	51
Number of eggs per litre	3900	4500	5200	6000	6800	7800	8800	9900

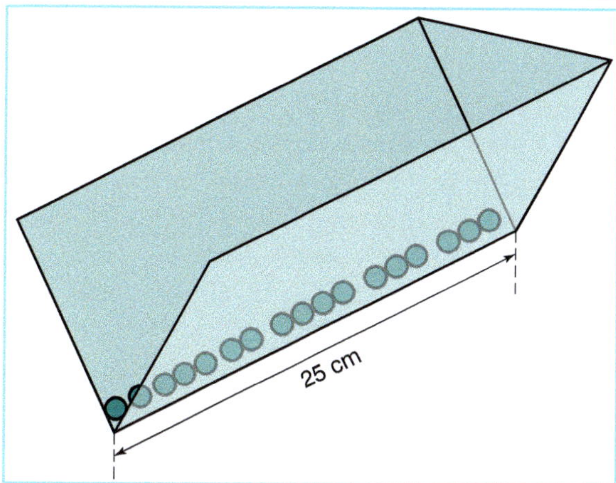

5.11.4 Transport of eyed eggs from the broodstock farm

The eggs are transported from the broodstock farms as eyed eggs, the stage when they can be handled without damaging the embryo. Eyed eggs are typically transported in specially designed extruded polystyrene foam boxes (Figure 5.11). The eggs are put in layers with ice on top. The ice will melt during transport and the water from the melted ice will drip down over the eggs. In

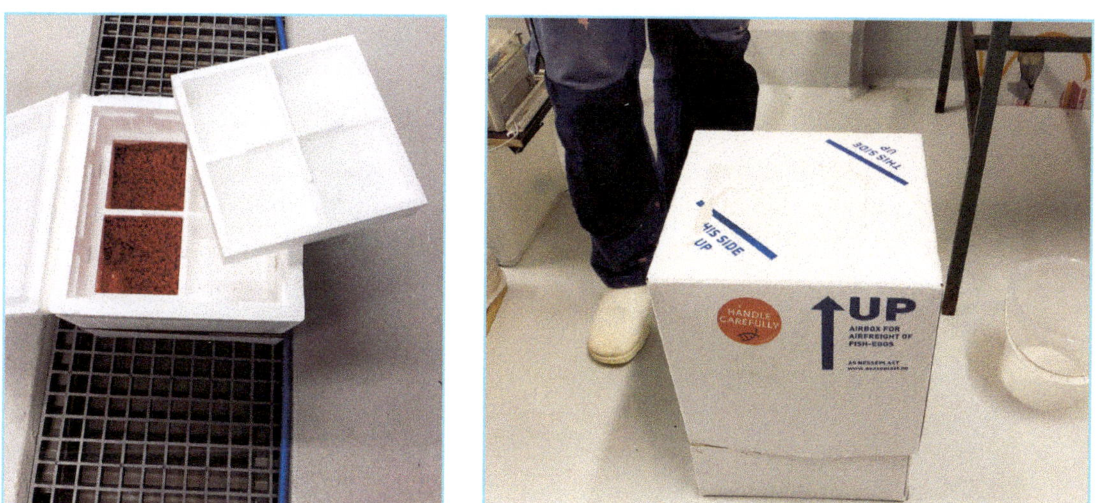

Figure 5.11 Boxes for shipping of eyed eggs.

this way the eggs will be kept at a high humidity and a low temperature in transit. The eggs can be transported for several days by this method.

Before dispatching the eggs from the broodstock farm, surface disinfection is carried out. A number of methods and chemicals are available for this [93], for example an iodine product (100 ml buffodine/10 l water for 10 minutes).

5.12 Tailor-made eggs

Today several of the broodstock farms that deliver eyed eggs to hatcheries/smolt farms can deliver a range of different types of eggs representing different traits. This is because breeding is included as part of the broodstock farm's activity, and therefore selective breeding for different traits can be performed. This also represents important progress for the industry because eggs can now be tailor-made for the environment that the salmon will meet. For example, eggs can be produced that have increased protection against sea lice or certain diseases (such as infectious pancreatic necrosis), or which are more suitable for special production systems, such as ongrowing in RAS systems where growth at high water temperatures with late maturation are desired traits. Triploid eggs are also of interest today, because salmon produced from these eggs will be sterile; this is an advantage regarding escape of salmon and possible interbreeding with wild stocks (see 5.13.1).

5.13 Controlling maturation and sex by genetic manipulation

It is possible to control maturation with genetic manipulation, which can also be used for manipulation and control of sex [94,95]. Several techniques are available, including hormonal treatment, production of hybrids and production of fish without functional germ cells (primordial germ cell [PGC] ablation). Many of these techniques are quite new and are based on knowledge of both advanced biotechnology and the salmon genome.

To avoid the problem of male early maturation in salmon for the consumer market, one solution is to produce all female salmon, while another is to produce sterile fish (fish that cannot reproduce) [96]. The latter will also provide a great benefit regarding the escape of farmed salmon, because a sterile farmed salmon cannot interbreed with wild salmon, causing genetic introgression [97–99]. Escaped farmed diploid salmon can migrate and when maturing they will enter rivers for spawning [100,101].

There are a number of ways to produce all female salmon, sterile salmon or almost sterile salmon [19,102,103]. Two methods for manipulation of sex include manipulation of the chromosomes (egg or sperm) and hormone treatment of the fry [103]. Male triploid salmon will still mature and if they escape may enter the wild spawning grounds and interfere, although they are not fertile; to avoid this, one solution is to produce all female triploids.

Sterile fish can be achieved by creating more than two sets of chromosomes in the egg (Figure 5.12), or by producing salmon that do not have active germ cells.

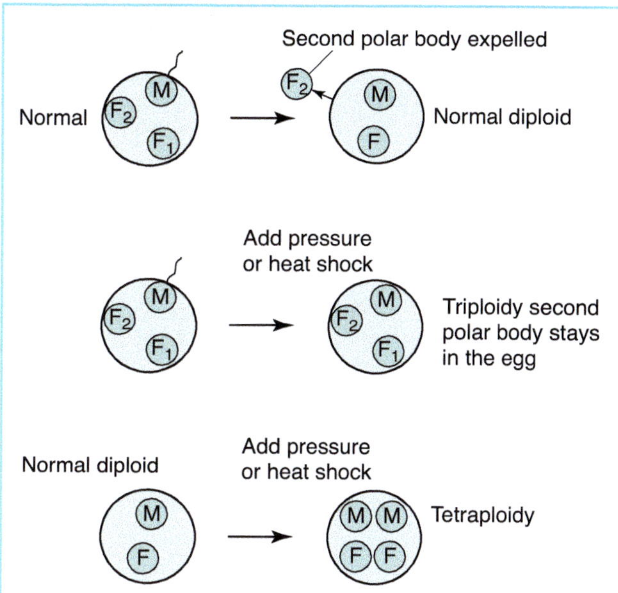

Figure 5.12 Production of normal (diploid), triploid or tetraploid salmon.

5.13.1 Triploid salmon

A triploid Atlantic salmon is an individual that has three sets of chromosomes in the cells rather than the normal two (diploid) [104,105]. A triploid female salmon will not mature and will not develop eggs. A triploid male will, however, produce milt, but the fertilization success of the sperm is very low, around 1%. These males will still compete with wild male salmon in the spawning grounds, so it is therefore a better alternative to produce all female triploid salmon to reduce any possibility for interaction between farmed and wild salmon.

A number of methods can be used to induce triploidy in fish [106,107]. The methods used can be divided into exposing the fish or eggs to a chemical or to expose the eggs to physical treatment (heat, cold or pressure). Triploid salmon may, for instance, be produced by adding large doses of androgens (natural or synthetic sex steroids) to the fry, either through feed or immersion of the fry in a bath containing androgens [103]. More common today, however, is to treat eggs just after fertilization with a thermal or pressure shock to induce the second polar body to remain within the egg (further described below). This can be achieved by exposing the eggs to high temperatures, up to 29°C for salmon, for a given period. One suggested temperature-time programme is to wait 10 minutes after fertilization and then expose the eggs to a temperature of 29°C for 10 minutes, or to 26°C for 20 minutes [103]. Exposing the eggs to a low-temperature shock may also be used, but this method is better suited to warm water species [106,107]. Exposing the fertilized eggs to a pressure shock after fertilization is the most common method used today, as this is more effective in achieving a repeated high triploid yield with low variation. Industrial processing lines handling this on a large scale have been developed. Pressure programmes that can be used at given temperatures have also been developed [108,109]. A combination of high pressure and time is needed. One proposal is to wait 300 minute degrees (time in minutes multiplied by incubation temperature) after fertilization and then apply a pressure of 655 bar that is released

350 minute degrees after fertilization [108,109]. An example of this would be to wait for a period of 42 minutes 15 seconds after fertilization if the temperature is 7°C, before adding pressure for 50 minutes. If the correct pressure-time programme is not used, the triploid yield will drop. The temperature at which pressure shock is performed used to be 10°C, but today 8°C or even lower is recommended [108,110,111].

When producing triploid salmon, the fertilized egg contains three sets of chromosomes, one from the male and two from the female. The second set of chromosomes from the female, the second polar body, is normally expelled from the egg, which is then left with two sets of chromosomes (diploid), as is the norm. In a triploid salmon, however, the pressure shock enables the second set of the female chromosomes to remain in the egg; together with the set coming from the male this means there are three sets of chromosomes in total.

Triploids may also be produced by cross breeding of normal diploid fish with tetraploid fish (which have two additional set of chromosomes); this is not common due to the low degree of survival of tetraploid salmonids [112–114].

When triploid salmon were first produced, reduced growth, health problems and higher mortality were observed. However, results improved when triploid salmon were treated as a new species with different nutrient and environmental requirements than ordinary salmon [104,105]. The fact that triploid salmon have different characteristics than diploid salmon has also resulted in a number of welfare challenges [115]. During egg incubation, temperatures above 6°C should be avoided [116]. Skeletal deformities can be avoided by using special feed with higher levels of phosphorous and histidine [117–120]. High seawater temperatures have been a problem in relation to development of cataracts. It also appears that triploid salmon have higher requirements for protein [120]. Compared to diploid salmon, triploid salmon in general show a better appetite at low temperatures (12°C), but a reduced appetite at high temperatures [121,122]. The upper lethal temperature is also lower than for diploids. Gut morphology has also been shown to differ between diploid and triploid [123]. Mortality rates until start feeding has finished are only slightly higher than for traditional salmon. Growth in the freshwater phase is the same as for diploids or higher, as is the case in seawater up to a size of 2–3 kg. The last phase of growth in seawater until slaughter seems to be slightly slower than normal salmon [124]. The oxygen requirement also seems to be a bit higher than traditional salmon, perhaps related to the larger cell size. Triploid salmon seem to be more susceptible to certain diseases, for example a higher degree of cataracts are observed, but this is not the case in general for diseases and parasites even though not much data from commercial triploid salmon production are available [122,124,125]. Triploid fish respond well to the traditional vaccines [124]. The quality of meat of triploid salmon seems to be similar to normal salmon, with a slightly lower number of fatty acids but a higher amount of long-chain omega-3 [126,127]. What can be seen is that with the trend in continuously optimizing husbandry conditions for triploids, the overall yield in triploid production seems to equal diploid salmon production [128,129].

Even though there are several ways to produce a sterile salmon, the only commercial method used today is triploid salmon production. These have been in commercial production for years in Tasmania where the aim has been to ensure delivery of salmon for slaughter all year round [130]. Early male maturation has been a hurdle for all year-round delivery of salmon of the correct size for the consumer market. In Norway triploid salmon are also used in commercial farming. Most of the farms producing triploid salmon have a special type of production licence which requires the use of triploid salmon. The major aim of these licences (green licences) has been to reduce the consequences of escaped salmon [131].

Triploid salmon is not a novel phenomenon and may also be found in wild salmon populations, though at a very low frequency [132]. Environmental factors such as high temperatures during the egg stage can cause triploidy in the wild. In the commercial salmon industry, there will also be a certain incidence of spontaneous triploidy, for example, around 2% of the overall Norwegian farmed salmon production was triploid in the period 2007 to 2014 [133].

5.13.2 All female production

Instead of normal salmon production, mono-sex production can be undertaken. By producing all female salmon (sex reversal), the problem of early maturation of males as parr or grilse can be avoided. Sex reversal of fish in aquaculture in general is quite commonly performed in other fish species to counteract early maturation of males, but until now has not been commonly used in salmon farming. Several techniques are available for all female production. The process is a manipulation of the chromosomes, resulting in the embryo only having maternal chromosomes. The most common techniques for all female production have involved hormone treatment but radiation of milt may also be used.

When exposing the milt to radiation with a dose of UV light or X-rays, the sperm will be inactivated. In one example, after first using a sperm extender solution the sperm were exposed to 400 microwatts/cm^2 UV light over 2–2.5 minutes (UV dose 12,000–15,000 ergs/mm^2) which seemed to inactivate all sperm [103]. After fertilization the eggs are exposed to heat or pressure shock to ensure that the second set of the female chromosome, the second polar body, does not leave the egg (as for triploid production). The result is that the egg has two sets of female chromosomes.

The method of using hormonal treatment is not currently used on Atlantic salmon. This can be performed by feeding the salmon fry with a feed that contains sex hormones, for example, oestradiol, in the form of either a natural hormone or a synthetic steroid [103,134]. Reported doses of 17 D-methyltestosterone are 3 mg/kg and 15 mg/kg feed [134,135]. Immersion of the salmon fry in a bath containing the sex steroid hormone is a further potential method [136]. 17 D-methyltestosterone has been used in bath doses at 400 Pg/l for 2 hours from 750–850 DD after fertilization, as the sex of the salmon is determined 700–800 DD after fertilization [137,138].

However, when using hormones the impact on the wider environment and on consumer health is important [139]. Supply of hormone-treated fish to the consumer market is also prohibited in a number of countries, such as EU member states, which may present a problem. This can, however, be handled by planning the production and sending the generation following the hormone-treated generation to the consumer market.

A normal female salmon has two X chromosomes (XX) and the male has one X and one Y chromosome (XY). If the offspring has inherited the X chromosome from the male together with the X from the female it will therefore be female, while it will be male if it has inherited the Y chromosome from the male. A neo-male is a genetically special male because it is a male that only has X chromosomes. It is actually a female that produces milt with active sperm: in other words, by genotype it is a female but by phenotype a male (a reversed female). Neo-males can be

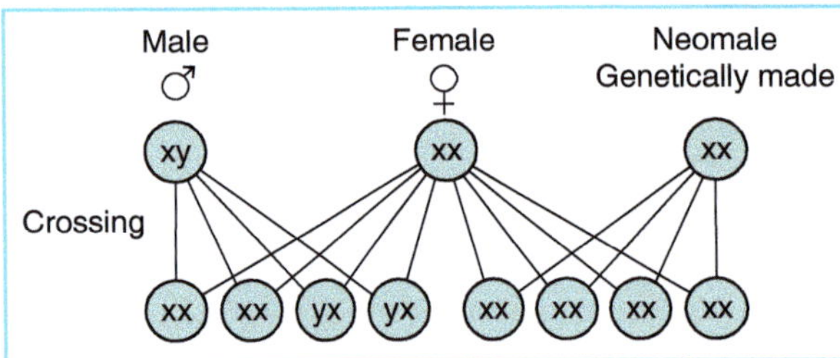

Figure 5.13 Normal production compared to all female production.

created when producing all females with hormonal treatment (Figure 5.13). When keeping fry that are hormone treated to maturation the normal males with the XY chromosome will differ from neo-males, and spawn normally, releasing milt. Neo-males, however, have no channel for releasing milt, and they will therefore not spawn normally [138]. They must be anaesthetized, killed and gutted to access their milt. When opening the abdomen and looking at the gonads, the neo-males have small non-developed eggs in addition to milt, and by this it can be observed that it is not a normal male but a neo-male. It is the milt from neo-males that is used further in production of all females because the milt contains only XX chromosomes. By manually removing the gonads from neo-males, the milt can be harvested and used to fertilize eggs which results in an all female population as no Y chromosomes will be present. The offspring of this process will not be hormone treated, so it is a way to produce a second generation of normal all female fish, thereby avoiding the problem of early male maturation without the use of hormones. It can be tricky to separate normal males from neo-males with 100% accuracy [136], but the use of genetic markers can aid this process [138].

5.13.3 Genes and genomics

Genomics, the study of the DNA sequence in the body, will certainly have enormous importance for the future development of the salmon industry. To understand a bit more about this rather complicated area of biology a very brief review is given below. It is greatly simplified and may therefore not be entirely accurate, and neither is the more advanced terminology within the field included. For further information, a number of textbooks, review articles and web pages are available, ranging from a quite basic level to very advanced (some examples [140–143]).

Breeding and genetics, the study of genetic properties, has until today been a major factor in the success of the Atlantic salmon farming industry (see also Chapter 12). The introduction of biotechnology and the 'omics' – genomics, proteomics and metabolomics – bring new tools that could improve the success of selective breeding further (example [144]). These are, however, also important standalone tools to increase basic biological knowledge [145,146]. Biotechnology and genomics require advanced laboratories, but new techniques that make it simpler to utilize advances are continuously being developed. Within genomics, which will be further described here, the sequencing of the salmon genome is one of the major steps driving all further development [147].

Going back to basic biology, the core of the cell (the nucleus) holds the chromosomes carrying the genetic information of an individual (Figure 5.14). Atlantic salmon have 27 or 29 pairs of chromosomes with 72 and 74 chromosome arms for North American and European strains,

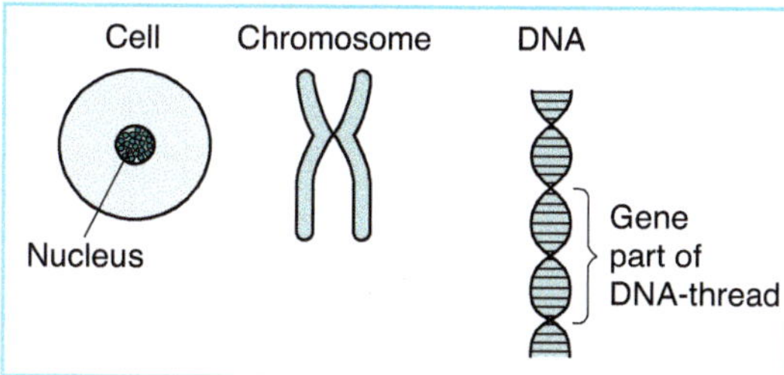

Figure 5.14 DNA is located in the chromosomes in the cell nucleus.

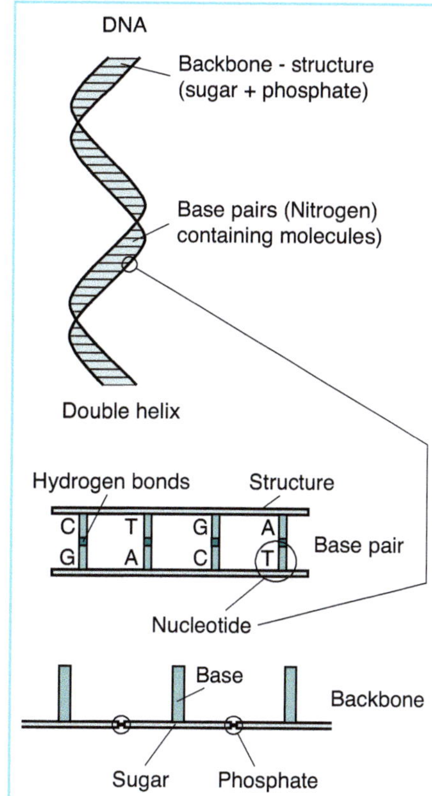

Figure 5.15 The structure of DNA.

respectively [148,149]. The chromosomes are presented in pairs with paternal chromosomes originating from the father and maternal chromosomes from the mother. One of the chromosome pairs is the allosome, or the sex chromosome, dictating the sex of the salmon. A female has, as mentioned, two X chromosomes while the male has one X and one Y chromosome. A chromosome is made up of a DNA molecule coiled around proteins called histones and packed in such a way that a long DNA thread can fit inside the cell nucleus.

The DNA molecule is again made up of a number of smaller units named nucleotides that are linked together to form the 'DNA thread'. Each nucleotide consists of three main parts: a sugar (pentose), a phosphate group that makes up the structure/backbone of the double helix, as typically illustrated, and a nitrogen base that makes up the 'pins' in the helix, like steps in a ladder (Figure 5.15). There are four nitrogen-containing molecules that constitute the base of DNA: adenine (A), cytosine (C), guanine (G) and thymine (T). They are always presented in pairs, A–T and C–G, linked by hydrogen bonds. The order of the base pairs in the helix is important, and is like a code for the individual. If following one of the 'backbones' in the double helix and the attached bases, it can produce a code like this: AACTGG and so on.

When sequencing the Atlantic salmon genome (further described below), 2.97 billion bases have been identified [147,150]. As an example, the Atlantic salmon mitochondrial genome, part of the overall salmon genome, has been sequenced and this contains 16,665 base pairs [151].

It is in the base pairs where the genetic information of the individual is located, and is what makes individuals genetically different. This may also be named the DNA fingerprint.

DNA contains many genetic instructions about protein assembly, and provides the information required to create a living organism. The genes are segments of the overall DNA, consisting of base pairs, and many genes represent codes for making specific proteins for special functions in the organism. Genes can therefore be compared to a manual, storing information about how to build and maintain an organism. Each gene can come in different forms named alleles, and the difference can be seen in the code for bases in the helix. If the gene gives the instruction, the allele represents the variation within the instruction. This is also what provides variation between individuals because the alleles may differ; for example, in humans this can be represented with brown and blue eyes. An individual will inherit alleles of a gene from mother and father, and they are therefore always in pairs. If they are the same, they are homozygotes and if they are different, they are heterozygotes. Alleles may be dominant or recessive, so for example, an allele from the mother may dominate over one from the father. A different form of the allele represents a small variation of the DNA base pairs (the code).

Each gene has a certain size, and this can be seen in the DNA thread of the chromosome where it can represent from hundreds to millions of base pairs. The gene will have a start and stop point on the DNA. The term locus or loci (plural) gives the location or 'address' of a specific gene or DNA sequence on the chromosome. The genotype of an individual represents the sum of all genes and gives the overall genetic composition, and will be different for each individual. The study of genes, the variation of genes and heritability is named genetics and is closely related to selective breeding (see Chapter 12).

The genome represents the total DNA, not only the genes, and each organism therefore has one genome which is unique for each individual (based on the order of base pairs, or the code). All information required for the body to function, grow and develop is given in the genome and it therefore represents the sum of the genes. But the genes are, as said, only a minor part of the entire DNA. Other information within the DNA includes how genes can be turned on and off, and it also contains what is called junk DNA or information of which the function is still unknown.

Genome sequencing or DNA sequencing is the process of identifying the order of the base pairs in the whole genome or parts of the genome, in other words the order of the four nitrogen compounds, A, C, G and T, in the base-pair code. It is a quite complex task to perform because first the DNA has to be broken down into several small fragments before it is assembled and annotated. The assembly process includes the organization of the nucleoids in the correct order: some, for instance, are overlapping, so it is put in a single 'line' and overlapping sequences removed. To simplify, it can then be visualized like a string of codes, such as ACCGGTT. The annotation process identifies the location of the genes and other coding regions. Much of this information is available in public databases, named genome browsers, for example, as provided for salmon [152]. Genome mapping (linkage or physical) is another term used and this describes the location of the genes on the chromosome and the distance between them, so is not as detailed as genome sequencing, where the entire code is identified.

Single nucleotide polymorphisms (SNPs, pronounced snips) are the smallest sequence of the genome and can be illustrated with the code of a single base: C, T, G or A. Also known as single DNA base variation. If the code occurs in more than 1% of the population but not in the majority it is named a SNP; if it occurs in less than 1% it is called a mutation. However, it is the same phenomenon: a difference in a single base. It can be illustrated with a normal code sequence of CTAGA, where a SNP occurring in over 1% of the population could be CTACA while a mutant occurring in less than 1% of the population is CTATA. Detection of SNPs is important because they can be used for identifying variations between individuals which could be related, for example, to a response to environmental factors or how susceptible they are to specific diseases. This may also be an important genetic tool to reduce transfer of hereditary diseases.

Natural mutations or changes in the DNA code may be the result of a mistake when copying or as a result of factors in the environment, for instance, exposure to radiation. Depending on the actual mutation, it can have a positive or negative impact on a species, but it will create an increased genetic variation in the species that later could be used in selective breeding.

When a full genome or a part of a genome is sequenced, assembled and annotated it allows for comparison, for instance, to see how individuals differ genetically by comparing the code of the base pairs. The genetic differences and similarities between species may be identified together with the occurrence of SNPs and mutations within one species.

After mapping and sequencing the genome it is possible to try to identify the locations of the different genes or other important areas in the DNA, and further identify which traits a gene or a sequence of the genome is involved in controlling: a genetic marker. Seldom is only one gene or a specific locus responsible for a trait, but sometimes this is the case. A good example here is the identification of the specific locus affecting the immunity against the disease infectious pancreatic necrosis (IPN). Here a variation in one single locus, a DNA sequence position, has been identified and one code has a higher resistance to IPN in seawater than the other. Such a locus is named a quantitative trait locus (QTL) and represents a located part of the genome that is quantitively affecting a trait. In practice it is used as a genetic marker linked to a specific locus. The coding here will differ, producing phenotypic variation as the alleles have different forms, so by selecting individuals that have a specific form such as higher IPN resistance, the problems with IPN can be greatly reduced, or almost eliminated, as has been happening within the salmon industry. Only individuals that have this code for resistance will be used in further selective breeding. Several other diseases, so-called DNA-related diseases or genetic diseases may also be identified by the DNA code once the genome or part of the genome is sequenced. One code, a SNP or mutation, may be responsible and by identifying this code in a salmon this individual can be removed from selective breeding to ensure that this code is not transferred to the next generation. The only way it could then return is through a new mutation, which is very rare. Identification of the DNA code will also give possibilities for early diagnostics and early prevention of diseases. This example also illustrates the possibilities with genetic methods related to genetically transferred diseases. QTL is therefore used more and more in selective breeding, as in addition to IPN such selection is used to give extra protection against diseases, such as pancreas disease (PD), cardiomyopathy syndrome (CMS) and heart and skeletal muscle inflammation (HSMI). It is also shown that some salmon function as lice collectors and here QTL is also used. QTL-selected ova for late maturation are also available on the market and a QTL related to early maturation has been identified [153]. Unfortunately, it is only in a few cases that a single locus controls so much of the variation as in the case of the resistance for IPN. Typically, this is not the case for many traits as a number of loci spread around the genome are responsible and are also difficult to locate. In addition, several of them may have only a small influence on the actual traits, but of course may still provide an improvement in the overall selective breeding. So far 50,080 genes have been identified in Atlantic salmon, in addition to 25,208 duplicated genes [152]. Genome mapping has therefore been of importance for selective breeding and development of future breeding programmes as more genome information is collected, because more precise and faster progress can be achieved by including gene mapping, also named molecular genetics, in the overall picture. This can be seen with the utilization of DNA markers linked to specific traits, such as the QTLs, in selection for breeding programmes; this is named marker-assisted selection (MAS) [154,155]. Genomic selection where a large number of genetic markers are taken into account is therefore increasingly included in traditional breeding programmes [156], such as the use of genome selection to predict breeding values related to resistance against sea lice [157].

5.13.4 Genetic modification

When the genome is sequenced, and the genes are mapped and the traits they are responsible for have been identified, modification of the structure of the genome or DNA may be directly performed. This is called gene editing or gene engineering, and the introduction of such techniques opens a new world. An organism exposed to genetic engineering can be called genetically modified (GM) and the organism created is called a genetically modified organism (GMO), but here the definitions are fluid based on the degree of engineering and whether foreign genes are included or not (described below). But even if a new world is opened by such techniques, there is also a great deal of legislation involved that represents obstacles for further development. The legislation regarding this topic varies a lot between countries; in some countries it is very restricted while in others it is more liberal. The preferences of the consumers, and therefore the market, must also be taken into consideration when starting activity within this area.

The area of genetic modification can be divided into two major categories: (1) introduction of 'foreign' genes or sequences of DNA into an individual, termed transgenics, and (2) modification of the existing DNA structure of an individual. The first might involve the addition of genes from tomatoes into apples while the second could involve changing the genes within an apple so all will be green. In salmon farming both categories have been introduced. An example of the first is the AquAdvantage salmon, while an example of the second is a salmon modified with Crispr/Cas9 technology (see section 5.13.6). Typically, it is easier to comply with the legislation dealing with the second category, where there is no introduction of new genes, but this is a debate for the future.

However, modification of the DNA structure of individuals is not new and continuously occurs in nature where what are called mutants are created. This can be illustrated by the small numbers of triploid salmon naturally occurring in nature. This type of modification is also what happens when exposing fish to hormonal treatment or radiation, methods that, for instance, are commercially used for production of triploids or all females. This is normally not categorized as genetic modification, showing that the definition is rather fluid. The large difference between this and performing engineering on the DNA structure is that engineering makes it possible to fully control the result with a high level of accuracy.

Genetic modification techniques are performed in advanced biotech laboratories and the process occurs at microscopic levels. An example of a possible intervention when a gene involved in a trait is identified is to turn it off, meaning the gene will be inactivated. A practical example of this is to delete a gene responsible for the production of a specific protein used for a certain purpose, for instance, a protein involved in initiating the maturation process in salmon. One way of deleting genes is to perform what is known as a gene 'knockout', meaning a permanent deletion of the gene. The genes may, however, also be 'knocked down' or silenced for only a certain time period. A lot of research within this area today is looking at locations of genes or DNA sequences controlling traits in salmon, and then starting to modify these. By using technology that knocks out or knocks down genes or DNA sequences, a lot of basic biological information can be gathered, such as the biological purpose of the genes or DNA sequences.

5.13.5 Genetically modified salmon

As discussed, today it is possible to genetically modify organisms by introducing new genes or fragments of foreign DNA into the organism. The purpose of this could be to improve performance or to increase disease tolerance. Genetic modification has had a major economic impact in today's plant breeding. Through genetic modification, plants that tolerate certain effective herbicides have been created, such as soy that tolerates the herbicide glyphosate. Spraying with this herbicide will kill normal soy but not the genetically modified soy. Today's boom in GM crops is within the area of producing crops containing the very long-chain omega-3 fatty acids. This is done by transferring genes from algae, the primary producer of these fatty acids, into the plants. This may also become very important for the salmon farming industry in the future because it could be possible to produce crops containing eicosapentaenoic acid (EPA) and docosahexaenoic acid (DHA), the important fatty acids which today are only available in fish oil and some marine algae. By using GM crops containing these fatty acids high volumes may become available at reasonable prices.

A genetically modified salmon has, as mentioned, already been created, and is the only GM animal available as food on the consumer market. The AquaBounty's AquAdvantage is a salmon that is genetically modified for improved growth in cold water (Figure 5.16). This was done by isolating specific genes/DNA sequences from two fish species, Chinook salmon (growth hormone gene) and ocean pout (promoter and terminator antifreeze protein) [158–160]. The ocean pout has a protein that inhibits the blood from freezing so it is able to live in very low water temperatures. The purpose of transferring the DNA sequence from ocean pout is to ensure maximum growth of the salmon in winter months, while the Chinook gene signals additional growth. In total, 4250 base pairs have been added to the 2.97 billion bases of salmon [161]. The development started with the creation of a transgene in a laboratory. This was further inserted into a plasmid that is placed inside bacteria for replication, producing large numbers. The next step was to inject the transgene into a normal Atlantic salmon egg (microinjection) at the same time as normal fertilization, so injected directly into the newly fertilized egg. Some of the hatched fry will then take up the inserted genes and become genetically modified, while some do not and will develop as normal Atlantic salmon. From this overall population the salmon where the genetic modification has been a success are then separated from the group to become a broodstock of genetically modified Atlantic salmon. This broodstock is then used as the base for brood fish for a new generation, so the genetic modification in the laboratory is only needed to be carried out once.

It is not, however, the offspring of these salmon that are used in commercial production for the consumer market. The next step is to produce all females, as described earlier, by hormone

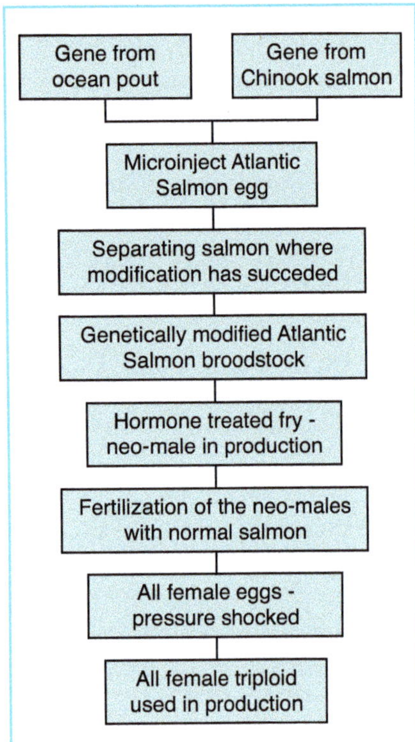

Figure 5.16 Production process for a genetically modified salmon, AquAdvantage salmon.

treatment of the transgenic fry. This produces neo-males (males that lack the Y chromosome), and these are then selected for use in further breeding. The milt from these neo-males is used for fertilization of eggs from normal female Atlantic salmon, not transgenic salmon. The fertilized eggs will then be exposed to high-pressure treatment ensuring triploidy, and the hatched fry will be used in commercial production for the consumer market. The GM salmon are therefore only female triploids. The last step is to ensure no interbreeding with wild salmon stocks if escape occurs. The production of this salmon today takes place in a land-based farm but with many additional efforts to prevent any possible escape of fish.

The GM salmon have improved growth, especially in the first stage, so they reach normal slaughter weight earlier than a comparable non-genetically modified salmon. The growth of the GM salmon will gradually reduce and stop after reaching slaughter weight, as does the growth of non-GM salmon. No "monster" salmon will be created. Feed utilization is also improved in comparison to non-GM salmon, by up to 25%. Optimal temperature for growing in RAS is recommended to be below 16.5°C, and experiments show 10.5°C or 13.5°C to be optimal, as even though the GM salmon show the highest growth rate at 16.5°C, feed utilization is reduced [162].

The development of transgenic salmon started in 1989, and the application for selling the produced salmon for human consumption was sent to the US Food and Drug Administration in 1995; in 2016 the first transgenic salmon was approved to be sold on the consumer market in Canada and in 2019 permission was given for the US market, 24 years after application [163–165]. This illustrates how difficult it is and how much time it has taken to get a new GM product on the market. Remember that this is the first and only transgenic animal approved for human consumption and it was pioneering work.

5.13.6 Genetically edited salmon

As mentioned, it is also possible to produce what is called a gene-edited salmon by editing the genome, modifying specific genes that are already within the Atlantic salmon rather than introducing foreign DNA into the salmon. Several methods can be used for this purpose such as ZFN, Talen and Crispr. The Crispr/Cas9 technology is a newly developed method for genome or gene editing that has proven to be very cost and labour effective compared to earlier methods [166,167]. The function of the genes that are already in the salmon are modified. Very simply described, a gene-editing molecule is sent into the cell (microinjection), and finds the nucleus and then the DNA. Then a specific RNA molecule, a guide molecule, locates the area in the genome that should be modified and identifies the base pairs; the cas9 protein (an enzyme that functions like a scissors) then cuts out this sequence of base pairs. This can then be totally removed or changed, or a new piece can be added (Figure 5.17). In aquaculture this method can be used by injecting the Crispr/Cas9 reagent into newly fertilized eggs [164]. Such methods have huge future possibilities; for example, it is known that only a few genes are responsible for initiation of maturation, so by elimination, knock out or silencing of those genes a salmon that will not mature can be produced. In this way, sterile salmon may also be produced. Another example is the problem with salmon lice. Coho salmon are resistant to sea lice; they attach but then drop off. An interesting future approach could therefore be to identify which DNA sequences control this in coho salmon and compare the sequences with those of Atlantic salmon [168,169]. It could then be seen if it is possible to edit this sequence in Atlantic salmon or maybe directly transfer it, although this would require that new genes are inserted. Another example could be to edit the genome so problems with genetic diseases could be eliminated. The deposition of polyunsaturated fatty acids (EPA, DHA) in the muscle could also be improved [170]. In the human sector, the use of Crispr technology is being evaluated for its potential for treating cancer. These examples illustrate some of the future potential of this method.

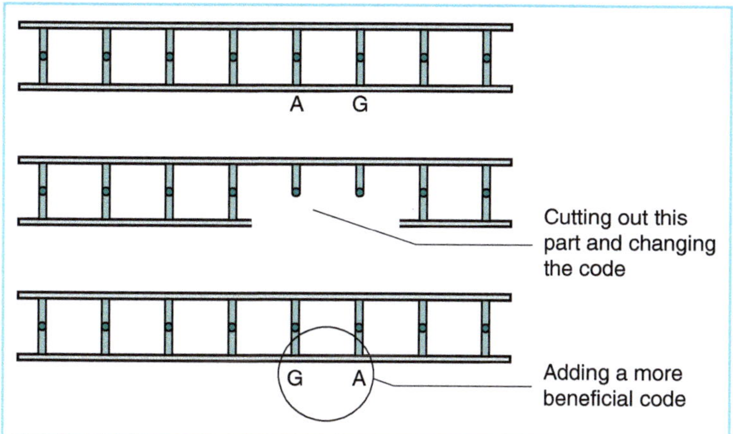

Figure 5.17 Gene editing is quite a new method where the genes can be changed in beneficial ways.

Gene editing will of course also be important for increasing basic knowledge of biology, because by knocking out genes it is possible to increase the knowledge of which traits the individual genes are involved in. It can also be taken further into separate DNA sequences.

5.13.7 Producing salmon without germ cells

When the genes that are involved in initiation of the maturation or development of germ cells and their purpose are known a sterile salmon or a salmon that does not develop functional germ cells can be produced. A locus that is involved in the age of maturity of Atlantic salmon (vgll1) has been identified [171–173], as have several proteins that must be present before maturation can commence.

Today at least two methods are followed to produce of germ-cell-free salmon: blocking of a gene necessary for fertilization, and the use of Crispr technology. The basis of the first method is to prevent the migration of the primordial germ cells into the gonads, as without this the gametes will not develop. The primordial germ cells (PGC) function as precursors for the development process. A gene (dead-end gene [dnd]) which codes for development of a protein required for survival and migration of PGC has been identified [174,175]. A technology under development is to bathe the eggs in a solution containing substances that block the development of the protein that is required very early in the development process of gametes [176]. In larger salmon the gene will become functional again so the fish appear to develop as normal, but the genitals develop without gametes so the salmon will therefore be infertile. This is also called gene silencing. A molecular transporter carries the silencing agent into the target cell in the egg, which inhibits the development of the PGC [177].

The eggs are bathed in the solution containing this silencing compound just before or at the time of fertilization. This is also called DNA vaccination. The vaccine contains one or more DNA sequences that inhibit maturation by a protein antigen which inhibits the development of the protein necessary for maturation [19]. Crispr technology may also be used to produce a salmon without germ cells [178], but then no germ cells at all will develop.

References

1. Erdahl, D.A. Inland salmonid broodstock management handbook. 1993. US Department of the Interior, Fish and Wildlife Service: Washington, DC.
2. Brännäs, E. et al. Arctic charr farming production of juveniles; a manual. 2011. Swed. Univ. Agricult. Sci. Rept. Dept. Wildlife Fish and Environm. Stud, 2011.
3. King, H.R. et al. Effect of elevated summer temperatures on gonadal steroid production, vitellogenesis and egg quality in female Atlantic salmon. *Journal of Fish Biology*. 2003. 63(1): 153–167.
4. Wootton, R.J. and Smith, C. *Reproductive Biology of Teleost Fishes*. 2014. Wiley: Hoboken, NJ.
5. Patiño, R. and Sullivan, C.V. Ovarian follicle growth, maturation, and ovulation in teleost fish. *Fish Physiology and Biochemistry*. 2002. 26(1): 57–70.
6. Lubzens, E. et al. Oogenesis in teleosts: how fish eggs are formed. *General and Comparative Endocrinology*. 2010. 165(3): 367–389.
7. Næve, I. et al. Ultrasound as a non-invasive tool for monitoring reproductive physiology in female Atlantic salmon (*Salmo salar*). *Physiological Reports*. 2018. 6(9): e13640.

8. Mattson, N.S. A new method to determine sex and gonad size in live fishes by using ultrasonography. *Journal of Fish Biology*. 1991. 39(5): 673–677.
9. Nagahama, Y. and Adachi, S. Identification of maturation-inducing steroid in a teleost, the amago salmon (*Oncorhynchus rhodurus*). *Developmental Biology*. 1985. 109(2): 428–435.
10. Næve, I. et al. Ultrasound as a non-invasive tool for monitoring reproductive physiology in male Atlantic salmon (*Salmo salar*). *Physiological Reports*. 2019. 7(13): e14167.
11. Taranger, G.L. et al. Control of puberty in farmed fish. *General and Comparative Endocrinology*. 2010. 165(3): 483–515.
12. Aksnes, A., Gjerde, B. and Roald, S.O. Biological, chemical and organoleptic changes during maturation of farmed Atlantic salmon, *Salmo salar*. *Aquaculture*. 1986. 53(1): 7–20.
13. Gjerde, B., Simianer, H. and Refstie, T. Estimates of genetic and phenotypic parameters for body weight, growth rate and sexual maturity in Atlantic salmon. *Livestock Production Science*. 1994. 38(2): 133–143.
14. Weltzien, F.-A. et al. The brain–pituitary–gonad axis in male teleosts, with special emphasis on flatfish (Pleuronectiformes). *Comparative Biochemistry and Physiology Part A: Molecular & Integrative Physiology*. 2004. 137(3): 447–477.
15. Kim, J., Hayton, W.L. and Schultz, I.R. Modeling the brain–pituitary–gonad axis in salmon. *Marine Environmental Research*. 2006. 62: S426–S432.
16. Antonopoulou, E. and Borg, B. The brain-pituitary-gonad axis in the Atlantic salmon, in *Evolutionary Biology of the Atlantic Salmon*. 2015. CRC Press: Abingdon. pp. 119–134.
17. Sang, H.M. et al. Steroid hormones in reproduction and roles of GnRH-a in gonadal maturation of marine fish: a review. *Annual Research & Review in Biology*. 2019: 1–12.
18. Okuzawa, K. Puberty in teleosts. *Fish Physiology and Biochemistry*. 2002. 26(1): 31–41.
19. Iversen, M., Myhr, A.I. and Wargelius, A. Approaches for delaying sexual maturation in salmon and their possible ecological and ethical implications. *Journal of Applied Aquaculture*. 2016. 28(4): 330–369.
20. Good, C. and Davidson, J. A review of factors influencing maturation of Atlantic Salmon, *Salmo salar*, with focus on water recirculation aquaculture system environments. *Journal of the World Aquaculture Society*. 2016. 47(5): 605–632.
21. Maugars, G. Endocrine regulation of early sexual maturation in male Atlantic salmon parr. 2007. Swedish University of Agricultural Sciences.
22. Skilbrei, O.T. and Heino, M. Reduced daylength stimulates size-dependent precocious maturity in 0+ male Atlantic salmon parr. *Aquaculture*. 2011. 311(1–4): 168–174.
23. Fjelldal, P.G., Hansen, T. and Huang, T.S. Continuous light and elevated temperature can trigger maturation both during and immediately after smoltification in male Atlantic salmon (*Salmo salar*). *Aquaculture*. 2011. 321(1–2): 93–100.
24. Fraser, T.W.K. et al. Termination of puberty in out-of-season male Atlantic salmon smolts. *Comparative Biochemistry and Physiology Part A: Molecular & Integrative Physiology*. 2019. 232: 60–66.
25. Imsland, A.K., Handeland, S.O. and Stefansson, S.O. Photoperiod and temperature effects on growth and maturation of pre-and post-smolt Atlantic salmon. *Aquaculture International*. 2014. 22(4): 1331–1345.
26. Brown-Peterson, N.J. et al. A standardized terminology for describing reproductive development in fishes. *Marine and Coastal Fisheries*. 2011. 3(1): 52–70.
27. Towers, L. Maturation and spawning in fish. The Fish Site. 2014.
28. Dahl, K. *Studier og forsøk over ørret og ørretvand*. 1917. Centraltrykkeriet: Kristiania.

29. Jonsson, B. and Matzow, D. *Fisk i vann og vassdrag: om økologien til aure, røyr og laks*. 1979. Aschehoug.
30. Agarwal, N.K. *Fish Reproduction*. 2008. APH Publishing Corporation: New Delhi.
31. Fleming, I.A. Reproductive strategies of Atlantic salmon: ecology and evolution. *Reviews in Fish Biology and Fisheries*. 1996. 6(4): 379–416.
32. Porter, M.J.R. et al. The use of cage lighting to reduce plasma melatonin in Atlantic salmon (*Salmo salar*) and its effects on the inhibition of grilsing. *Aquaculture*. 1999. 176(3): 237–244.
33. Taranger, G.L. et al. Control of puberty in farmed fish. *General and Comparative Endocrinology*. 2010. 165(3): 483–515.
34. Hansen, T.J. et al. Effects of light source and intensity on sexual maturation, growth and swimming behaviour of Atlantic salmon in sea cages. *Aquaculture Environment Interactions*. 2017. 9: 193–204.
35. Schulz, R.W., Andersson, E. and Taranger, G.L. Photoperiod manipulation can stimulate or inhibit pubertal testis maturation in Atlantic salmon (*Salmo salar*). *Animal Reproduction (AR)*. 2018. 3(2): 121–126.
36. Jensen, P.M. Forskere advarer: Laksen kan nå bråmodne av høye sjøtemperaturer (in Norwegian). 2018. https://www.kyst.no/article/forskere-advarer-laksen-kan-bli-braa-kjoennsmoden/.
37. Thorpe, J.E. et al. Control of maturation in cultured Atlantic salmon, *Salmo salar*, in pumped seawater tanks, by restricting food intake. *Aquaculture*. 1990. 86(2–3): 315–326.
38. Thorpe, J.E. Reproductive strategies in Atlantic salmon, *Salmo salar* L. *Aquaculture Research*. 1994. 25(1): 77–87.
39. Arge, R. et al. Reduction of early sexual maturation in male S0 Atlantic salmon (*Salmo salar* L.) by dietary supplementation of tetradecylthioacetic acid (TTA). *Aquaculture Research*. 2014. 45(5): 922–933.
40. Fjelldal, P.G., Imsland, A. and Hansen, T. Vaccination and elevated dietary phosphorus reduces the incidence of early sexual maturation in Atlantic salmon (*Salmo salar* L.). *Aquaculture*. 2012. 364: 333–337.
41. Gjedrem, T. and Baranski, M. *Selective Breeding in Aquaculture: an Introduction. Vol. 10*. 2010. Springer: Dordrecht.
42. Zohar, Y. and Mylonas, C.C. Endocrine manipulations of spawning in cultured fish: from hormones to genes, in *Reproductive Biotechnology in Finfish Aquaculture*, C.-S. Lee and E.M. Donaldson (eds). 2001. Elsevier: Amsterdam. pp. 99–136.
43. Mylonas, C.C., Fostier, A. and Zanuy, S. Broodstock management and hormonal manipulations of fish reproduction. *General and Comparative Endocrinology*. 2010. 165(3): 516–534.
44. Mehdi, Y. and Mousavi, S.E. A review of the control of reproduction and hormonal manipulations in finfish species. *African Journal of Agricultural Research*. 2011. 6(7): 1643–1650.
45. Mittelmark, J. and Kapuscinski, A.R. *Induced Reproduction in Fish*. 1993. Minnesota Sea Grant Extension.
46. Mylonas, C.C. and Zohar, Y. Use of GnRHa-delivery systems for the control of reproduction in fish. *Reviews in Fish Biology and Fisheries*. 2000. 10(4): 463–491.
47. Taranger, G.L. et al. Effects of photoperiod, temperature and GnRHa treatment on the reproductive physiology of Atlantic salmon (*Salmo salar* L.) broodstock. *Fish Physiology and Biochemistry*. 2003. 28(1–4): 403–406.
48. Taranger, G.L., Stefansson, S.O. and Hansen, T. Advancement and synchronization of ovulation in Atlantic salmon (*Salmo salar* L.) following injections of LHRH analogue. *Aquaculture*. 1992. 102(1): 169–175.

49. OECD. *Harmonization of Regulatory Oversight in Biotechnology Safety Assessment of Transgenic Organisms in the Environment.* Volume 7 OECD Consensus Documents: OECD Consensus Documents. 2017: OECD Publishing.
50. Springate, J.R.C. et al. The timing of ovulation and stripping and their effects on the rates of fertilization and survival to eying, hatch and swim-up in the rainbow trout (*Salmo gairdneri* R.). *Aquaculture.* 1984. 43(1–3): 313–322.
51. Fitzpatrick, M.S. et al. Viabilities of eggs stripped from coho salmon at various times after ovulation. *The Progressive Fish-Culturist.* 1987. 49(3): 177–180.
52. Gjerde, B. Variation in semen production of farmed Atlantic salmon and rainbow trout. *Aquaculture.* 1984. 40(2): 109–114.
53. Lahnsteiner, F., Weismann, T. and Patzner, R.A. Composition of the ovarian fluid in 4 salmonid species: *Oncorhynchus mykiss, Salmo trutta f lacustris, Saivelinus alpinus* and *Hucho hucho. Reproduction Nutrition Development.* 1995. 35(5): 465–474.
54. Zadmajid, V. et al. Ovarian fluid and its impacts on spermatozoa performance in fish: a review. *Theriogenology.* 2019. 132: 144–152.
55. Rosengrave, P. et al. Chemical composition of seminal and ovarian fluids of chinook salmon (*Oncorhynchus tshawytscha*) and their effects on sperm motility traits. *Comparative Biochemistry and Physiology Part A: Molecular & Integrative Physiology.* 2009. 152(1): 123–129.
56. Hwang, P.C. and Idler, D.R. A study of major cations, osmotic pressure, and pH in seminal components of Atlantic salmon. *Journal of the Fisheries Board of Canada.* 1969. 26(2): 413–419.
57. Yanagimachi, R. et al. Sperm attractant in the micropyle region of fish and insect eggs. *Biology of Reproduction.* 2013. 88(2): 47–1.
58. Johnson, S.L. et al. Proteomic analysis of chinook salmon (*Oncorhynchus tshawytscha*) ovarian fluid. *PLoS One.* 2014. 9(8).
59. Thorpe, J.E., Miles, M.S. and Keay, D.S. Developmental rate, fecundity and egg size in Atlantic salmon, *Salmo salar* L. *Aquaculture.* 1984. 43(1–3): 289–305.
60. Kaushik, S.J. Feed management and on-farm feeding practices of temperate fish with special reference to salmonids. FAO Fisheries and Aquaculture Technical Paper. 2013(583): 519–551.
61. Lush, L. et al. Size and ATP content of unfertilized eggs from farmed and wild Atlantic salmon in Newfoundland. *North American Journal of Aquaculture.* 2014. 76(2): 138–142.
62. Heinimaa, S. and Heinimaa, P. Effect of the female size on egg quality and fecundity of the wild Atlantic salmon in the sub-arctic River Teno. *Boreal Environment Research.* 2004. 9(1): 55–62.
63. Halseth, V. and Haus, E.O. *Beregning av rognvolum og rognantall hos laks og regnbueoerret.* 1984. Norsk Fiskeoppdrett.
64. Institute, N.v., Strykeplakaten (In Norwegian). Norwegian veterinary institute.
65. Morisawa, M., Suzuki, K. and Morisawa, S. Effects of potassium and osmolality on spermatozoan motility of salmonid fishes. *Journal of Experimental Biology.* 1983. 107(1): 105–113.
66. Vílchez, M.C. et al. Role of potassium and pH on the initiation of sperm motility in the European eel. *Comparative Biochemistry and Physiology Part A: Molecular & Integrative Physiology.* 2017. 203: 210–219.
67. Alavi, S.M.H. and Cosson, J. Sperm motility in fishes. I. Effects of temperature and pH: a review. *Cell Biology International.* 2005. 29(2): 101–110.
68. Alavi, S.M.H. and Cosson, J. Sperm motility in fishes.(II) Effects of ions and osmolality: a review. *Cell Biology International.* 2006. 30(1): 1–14.
69. Billard, R. and Jensen, J.J.O.T. Gamete removal, fertilization and incubation, in *Developments in Aquaculture and Fisheries Science.* 1996. Elsevier: Amsterdam, pp. 291–364.

70. Cosson, M.P., Billard, R. and Letellier, L. Rise of internal Ca2+ accompanies the initiation of trout sperm motility. *Cell Motility and the Cytoskeleton*. 1989. 14(3): 424–434.
71. Alderson, R. and MacNeil, A.J. Preliminary investigations of cryopreservation of milt of Atlantic salmon (*Salmo salar*) and its application to commercial farming. *Aquaculture*. 1984. 43(1–3): 351–354.
72. Dziewulska, K., Rzemieniecki, A. and Domagała, J. Motility and energetic status of Atlantic salmon (*Salmo salar* L.) sperm after refrigerated storage. *Journal of Applied Ichthyology*. 2010. 26(5): 668–673.
73. Scott, A.P. and Baynes, S.M. A review of the biology, handling and storage of salmonid spermatozoa. *Journal of Fish Biology*. 1980. 17(6): 707–739.
74. Daye, P.G. and Glebe, B.D. Fertilization success and sperm motility of Atlantic salmon (*Salmo salar* L.) in acidified water. *Aquaculture*. 1984. 43(1–3): 307–312.
75. Billard, R. Spermatogenesis and spermatology of some teleost fish species. *Reproduction Nutrition Développement*. 1986. 26(4): 877–920.
76. Stefansson, S.O., Holm, J.C. and Taranger, G.L. Oppdrett av laks og aure i Norge (in Norwegian). 2002. Universitet i Bergen: Institutt for fiskeri- og marinbiologi.
77. Baeverfjord, G. et al. Feilutvikling og deformiteter hos laks. 1998. Aas, Norway.
78. Baeverfjord, G. et al. High temperatures during egg incubation may induce malformations in Atlantic salmon (*Salmo salar* L.). *Aquaculture and Water*, EAS special publication. 1998(26): 24–25.
79. Vernier, J.M. Table chronogique du development embryonnaire de la truite arc-en-ciel, Salmo gairdneri Rich. *Ann. Embryol. Morphogenet.* 1969. 2: 495–520.
80. Velsen, F.P.J. Embryonic development in eggs of sockeye salmon, *Oncorhynchus nerka*. Vol. 49. 1980. Department of Fisheries and Oceans.
81. Boyd, J.W., Oldenburg, E.W. and McMichael, G.A. Color photographic index of fall Chinook salmon embryonic development and accumulated thermal units. *PLoS One*. 2010. 5(7).
82. Lekang, O.-I. *Aquaculture Engineering*. 2020. Wiley; Hoboken, NJ.
83. Piper, R.G. et al. *Fish Hatchery Management*. 1982. United States Department of the Interior: Washington, DC.
84. Thoen, E., Evensen, Ø. and Skaar, I. Pathogenicity of *Saprolegnia* spp. to Atlantic salmon, *Salmo salar* L., eggs. *Journal of Fish Diseases*. 2011. 34(8): 601–608.
85. Songe, M.M. et al. Saprolegnia diclina IIIA and S. parasitica employ different infection strategies when colonizing eggs of Atlantic salmon, *Salmo salar* L. *Journal of Fish Diseases*. 2016. 39(3): 343–352.
86. Sudova, E. et al. Negative effects of malachite green and possibilities of its replacement in the treatment of fish eggs and fish: a review. *Veterinarni Medicina-Praha*. 2007. 52(12): 527.
87. Srivastava, S., Sinha, R. and Roy, D. Toxicological effects of malachite green. *Aquatic Toxicology*. 2004. 66(3): 319–329.
88. Culp, S.J. and Beland, F.A. Malachite green: a toxicological review. *Journal of the American College of Toxicology*. 1996. 15(3): 219–238.
89. Oono, H. et al. The use of bronopol to control fungal infection in rainbow trout eggs. *Biocontrol Science*. 2007. 12(2): 55–57.
90. Ministry of Natural Resources and Forestry. Egg disinfection and incubation procedures for salmonids (salmon, troout and whitefish). Fish Culture Technical Bulletin Best Management Practices. 2014. http://www.communityhatcheries.com/wp-content/uploads/2014/07/BMP_2009–01-Salmonid_Egg_Disinfection-Updated.pdf.
91. Brofeldt, P. En enkel metod att beräkna antalet romkorn per litre [A simple method for calculating the number of fish eggs per litre]. *Svensk Fiskeri Tidsskrift*. 1935. 44: 158–160.

92. Hagala, P. *Drift av stamlaksbasseng*. 1978. Norges vassdrags- og elektrisitetsvesen, Statskraftverkene: Direktoratet for vilt og ferskvannsfisk: Trondheim.
93. De Swaef, E. et al. Disinfection of teleost eggs: a review. *Reviews in Aquaculture*. 2016. 8(4): 321–341.
94. Hulata, G. Genetic manipulations in aquaculture: a review of stock improvement by classical and modern technologies. *Genetica*. 2001. 111(1–3): 155–173.
95. Budd, A.M. et al. Sex control in fish: approaches, challenges and opportunities for aquaculture. *Journal of Marine Science and Engineering*. 2015. 3(2): 329–355.
96. Wang, H. et al. *Sex Control in Aquaculture*. 2018: Wiley: Hoboken, NJ.
97. Yeates, S.E. et al. Assessing risks of invasion through gamete performance: farm Atlantic salmon sperm and eggs show equivalence in function, fertility, compatibility and competitiveness to wild Atlantic salmon. *Evolutionary Applications*. 2014. 7(4): 493–505.
98. Thorstad, E.B. et al. Incidence and impacts of escaped farmed Atlantic salmon *Salmo salar* in nature, in *NINA,* special report. 2008.
99. Glover, K.A. et al. Atlantic salmon populations invaded by farmed escapees: quantifying genetic introgression with a Bayesian approach and SNPs. *BMC Genetics*. 2013. 14(1): 74.
100. Glover, K.A. et al. Three decades of farmed escapees in the wild: a spatio-temporal analysis of Atlantic salmon population genetic structure throughout Norway. *PLoS One*. 2012. 7(8).
101. Glover, K.A. et al. Half a century of genetic interaction between farmed and wild Atlantic salmon: status of knowledge and unanswered questions. *Fish and Fisheries*. 2017. 18(5): 890–927.
102. Donaldson, E.M. and Hunter, G.A. Sex control in fish with particular reference to salmonids. *Canadian Journal of Fisheries and Aquatic Sciences*. 1982. 39(1): 99–110.
103. Feist, G., Schreck, C.B. and Gharrett, A.J. *Controlling the Sex of Salmonids*. 1996. Oregon Sea Grant.
104. Benfey, T.J. Use of sterile triploid Atlantic salmon (*Salmo salar* L.) for aquaculture in New Brunswick, Canada. *ICES Journal of Marine Science*. 2001. 58(2): 525–529.
105. Benfey, T.J. Effectiveness of triploidy as a management tool for reproductive containment of farmed fish: Atlantic salmon (*Salmo salar*) as a case study. *Reviews in Aquaculture*. 2016. 8(3): 264–282.
106. Tiwary, B.K., Kirubagaran, R. and Ray, A.K. The biology of triploid fish. *Reviews in Fish Biology and Fisheries*. 2004. 14(4): 391–402.
107. Maxime, V. The physiology of triploid fish: current knowledge and comparisons with diploid fish. *Fish and Fisheries*. 2008. 9(1): 67–78.
108. Hansen, T. et al. *Oppdrett av steril fisk*. 2012. Havforskningsinstituttet.
109. Benfey, T.J. et al. Effectiveness of a commercial-scale pressure shocking device for producing triploid salmonids. *Aquacultural Engineering*. 1988. 7(3): 147–154.
110. Taylor, J.F. et al. Ploidy effects on hatchery survival, deformities, and performance in Atlantic salmon (*Salmo salar*). *Aquaculture*. 2011. 315(1–2): 61–68.
111. Johnstone, R. and Stet, R.J.M. The production of gynogenetic Atlantic salmon, *Salmo salar* L. *Theoretical and Applied Genetics*. 1995. 90(6): 819–826.
112. Myers, J.M., Hershberger, W.K. and Iwamoto, R.N. The induction of tetraploidy in salmonids. *Journal of the World Aquaculture Society*. 1986. 17(1-4): 1–7.
113. Allendorf, F.W. and Thorgaard, G.H. Tetraploidy and the evolution of salmonid fishes, in *Evolutionary Genetics of Fishes*. 1984. Springer: Dordrecht, pp. 1–53.
114. Donaldson, E.M. and Devlin, R.H. Uses of biotechnology to enhance production, in *Developments in Aquaculture and Fisheries Science*, W. Pennell and B.A. Barton (eds). 1996. Elsevier: Amsterdam, pp. 969–1020.

115. Fraser, T.W.K. et al. Welfare considerations of triploid fish. *Reviews in Fisheries Science*. 2012. 20(4): 192–211.
116. Fraser, T.W.K. et al. The prevalence of vertebral deformities is increased with higher egg incubation temperatures and triploidy in Atlantic salmon *Salmo salar* L. *Journal of Fish Diseases*. 2015. 38(1): 75–89.
117. Taylor, J.F. et al. Adult triploid Atlantic salmon (*Salmo salar*) have higher dietary histidine requirements to prevent cataract development in seawater. *Aquacutulre Nutrition*. 2015. 21(1): 18–32.
118. Fjelldal, P.G. et al. Increased dietary phosphorous prevents vertebral deformities in triploid A tlantic salmon (*Salmo salar* L.). *Aquaculture Nutrition*. 2016. 22(1): 72–90.
119. Burke, H.A. et al. The effect of triploidy on juvenile Atlantic salmon (*Salmo salar*) response to varying levels of dietary phosphorus. *Aquaculture*. 2010. 306(1–4): 295–301.
120. Smedley, M.A. et al. Dietary phosphorous and protein supplementation enhances seawater growth and reduces severity of vertebral malformation in triploid Atlantic salmon (*Salmo salar* L.). *Aquaculture*. 2016. 451: 357–368.
121. Sambraus, F. et al. Water temperature and oxygen: the effect of triploidy on performance and metabolism in farmed Atlantic salmon (*Salmo salar* L.) post-smolts. *Aquaculture*. 2017. 473: 1–12.
122. Sambraus, F. et al. Changes in water temperature and oxygen: the effect of triploidy on performance and metabolism in large farmed Atlantic salmon. *Aquaculture Environment Interactions*. 2018. 10: 157–172.
123. Peruzzi, S., Hagen, Ø. and Jobling, M. Gut morphology of diploid and triploid Atlantic salmon (*Salmo salar* L.). *Aquaculture International*. 2015. 23(4): 1105–1108.
124. Chalmers, L. A comparative assessment of health and immune response between triploid and diploid Atlantic salmon (*Salmo salar*). 2017. Thesis, University of Stirling. http://hdl.handle.net/1893/26609.
125. Chalmers, L. et al. Response of triploid Atlantic salmon (*Salmo salar*) to commercial vaccines. *Fish & Shellfish Immunology*. 2020. 97: 624–636.
126. Murray, D.S. et al. Comparisons of reproductive function and fatty acid fillet quality between triploid and diploid farm Atlantic salmon (*Salmo salar*). *Royal Society Open Science*. 2018. 5(8): 180493.
127. Taylor, J.F. et al. Ploidy and family effects on Atlantic salmon (Salmo salar) growth, deformity and harvest quality during a full commercial production cycle. *Aquaculture*. 2013. 410: 41–50.
128. Nuez-Ortín, W.G. et al. Triploid Atlantic salmon shows similar performance, fatty acid composition and proteome response to diploids during early freshwater rearing. *Comparative Biochemistry and Physiology Part D: Genomics and Proteomics*. 2017. 22: 67–77.
129. Peruzzi, S. et al. Growth and development of skeletal anomalies in diploid and triploid Atlantic salmon (*Salmo salar*) fed phosphorus-rich diets with fish meal and hydrolyzed fish protein. *PLoS One*, 2018. 13(3).
130. Benfey, T.J. Producing sterile and single-sex populations of fish for aquaculture. *New Technologies in Aquaculture*. 2009: 143–164.
131. Lekang, O.-I., Salas-Bringas, C. and Bostock, J.C. Challenges and emerging technical solutions in ongrowing salmon farming. *Aquaculture International*. 2016. 24(3): 757–766.
132. Jørgensen, K.M. et al. Investigating the frequency of triploid Atlantic salmon in wild Norwegian and Russian populations. *BMC Genetics*. 2018. 19(1): 1–9.
133. Glover, K.A. et al. The frequency of spontaneous triploidy in farmed Atlantic salmon produced in Norway during the period 2007–2014. *BMC Genetics*. 2015. 16(1): 37.
134. Johnstone, R., Simpson, T.H. and Youngson, A.F. Sex reversal in salmonid culture. *Aquaculture*. 1978. 13(2): 115–134.

135. Wilkins, N.P., Cotter, D. and O'Maoiléidigh, N. Ocean migration and recaptures of tagged, triploid, mixed-sex and all female Atlantic salmon (*Salmo salar* L.) released from rivers in Ireland. *Genetica*. 2001. 111(1–3): 197–212.
136. Lee, P., King, H. and Pankhurst, N. Preliminary assessment of sex inversion of farmed Atlantic salmon by dietary and immersion androgen treatments. *North American Journal of Aquaculture*. 2004. 66(1): 1–7.
137. Johnstone, R. and MacLachlan, P.M. Further observations on the sex inversion of Atlantic salmon, *Salmo salar* L., using 17α methyl testosterone. *Aquaculture Research*. 1994. 25(8): 855–859.
138. Hansen, T., Wargelius, A., Taranger, G.L. and Fjelldal, P.G. *Oppdrett av steril fisk in Rapport fra Havforskningen*. 2012. Havforskningsinstituttet: Bergen.
139. Hoga, C.A., Almeida, F.L. and Reyes, F.G.R. A review on the use of hormones in fish farming: analytical methods to determine their residues. *CyTA-Journal of Food*. 2018. 16(1): 679–691.
140. Fletcher, G.L. and Rise, M.L. *Aquaculture Biotechnology*. 2012. Wiley: Hoboken, NJ.
141. DNA Learning Center. https://dnalc.cshl.edu/.
142. Genome Network. http://www.genomenewsnetwork.org/.
143. Abdelrahman, H. et al. Aquaculture genomics, genetics and breeding in the United States: current status, challenges, and priorities for future research. *BMC Genomics*. 2017. 18(1): 1–23.
144. Houston, R.D. et al. Harnessing genomics to fast-track genetic improvement in aquaculture. *Nature Reviews Genetics*. 2020. 21(7): 389–409.
145. Yáñez, J.M., Houston, R. and Newman, S. *Genomics in Aquaculture to Better Understand Species Biology and Accelerate Genetic Progress*. 2016. Frontiers Media SA: Lausanne.
146. Houston, R.D. and Macqueen, D.J. Atlantic salmon (*Salmo salar* L.) genetics in the 21st century: taking leaps forward in aquaculture and biological understanding. *Animal Genetics*. 2019. 50(1): 3–14.
147. Lien, S. et al. The Atlantic salmon genome provides insights into rediploidization. *Nature*. 2016. 533(7602): 200–205.
148. Roberts, F.L. Atlantic salmon (*Salmo salar*) chromosomes and speciation. *Transactions of the American Fisheries Society*. 1970. 99(1): 105–111.
149. Brenna-Hansen, S. et al. Chromosomal differences between European and North American Atlantic salmon discovered by linkage mapping and supported by fluorescence in situ hybridization analysis. *BMC Genomics*. 2012. 13(1): 432.
150. National Center for Biotechnology Information (nih.gov). https://www.ncbi.nlm.nih.gov/.
151. Hurst, C.D. et al. The complete mitochondrial DNA sequence of the Atlantic salmon, *Salmo salar*. *Gene*. 1999. 239(2): 237–242.
152. Salmobase. https://salmobase.org/.
153. Gutierrez, A.P. et al. Detection of quantitative trait loci (QTL) related to grilsing and late sexual maturation in Atlantic salmon (*Salmo salar*). *Marine Biotechnology*. 2014. 16(1): 103–110.
154. Houston, R.D. et al. Major quantitative trait loci affect resistance to infectious pancreatic necrosis in Atlantic salmon (*Salmo salar*). *Genetics*. 2008. 178(2): 1109–1115.
155. Moen, T. et al. Confirmation and fine-mapping of a major QTL for resistance to infectious pancreatic necrosis in Atlantic salmon (*Salmo salar*): population-level associations between markers and trait. *BMC Genomics*. 2009. 10(1): 368.
156. Hayes, B.J. and Goddard, M.E. Prediction of total genetic value using genome-wide dense marker maps. *Genetics*. 2001. 157(4): 1819–1829.
157. Ødegård, J. et al. Genomic prediction in an admixed population of Atlantic salmon (*Salmo salar*). *Frontiers in Genetics*. 2014. 5: 402.

158. Du, S.J. et al. Growth enhancement in transgenic Atlantic salmon by the use of an "all fish" chimeric growth hormone gene construct. *Bio/technology*. 1992. 10(2): 176–181.
159. Hew, C.L., Fletcher, G.L. and Davies, P.L. Transgenic salmon: tailoring the genome for food production. *Journal of Fish Biology*. 1995. 47: 1–19.
160. Yaskowiak, E.S. et al. Characterization and multi-generational stability of the growth hormone transgene (EO-1α) responsible for enhanced growth rates in Atlantic Salmon. *Transgenic Research*. 2006. 15(4): 465–480.
161. DFO. Summary of the Environmental and Indirect Human Health Risk Assessment of AquAdvantage® Salmon. 2013.
162. Ignatz, E.H. et al. Growth performance and nutrient utilization of growth hormone transgenic female triploid Atlantic salmon (*Salmo salar*) reared at three temperatures in a land-based freshwater recirculating aquaculture system (RAS). *Aquaculture*. 2020. 519: 734896.
163. Waltz, E. GM salmon declared fit for dinner plates. *Nature Biotechnology*. 2016. 34(1): 7.
164. Cleveland, B.M. A perspective of the future value and challenges of genetic engineering in aquaculture. *Journal of the World Aquaculture Society*. 2019. 50(5): 890–893.
165. Bodnar, A. AquAdvantage Salmon Regulatory Timeline. Version 1.0. Biology Fortified. Inc. Mar, 2019. 12.
166. Kisak, P.F. *Crispr Technology: The Revolutionary Breakthrough for Genetics and Evolution*. 2017. CreateSpace Independent Publishing Platform.
167. Ridge, Y. and Boersma, A. *Crispr: A Powerful Way to Change DNA*. 2020. Annick Press: Toronto.
168. Robledo, D. et al. Gene expression response to sea lice in Atlantic salmon skin: RNA sequencing comparison between resistant and susceptible animals. *Frontiers in Genetics*. 2018. 9: 287.
169. Moore, G. Coho biomarkers offer hope of progress in lice fight, in *Fishfarmingexpert*. 2018.
170. Datsomor, A.K. The molecular basis of long-chain polyunsaturated fatty acid (LC-PUFA) biosynthesis in Atlantic salmon (*Salmo salar* L): In vivo functions, functional redundancy and transcriptional regulation of LC-PUFA biosynthetic enzymes. 2019, NTNU.
171. Ayllon, F. et al. The vgll3 locus controls age at maturity in wild and domesticated Atlantic salmon (*Salmo salar* L.) males. *PLoS Genetics*. 2015. 11(11).
172. Barson, N.J. et al. Sex-dependent dominance at a single locus maintains variation in age at maturity in salmon. *Nature*. 2015. 528(7582): 405–408.
173. Ayllon, F. et al. The influence of vgll3 genotypes on sea age at maturity is altered in farmed mowi strain Atlantic salmon. *BMC Genetics*. 2019. 20(1): 44.
174. Vasconcelos, A.C.N. et al. The germ cell marker dead end reveals alternatively spliced transcripts with dissimilar expression. *Scientific Reports*. 2019. 9(1): 1–10.
175. Weidinger, G. et al. Dead end, a novel vertebrate germ plasm component, is required for zebrafish primordial germ cell migration and survival. *Current Biology*. 2003. 13(16): 1429–1434.
176. Wong, T.-T. and Y. Zohar, Production of reproductively sterile fish by a non-transgenic gene silencing technology. *Scientific Reports*. 2015. 5(1): 15822.
177. Pharma, A. Production of reproductively sterile Atlantic salmon for environmentally responsible aquaculture in Norway. https://acdpharma.com/en/products/sterile-fish/.
178. Wargelius, A. et al. Dnd knockout ablates germ cells and demonstrates germ cell independent sex differentiation in Atlantic salmon. *Scientific Reports*. 2016. 6(1): 1–8.

CHAPTER 6

Eyed eggs and start feeding

6.1 Introduction

The eggs are normally sent from the broodstock farm to the hatchery or smolt farm at eyed egg stage. Here the eyed eggs go through hatching, start feeding, fry and smoltification so they are ready for delivery to ongrowing farms in the sea. In rare cases there may be a subdivision whereby one farm produces fry up to around 5 g and another takes them further to smolt, where they can be kept in cages in lakes or rivers.

The first stage at the smolt production plant is hatching and start feeding, which is described in this chapter.

6.2 The hatchery

The hatchery at the smolt farm is where the eyed eggs are taken in and yolk sac fry ready for start feeding are sent out; this is typically done in a separate area of the smolt farm. This area is only used for a limited period of the year but compared to fish-producing units the space required is limited. The area should have its own entrance through a disinfection sluice to avoid possible disease transfer from incoming personnel. Consideration of biosecurity (see Chapter 11) is of major importance in this area because this is the only route of entry of biological material to the farm. Between batches of eggs the area is not used, and is dried out, which is another important biosecurity precaution.

6.2.1 Units for storing eyed eggs

The storing units used for incubation of eyed eggs up until hatching include trays and cabinets (see more in Chapter 5). Cylinders like those used on the broodstock farms cannot be used because the eggs will hatch in these smolt farm hatchery units (Figure 6.1). Combined solutions may also be used, in which the storing of eggs, hatching and start feeding are all performed in the same unit; however, this is only used in smaller-sized farms, and then the hatchery will not be a separate area, but combined with the start feeding area.

Figure 6.1 Units for storing eggs in the period from eyed egg stage to hatching.

The size of the area is dependent on the amount of eggs and the incubation system used; values range from 10,000 to 80,000 eggs per square metre [1].

A hatching substrate is recommended. When the eggs hatch and the yolk sac fry emerge, they will, based on instinct, swim down into the substrate and lie there (Figure 6.2). 'Pillars' in the substrate will help the fry to stay in an upright position and prevent them from swimming around. This ensures that the nutrients in the yolk sac are used for growth, rather than as energy for swimming. Better performance has been documented when using substrates [2–4]. A number of different substrate designs are available, but the major aim is to keep the fry in an upright position.

The water supply to the hatchery unit is either a flow through system or a system that reuses the water several times (RAS) (Figure 6.3). Since there is no feeding, the only release from the eggs is gas so it will be a much simpler RAS compared to what is necessary for fish. A protein skimmer may, however, be added in the RAS to remove released protein and eggshells. There are indications that the eggs are more likely to hatch at the same time when using RAS in the hatchery, as substances released into the water from the first eggs to hatch may stimulate more hatching as the water is reused in the RAS. It is also easier to keep a stable water quality and a water temperature with limited fluctuations if using RAS.

Figure 6.2 Hatching substrate improves growth and survival of the yolk sac fry.

6.2.2 Light conditions

It is of major importance to use the correct wavelength of light in the hatchery. Direct sunlight damages the eggs and must be avoided. In nature this is not a problem because the eggs are lying down in the river gravel. Under farming conditions, low intensity light, perhaps with a cover over the trays, can be used. When adequate working light is needed, for instance, when removing dead eggs, light without the blue-violet wavelength (short wavelength of the visual spectrum), the harmful part, should be used. Warm yellow or red light is suitable for use. It is recommended to have dark rooms or at least dimmed light in general during this period and until the time when start feeding is reached.

6.2.3 Preparation of the hatchery to receive the eyed eggs

Before the smolt farm receives the eyed eggs, the hatchery must be prepared. It must be cleaned, disinfected and dried for a period to avoid potential disease transfer from one intake of eggs to the next. It is important to check the water system for any leakages and there must be no places where eggs or fry can disappear or escape. It is obviously of major importance that all is ready and tested before the eggs arrive and that experienced personnel are responsible for this operation.

Figure 6.3 System for recycling of water in the hatchery.

6.2.4 Receiving the eyed eggs

When the eyed eggs enter the hatchery of the smolt farm, they should be disinfected. This adds another step to ensure no transfer of disease on the eggs, even if they have already been disinfected when leaving the broodstock farm. The disinfection is carried out with a permitted substance, for instance, an iodophor [5] such as buffodine (buffered iodine). The eggs are dipped into this solution at the concentration recommended by the manufacturer, for instance, 100 ml per 10 l of water (see also disinfection in Chapter 5). This will disinfect the surface of the eggs, but it is important to remember that what is contained within the eggs is not disinfected. Vertical transfer of disease agents from generation to generation can therefore not be eliminated. A veterinarian status report on the broodstock is required to reduce the risk of vertical contamination. The disinfection bath will also reduce later fungal problems to some degree. It is also important that all handling of the eggs is done with care, as stress on the eggs may result in premature hatching.

After cleaning of the hatchery, it will be a 'clean zone' and no direct traffic into the hatchery should occur. The extruded polystyrene foam boxes in which the eyed eggs enter the smolt farm should under no circumstances be taken into the hatchery without proper external disinfection. All excess packing material should be discarded. The eggs may also be transferred into the farm's own buckets and water should be taken into the hatchery, or just the individual plates in the box can be taken in after disinfection.

Some viruses and bacteria may, as mentioned, be transferred with the incoming eyed eggs: vertical transfer. Infectious pancreatic necrosis (IPN) is a virus that is very difficult to destroy, and at one point was present in up to 80–100% of ongrowing farms and 40–60% of smolt farms. Fish that have this infection carry the virus for the rest of their lives. Both horizontal (between individuals) and vertical (between generations) spread of the virus occurs. IPN is present inside the eggs and therefore impossible to control. Other diseases can be transferred in the same way. When almost all broodstock farms had IPN, there was no way to guarantee IPN-free eggs. Different strains result in different levels of disease and are described as being of low or high virulence: high virulence indicates increased severity or harmfulness of a disease. There are also large differences in mortality between the strains, from almost 0 (the virus can be present but no deaths are seen) up to almost 100% at fry stage. Inside the farm, different strains of IPN can develop, which can then be transferred between the different egg intakes. Owing to this, it is very important to use disease-preventing efforts such as proper disinfection and drying. Today the gene position (QTL) representing IPN has been identified and this makes it possible to source IPN-free eggs from broodstock farms; the same technology is also being tested on other pathogens (further described in Chapter 5).

6.2.5 Counting eggs

The eggs should be counted or weighed when taken into the hatchery to enable production control on the farm. This must be done for each individual incubation unit. It can be based on volume (when the number of eggs per litre is known), by a counting plate or by weighing scales (when the weight of the eggs is known) (for further description see Chapter 5). Compared to a newly fertilized egg, the weight is slightly lower at eyed egg stage. The size of the eyed eggs may vary to a degree, but there seems to be no correlation between this and the survival rate. However, if the eggs are large then the yolk sac fry are larger when hatched. If different batches are taken into the hatchery, it is advantageous to keep them separately to maintain control of their performance. It is important to fill out a log of the eggs that were received and to track the eggs properly through production so any faults can be discovered.

6.2.6 Water quality and temperature

It is crucial to maintain good and stable water quality during incubation and hatching at the smolt farm. A stable pH of around 6.3–7 is recommended. Below pH 5 there will be greatly reduced hatching and growth [6]. Between pH 5 and just above pH 6 there might also be problems if the water contains metals such as aluminium or iron, which can occur in the snow-melting period, so additional care should be taken to avoid this. The gases in the water and air must be in equilibrium, with no supersaturation of gases. Gas bubbles within the inlet water must be avoided because they may clog systems or disturb the eggs. This includes oxygen, so oxygenated water should not be used. In nature, oxygen saturation has to be above 66% with a temperature above 12.5°C [7]. Under farming conditions, over 7 mg oxygen per litre or over 80% saturation in the incubator is recommended. The recommended carbon dioxide level of the incubator is below 6 mg/l. A fine particle filter ought to be installed on the water supply. The eggs will attract small particles which will settle on the surface of the eggs. Disinfection of the water may also be carried out, and this has the advantage of reducing fungal problems, at least to some degree.

When the eggs arrive on the farm the water temperature is typically in the range 2–5°C, but it is important to measure this to maintain control. To avoid unnecessarily stressing of the eggs at arrival it is important to start with similar temperatures in the production water, then it can be increased gradually to the desired incubation temperature.

It is important that the water temperature is stable during the incubation period before reaching the hatching stage; temperatures between 4°C and 8°C seem appropriate. The typical temperature is between 6–8°C. Temperatures above 8°C degree may result in deformities in the salmon [8,9]. As incubation mostly occurs in late winter and spring, heating of the water to be used for the eggs is normally required but of course depends on the temperature of the water supply. For triploid eggs, temperatures above 6°C are not recommended [10]. Good results have been produced with reuse of water at this stage of production.

6.2.7 Water flow

Water supplies eggs with oxygen and removes metabolic products such as carbon dioxide, so a supply of water is necessary. Generally, a supply of 1–1.7 l/minute per litre of eggs is suggested, but this depends on the egg storing system used; recommendations from the suppliers should be followed. A typical trough with four trays (California system) requires 7 l/minute. It is important to ensure correct water flow. If it is too low the oxygen supply will be too low, and if it is too high then water flow may result in movement of the eggs, which is to be avoided. The oxygen concentration of the production unit should be above 7 mg/l. Gas bubbles must not be allowed to inhibit water flow, and which in turn may also result in early hatching because an oxygen drop, caused by insufficient flow, can induce hatching. It is important to ensure proper distribution of the water, so the water flow is equal around all eggs. A velocity of 100 cm/hour has shown good survival rates [7].

6.2.8 Fungal problems

Fungi present a problem in the hatchery (Figure 6.4). Dead eggs, as described in the Chapter 5, are a very good substrate for fungus growth (*Saprolegnia* spp.). A dead egg can be covered by fungus within a day and the oxygen supply to nearby eggs will be reduced. The fungi will continue to grow and will eventually cover nearby live eggs and totally prevent oxygen supply to them, resulting in their death. The process will continue and if nothing is done the entire hatching unit may be covered with fungi and all the eggs may die.

There are several strategies that are used to deal with fungi, because it is almost impossible to avoid the presence of fungal spores within the water supplied to the hatching units. However, the incidence and severity of fungal problems vary from site to site, and some sites have minimal or no fungal problems. This depends on the water source. The main control strategy used is daily removal of dead eggs. This reduces the problem because the substrate for fungal growth is removed. Addition of chemicals to the water, with flushing or bathing, can be another solution. In the past, malachite green was used (1:200,000, two to three times a week), and it was very effective. However, it is carcinogenic, and also presented a number of other health risks to the

Figure 6.4 Fungi is a problem in hatcheries when eggs die, and if not treated widespread mortalities may be the result.

operators [11]. When eggs or fry were treated with this substance, residues could be found even in market size salmon. Therefore, use of this substance is now prohibited.

A number of other chemical substances may still be used to reduce fungal problems [12]. One substance used today is bronopol (20 mg/l for 30 minutes) [13–15], but other effective substances include hydrogen peroxide (H_2O_2), formalin or salt [16–18]. Disinfection of the inlet water, for example, with UV or ozone, may also have some effect on the concentration of spores [12,19,20]. Fungal spores are in general tricky to remove with disinfection, and high doses are required.

The potential for fungal problems also highlights the importance of optimal water quality and storing conditions to keep mortality as low as possible, since the fungi need organic matter to develop. For the same reason, it is important to disinfect equipment prior to egg intake, and to disinfect the surface of eggs before they are taken into the hatchery.

Use of chemicals or other methods for treatment against fungi, or in general for disinfection, may affect the development of the eggs negatively. It is therefore of major importance that the chemicals or method used do not negatively affect the hatchability, nor has any impact on post-hatching development, such as malformations, growth rate and mortality [12].

6.2.9 Daily work in the hatchery

The daily work in the hatchery includes control of water flow, and control and removal of dead eggs and eggshells at least twice a week. In addition, the fungal situation must be monitored regularly. It is important to maintain good working routines and procedures, and to have dedicated operators for different operations. Handling of eggs at this stage ideally should be avoided because it is best for the eggs to remain undisturbed; however, dead eggs must be removed to avoid increasing any fungal problems, so a difficult balance much be struck.

It is important to have protocols from the stage at which eyed eggs enter the farm. For the eggs, this should include daily recording of mortality. Water temperature and water quality parameters should also be included. The total mortality rate from eyed egg stage to hatching can be around 1–2%, including both unhatched eggs and mortality of hatched fry.

6.2.10 Hatching

The time of hatching can be estimated based on the number of day degrees after fertilization or from when the eyed egg stage was reached. The egg supplier will supply this information when the eggs are delivered. Hatching occurs approximately 280 day degrees (D°C) after eyed egg stage or around 500 D°C after fertilization, with lower values when using lower water temperatures. When taking the eggs into the hatchery the eggs should be at eyed egg stage and at least 50–100 D°C before expected hatching. If taken in too close to hatching time, preliminary hatching may occur as a result of handling. Typically, intake of eggs is at 320 to 380 D°C. If incubation has been performed at reduced water temperatures (2–5°C), calculation of day degrees will overestimate the hatching time, so in general the day degrees should be reduced by 50, meaning hatching will occur around 450 D°C. Hatching will not occur simultaneously for all eggs, but will be within what is known as a hatching window. The hatching window varies but can be quite wide (from 2–20 days). When hatching occurs, the majority will hatch over 2 to 7 days [21]. There will be some premature hatching and some late hatching; typically, the fry from those hatchings will be of reduced quality. When the hatching window passes 7 days it is recommended to destroy the remaining eggs, because higher mortality will occur and they may function as fungus collectors [22].

When close to the hatching point, the embryo gets more and more active, and there will be much movement within the egg. This illustrates the necessity of a good distribution of the incoming water so the egg can take up sufficient oxygen from the surrounding water. When it is close to the point of hatching there will not be enough oxygen inside the egg for embryo development, and this lack of oxygen is a signal for the embryo to start releasing enzymes that weaken the strength of the eggshell. This together with the increasing movement of the embryo results in bursting of the shell so the yolk sac fry can emerge. During hatching it is very important to maintain proper control of the water flow (Figure 6.5). It is easy for the outlet grating in the production unit to become clogged with eggshells from the hatched eggs. Adding an L-shaped grate or sieve in hatching trays can increase the grate/sieve area. Eggshells can also be removed regularly with a siphon. It is also typical for a white layer to be created on the water surface in the incubation unit, due to protein from the shells creating foam. Increasing the water flow prior to hatching can be carried out, but it is not the best solution. Increasing temperature when there is slow hatching is neither an optimal solution because the quality of the hatched fry can be reduced. Some farms increase the temperature by 0.2–0.5°C when hatching starts to achieve a more uniform hatching. For further production planning, the start time, or zero point, can be set when approximately 50% of the eggs have hatched (see also Chapter 9 on production planning).

6.2.11 Storage of yolk sac fry – alevin

When the yolk sac fry emerge from the eggs they are not fully developed fish, and during the post-hatching period they must lie still and use all energy stored in the yolk sac for further

Figure 6.5 Hatching of eggs.

development into a fry that can move and eat (Figure 6.6). This includes full development of the digestive system, organs, fins and so on. A hatching substrate should be included in the incubation unit, as mentioned earlier. The yolk sac fry will instinctively swim down seeking shelter after hatching, and will reach the hatching substrate. They will then lie still while they grow; therefore, improved growth is achieved by use of substrate. It is, however, of major importance that the yolk sac fry are not exposed to any stress during this period, such as handling, high water flow or suboptimal water quality. This may result in a ligated yolk sac, a broken yolk sac, coagulation of the yolk sac (white spots inside) or water in the yolk sac (water shot). It is also important to remove dead fry in this period to avoid fungal problems, but again, this will be a problem because the live fry ought to be disturbed as little as possible.

The water temperature can be at 8°C for hatching and further storage of the yolk sac fry until start feeding, but can be reduced to 2°C to hold the fry longer at this developmental stage. When close to the time for start feeding, the temperature can be raised to 10°C but this must be done gradually over a period of time.

Fry weight is within the range 0.2–0.3 g. The size of the yolk sac fry depends on the egg size: larger eggs produce larger yolk sac fry. At hatching, most of the weight is in the yolk sac, not in the embryo (70% : 30%). Their density is higher than water so the yolk sac fry will naturally sink if not actively swimming. This is related to the natural behaviour where the hatched fry remain within river gravel.

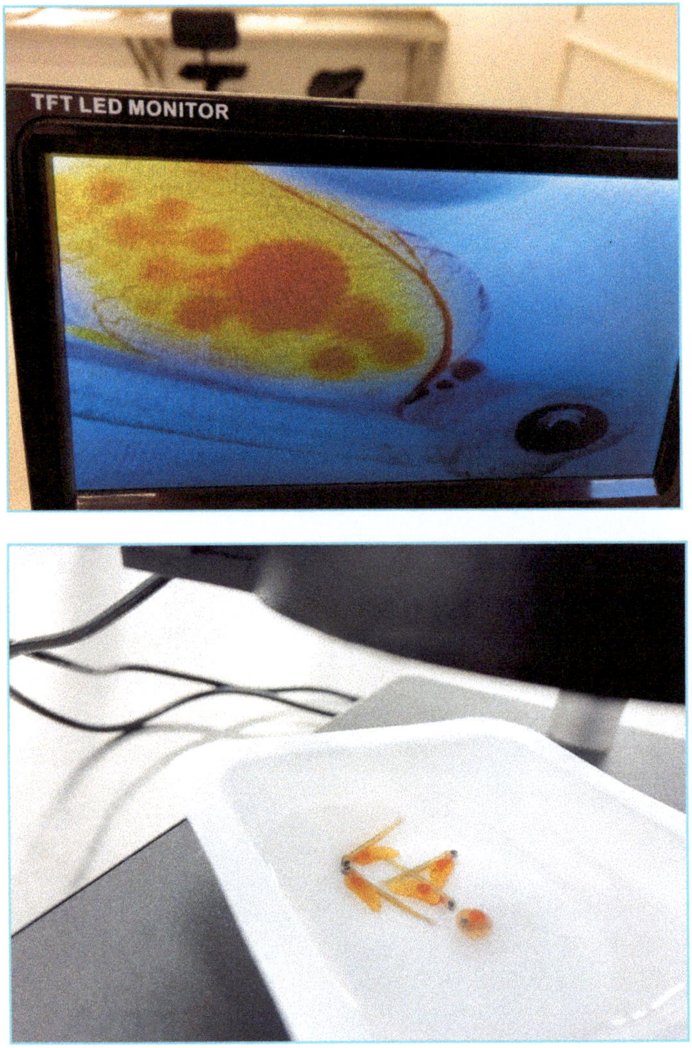

Figure 6.6 Yolk sac fry.

The same light conditions as for the eggs can be used for the stored fry, so either no light or low intensity light. In nature, the fry now lie in the gravel with no light during this period.

6.3 Start feeding

6.3.1 The start-feeding area

Before start feeding, the fry have to be moved from the hatchery and into the start-feeding area. If combined units are used, which is not common, no move is required because the entire process occurs in the same area.

The start-feeding area, including the tanks, must be thoroughly cleaned and disinfected before the fry are taken in. The correct size of outlet grating (typically 2 mm holes) is installed in the start-feeding tanks. This is to avoid the fry being sucked out of the tank with the outlet water. The grating is fixed with a non-toxic sealant.

6.3.2 Intake of sac fry to start feeding

When the fry are taken into the start feeding area, they typically have a weight of around 0.2 g and a length of around 2.5 cm. The yolk sac fry can be moved into the start-feeding tanks when the yolk sac has been consumed by most of the fry. It may be easier to use their behaviour to see when they are ready to be moved, as the fry will gradually move up from the hatching substrate at the bottom and into the water column to search for food: the swim-up stage. In general, once behaviour becomes more active it is time to move fry as a group. This mirrors the process that occurs in the wild when the fry are coming up from the river gravel. This 'swim-up behaviour' results in the fry going to the water surface and filling their swim bladders for the first time. They then become more weightless in water and need to use less energy to stay in the water column searching for food. This behaviour occurs approximately 350–400 D°C after hatching or approximately 900 D°C after fertilization. Some farms prefer to wait until the fry are almost ready for start feeding before moving to the start-feeding tanks, which is when all fry are swimming actively at the water surface of the hatching system.

One to two days after moving, it is usual for the fish to start to take feed. Start feeding is easier in smaller tanks because this simplifies visual control and monitoring of individuals. In the past, tanks with 1–4 m^2 surface area were typically used. Today, start feeding is performed in much larger tanks (>10 m^2) with good results, mainly due to better start feed and increased knowledge of the optimal environment.

Previously, tanks with a reduced water height were used, as the yolk sac fry are not distributed vertically in the water column. Therefore, it was typical to lower the water level during this stage to 15–30 cm. Deep tanks were not considered necessary and had the disadvantage of being more difficult to operate, with respect to catching dead fish and with cleaning. However, it is far more cost effective to use larger tanks and volumes, and therefore it is normal today to use full volume tanks for start feeding. Although when moving the yolk sac fry into the start-feeding tanks they typically remain near the bottom, they will gradually adapt and distribute well in the water column.

It is important to have low enough water velocity at this stage, with a maximum of 3–4 cm/second. A general rule is 1.2–1.5 BL (body length/second). The fry are not adapted for faster swimming. If the water flow is forcing the fry backwards, it is too high, and may also drag the fry backwards resulting in them becoming stacked on the outlet screen. Low velocity may increase the likelihood of gill diseases because the number of suspended solids can increase when the retention time in the tanks increases. Hydraulic self-cleaning is not possible at this stage, as it requires too high a water flow, so manual cleaning is very important for removal of faeces and waste feed.

The water flow has to be high enough to ensure oxygen in the outlet water is above 7 mg. Some approximate values are 0.7–1 l/minute per 1000 yolk sac fry.

6.3.3 Water quality and temperature

The water quality should in general be very good at this stage, because the fry are in a vulnerable period and are going through fast biological development. The total gas concentration needs to be below 102% and nitrogen is recommended to be below 100.5%, so only a very low degree of supersaturation is allowed.

Water temperature is very important at this stage and it is of major importance for the fry to take the artificial food (dry feed). This seems difficult below 7–8°C; the optimum seems to between 12°C and 14°C. In this period wild salmon live in the rivers and therefore will tolerate quite a large temperature fluctuation (0°C to 20°C [23]). It can be an advantage to raise the temperature by some degrees when moving the fry to the start-feeding area. This will ensure a good distribution of the fry in the tank and increase their activity searching for food. If the temperature is too low, overfeeding is easier and will result in an increased number of suspended solids in the water which may create gill inflammation. During this period, the fry will also be more susceptible to gill problems due to feed being supplied with small-sized particles. The problem with higher temperatures is that they can increase any fungal problems. It is important to have a stable water temperature during this period; temperature variations during start feeding are not wanted because they 'disturb' the fish and it becomes more difficult for them to take feed. Typically, the water does need to be artificially heated during the period of start feeding because it normally takes place in wintertime with naturally low water temperatures. By having heaters that control the water temperature, a more stable water temperature can be kept. When the fish are moved to the start-feeding area the water temperature therefore needs to be higher compared to hatching. Exposing the fry to large temperature variations is not desirable, so increasing the temperature by 0.5°C to 1°C per day from the yolk sac fry stage up to the required start-feeding temperature can be a solution. Start feeding is, as said, typically done within a water temperature range from 12°C to 14°C.

Light is important during this stage and correct light intensity will increase both survival and growth. It is important that the fry can see the feed. Light intensity of 10–30 lux at the surface gives good results, but up to 1000 lux is also acceptable [24]. The intensity seems not to be so important, but there must be some light. Too high an intensity may, however, result in increased mortality, so it can be an advantage to have a light control system with the possibility of increasing the light intensity only when working. By having the light source close to the surface, it reduces the creation of shadows when working which disturb the fish. Equal light over the entire tank surface is preferable. It appears that 24-hour light gives the best production results in this period.

The density in kg/m^3 during this phase will be low, up to 10 kg/m^3. The most important requirement is that the fish get easy access to feed. An old rule of thumb was a density of 10,000 yolk sac fry for each square metre of surface area during this period, but that was before the volume was increased.

6.3.4 Start-feeding behaviour

After moving the yolk sac fry to the start-feeding tanks, the fry typically lie on the bottom for the first 1 to 3 days, depending on when they are moved. During this period the fry are vulnerable and care should be taken to avoid them being dragged against the outlet grating. One way to control this is to direct the inlet flow directly against the outlet. Monitoring of the remaining yolk sac is a possibility, as all the yolk sac will be consumed before start-feeding behaviour commences, but this can be difficult to observe in a large tank. One way to check is to take a couple of fry into a small glass, in which it will be easier to examine them.

6.3.5 The feed

Until now the fry have been using nutrients from the yolk sac (endogen) for growth and development. These nutrients are well adapted to the needs of the fish. Now the fry will start on an artificial diet (exogen) and of course this is a great challenge for the fish. However, today's start feed is very well adapted and tasty for the fry, so low mortality associated with start feeding should be expected.

6.3.6 Routines for feeding

Good start feeding results are essential for the final smolt production, so effort ought to be put into this production stage. Specially formulated start feed adapted to the requirements of the fish is used. The mouth of the fish is small so feed particles of a small size, such as 0.5–0.6 mm diameter crumbs, must be used. Hand feeding should be used in addition to automatic feeding, in order to control feeding according to the appetite of the fry. The following feeding behaviour can be observed: a fish starts to take feed into the mouth then spits it out and catches it again until it has learned to eat. The fry prefer to take the feed close to the bottom at first but will gradually go higher in the water column. The fry have small stomachs and need food often, therefore regularly feeding small batches each time, at a frequency of 5–10 times every hour, can be used. Feeding is required for 24 hours a day, so this also means 24-hour light because salmon prefer to have light when eating. The feeders need to be placed close to the water surface to avoid too high a free fall of the crumbs which results in too high a sinking velocity. It is important that the feed has a low sinking velocity so the fry can catch the feed particles. Good distribution of the feed in the tank is important, even though the water flow will help to transport feed around in the tank. Aggressive behaviour with damage to eyes and fins may be a result of poor feed distribution. Use of feeding tables alone to control feeding is not sufficient, as the accumulation of lost feed must be considered. The great advantage for salmon that have used their yolk sac is that their digestive systems are well developed, and they can start eating dry formulated feed at once. This is not the case for most other farmed species, such as marine fish species.

6.3.7 Sheltering

Some farms use small pieces of hatching substrate at the bottom of the start-feeding tank to create shelter for the fry (Figure 6.7). This will inhibit water flow slightly, so it is important to carry out proper cleaning under the substrate, which creates additional work. In the larger tanks that are more commonly used today this is not possible. The velocity of the water must not be so high that the fry are sucked against the grating at the outlet.

Figure 6.7 Sheltering can be used in the tanks for the fry but inhibit the water flow; it is also important to avoid the fry becoming sucked against the outlet grating.

6.3.8 Mortality

Start feeding represents the most difficult stage in the production of salmon and higher mortality than at later stages must be expected. Even with increased knowledge about optimal farming conditions and improved start feed, a total mortality of up to 10–15% can occur during this period, although very low mortality of a few percent is more typical. After 2–3 weeks, dependent on water temperature, there is normally a peak in mortality. This is because the fry have not yet started to eat and the yolk sac has been fully consumed.

6.3.9 Hygiene

Hygiene is important at this stage because the fry are small and weak. The tanks must be brushed regularly due to the lack of hydraulic self-cleaning. Use of a separate brush for each tank to avoid any cross contamination is recommended. Dead fry must be regularly removed, and it is preferable to use a separate net for each fish tank for removal of dead fish if there is no system on the tank outlet where dead fish are collected and removed. Overfeeding and faeces accumulation must be controlled, and the unit has to be regularly drained to avoid blocking of the outlet pipe which can occur because the water velocity is so low that particles settle in the pipe. For disinfection of surfaces and equipment, many chemicals can be used, for instance, ethanol, potassium peroxymonosulfate (oxidation), iodophors and acids/bases [25].

6.3.10 Parasites, diseases and fungus

The yolk sac is sensitive to parasites. Costia (*Ichthyobodo necator*) and Trichodina are typically seen. These parasites enter with the inlet water and are quite small in size. However, they can be eliminated using a fine mesh filter in the inlet. If they do enter the tank, a common treatment if too high a concentration occurs is with an antiparasitic agent, such as a formalin or hydrogen peroxide bath. Using a formalin bath with a concentration of 1:4000 for 30 minutes is one solution. One procedure for this is to reduce the water level in the tank, add formalin at the required amount and then refill the tank with fresh water [24]. It is important to remember that yolk sac fry have a lower tolerance of formalin and chemicals in general than the eggs.

Gill inflammation is also quite typical at this stage, due to the potential for high particle concentration in the water in addition to bacteria that will always be present in the water. Possible reasons for an outbreak can be overfeeding, insufficient cleaning of inlet water, suboptimal water quality and improper cleaning of tanks. The usual treatment for this is again a formalin bath.

Fin rot is another bacterial infection that can occur. Bacteria (*Flexibacter* spp.) cause the infection which results from damage to the skin protection of the fins. It is mainly the back fin that is affected. Reasons are again related to bad water quality, too little feed followed by aggression, too high a fish density and other factors which result in stress on the fish. Again, the treatment is a formalin bath but antibiotics can also be used.

Other diseases, such as bacterial kidney disease (BKD), may also occur but these require veterinary treatment (see Chapter 11).

References

1. Benchmark. *Atlantic Salmon Hatchery Manual 2019*. Benchmark: Sheffield.
2. Peterson, R.H. and Martin-Robichaud, D.J. Yolk utilization by Atlantic salmon (*Salmo salar* L.) alevins in response to temperature and substrate. *Aquacultural Engineering*. 1995. 14(1): 85–99.
3. Hansen, T.J. and Møller, D. Yolk absorption, yolk sac constrictions, mortality, and growth during first feeding of Atlantic salmon (*Salmo salar*) incubated on astro-turf. *Canadian Journal of Fisheries and Aquatic Sciences*. 1985. 42(6): 1073–1078.
4. Hansen, T. and Torrissen, K.R. Artificial hatching substrate and different times of transfer to first feeding: effect on growth and protease activities of the Atlantic salmon (*Salmo salar*). *Aquaculture*. 1985. 48(2): 177–188.
5. Alderman, D.J. The toxicity of iodophors to salmonid eggs. *Aquaculture*. 1984. 40(1): 7–16.
6. Buckler, D.R.et al. Survival, sub-lethal responses, and tissue residues of Atlantic salmon exposed to acidic pH and aluminum. *Aquatic Toxicology*. 1995. 31(3): 203–216.
7. Crisp, D.T. Environmental requirements of common riverine European salmonid fish species in freshwater with particular reference to physical and chemical aspects. *Hydrobiologia*. 1996. 323(3): 201–221.
8. Ørnsrud, R., Gil, L. and Waagbø, R. Teratogenicity of elevated egg incubation temperature and egg vitamin A status in Atlantic salmon, *Salmo salar* L. *Joural of Fish Disease*. 2004. 27(4): 213–223.
9. Baeverfjord, G. et al. High temperatures during egg incubation may induce malformations in Atlantic salmon (*Salmo salar* L.). *Aquaculture and Water*, EAS special publication, 1998(26): 24–25.
10. Fraser, T.W.K. et al. The prevalence of vertebral deformities is increased with higher egg incubation temperatures and triploidy in Atlantic salmon *Salmo salar* L. *Joural of Fish Disease*. 2015. 38(1): 75–89.
11. Sudova, E. et al. Negative effects of malachite green and possibilities of its replacement in the treatment of fish eggs and fish: a review. *Veterinari Medicina Praha*. 2007. 52(12): 527.
12. De Swaef, E. et al. Disinfection of teleost eggs: a review. *Reviews in Aquaculture*. 2016. 8(4): 321–341.

13. Oono, H. et al. The use of bronopol to control fungal infection in rainbow trout eggs. *Biocontrol Science*. 2007. 12(2): 55–57.
14. Braidwood, J.C. Use of bronopol for the treatment of diseases in fish. 2000. Google Patents.
15. EMEA, Bronopol. 2001. The European Agency for the Evaulation of Medicinal Products, Veterinary Medicines and Inspections, EMEA.
16. Yamamoto, A. et al. Control of fungal infection of salmonid eggs by hydrogen peroxide. *Fish Pathology*. 2001. 36(4): 241–246.
17. Lee, B.-Y. and Lee, D.-C. Comparative efficacy of antifungal agents for aquaculture fish and their eggs. *Korean Journal of Fisheries and Aquatic Sciences*. 2009. 42(1): 34–40.
18. Marking, L.L., Rach, J.J. and Schreier, T.M. American fisheries society evaluation of antifungal agents for fish culture. *The Progressive Fish-Culturist*. 1994. 56(4): 225–231.
19. Forneris, G. et al. The use of ozone in trout hatchery to reduce saprolegniasis incidence. *Aquaculture*. 2003. 221(1–4): 157–166.
20. Kasai, H., Yoshimizu, M. and Ezura, Y. Disinfection of water for aquaculture. *Fisheries Science*. 2002. 68(sup1): 821–824.
21. Aqua-Gen. Consistent quality and good hatching results for early eggs 2012. AquaGen egg update, 2013(1/2013).
22. Aqua-Gen. Optimal egg quality – the basis for a good start on a new salmon generation. AquaGen Informtion letter. 2006.
23. Elliott, J. and Elliott, J.A. Temperature requirements of Atlantic salmon *Salmo salar*, brown trout *Salmo trutta* and Arctic charr *Salvelinus alpinus*: predicting the effects of climate change. *Journal of Fish Biology*. 2010. 77(8): 1793–1817.
24. Kittelsen A.et al. *Drift og røkt, akvakultur*. 2002. Oslo: Gan Forlag.
25. OIE. Disinfection of aquaculture establishment and equipment, in *Aquatic Animal Health Code 2019*. The World Organisation for Animal Health (OIE).

CHAPTER 7

Fry to smolt: ongrowing at the smolt farm

7.1 Introduction

When fry reach 1 g in size, the normal ongrowing period on the smolt farm starts. The fry are now familiar with eating formulated feed. They will gradually develop to what is called the parr stage, which is reached at a size of 1–2 g. They now develop the characteristic parr marks: finger-like marks on the side of the fish (Figure 7.1).

At the outset of the ongrowing stage, it must be decided whether the 0^+, 1^+ or 2^+ production model will be followed (Figure 7.2). A 0^+ or ½ year smolt is a salmon that is ready for transfer to seawater as smolt in the autumn, less than 1 year after hatching. A 1^+ smolt is ready for the sea around 1 year after hatching, and is transferred to seawater in the spring or early summer. A 2^+ smolt is transferred to the sea around 2 years after hatching. This last model is generally not used in commercial aquaculture anymore, but can be used in cultivation (salmon released for restocking of natural strains). However, some facilities are now returning to this model, testing it out in commercial farming in order to create a strong smolt that can better handle sea conditions, and in addition one that might achieve the same fast growth in the sea as is seen in wild salmon. Today though the majority of smolt production is 0^+ and 1^+. With the introduction of more RAS on smolt farms, utilizing higher water temperatures, more and more smolt are 0^+ fish and smolt production can be independent of season. This is possible since production is indoors, and light and water temperature can be controlled. With this seasonally independent smolt production, the ongrowing farms have access to smolt all year round. The result is that it is easy to deliver slaughter-size fish to the consumer market all year round. In addition, it will be possible to optimize the maximum allowed biomass (MAB) on a site, if this is a limitation for the ongrowing phase in the sea cage farm.

Each production model or regime requires different preparation. To produce 0^+ smolt, both heating of the water and light manipulation (to create artificial day length) is necessary. To produce 1^+ smolt, heating of the natural water supply will normally be necessary. Depending on the initial temperature of the water to produce 2^+ smolt, natural water temperatures may be used as there is normally no need to heat the water to obtain smolt ready for seawater within 2 years after hatching.

Figure 7.1 Parr of salmonids can be identified by characteristic finger marks.

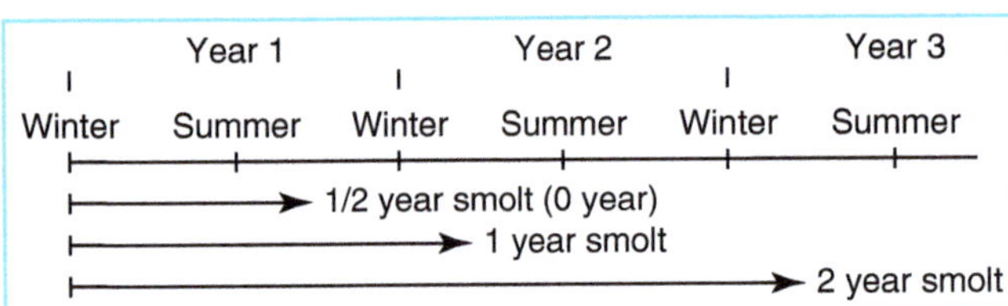

Figure 7.2 Different production models.

Other factors that have to be decided are the size of the smolt ordered, and number in each of the size categories. Estimated delivery dates also need to be known. All these factors will dictate the production plan for the smolt farm.

Another production model that has been used, although to a very limited degree, is to produce fry inside and then place the fry in cages in lakes or large, slow-flowing rivers. The fry are transferred to the cages in freshwater at a weight of 5–10 g. One problem with such a production model is controlling the discharge from the cages. It is also not possible to control the process of smoltification, as it will follow the natural rhythm of smoltification in the spring when natural water temperatures and light conditions increase.

7.2 Water temperature, quality and supply

7.2.1 Water temperature

Similar to incubation and start-feeding, the water supply, temperature and water quality are important. The water temperature needed for the fry is dependent on the chosen production regime. It ought to be within the range 1–20°C. However, at low temperatures the growth rate of the fish will be very limited. Close to 20°C it will be difficult to supply enough oxygen to the

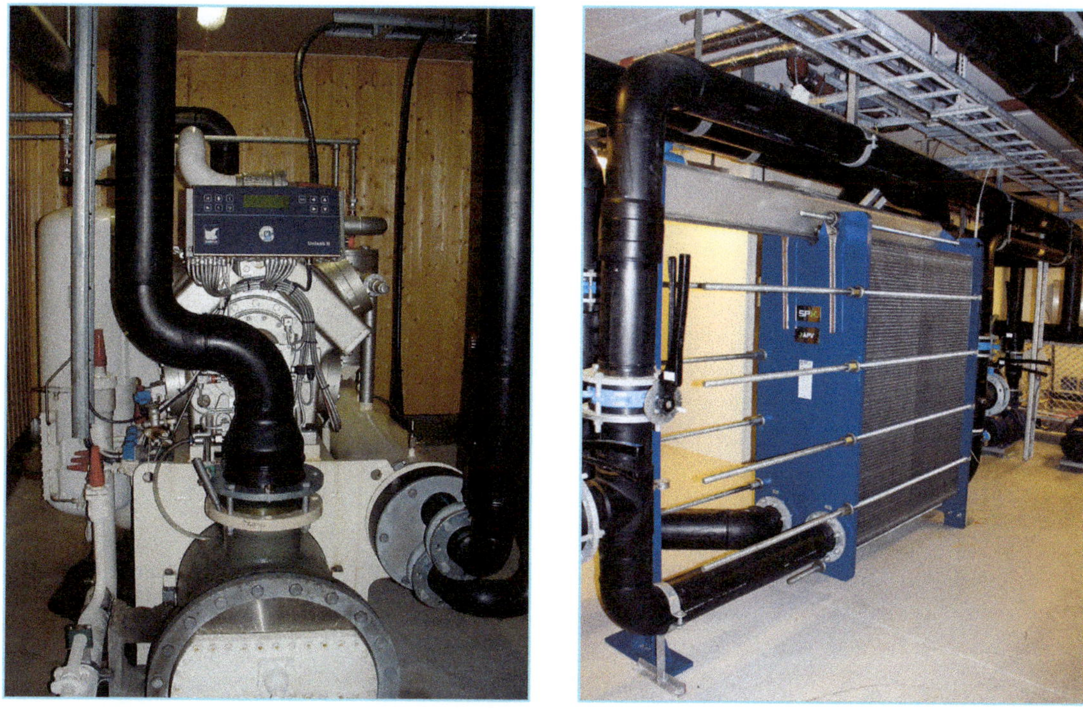

Figure 7.3 Heating systems, including heat exchangers and heat pumps, are used to heat and obtain a more stable water temperature.

fish, as the fish have high oxygen requirements and the amount of oxygen gas that the water naturally contains drops rapidly as the temperature increases. Production temperatures above 16°C are therefore not recommended. However, at this temperature the growth rate will be very high and so a fast turnover is possible; as an example, the required time for the fry to grow from 30 g to 70 g will vary from 25 to 39 days when the temperature is raised from 12°C to 16°C, based on normal growth rates. A stable water temperature is preferable, since temperature fluctuations may stress the fish and will also make production planning more difficult (Figure 7.3).

A drawback of increasing the water temperature above the recommended levels is that it is more likely to result in malformations in the salmon [1,2]. Temperatures above 16°C have experimentally resulted in increased vertebral malformations [1]. This is also a highlighted problem when using RAS, where it is easy to obtain high temperatures and fast growth, especially in summer. To achieve higher water temperatures than those of the water source requires a water heating plant, and a water aeration plant will also be required to ensure safe water quality, with no supersaturation of nitrogen gas. If the water temperature gets too high, a water chilling plant will be required.

7.2.2 Water quality

To achieve optimal growth and maintain good fish welfare within the given temperature regime the water quality has to be excellent. A number of toxic elements can make the water quality suboptimal. The pH of the water may range from 5.5–8.0 (tolerance depends on life stage); the optimum range is from pH 6.2–7. High metal concentrations must be avoided, and special precautions must be taken to avoid precipitation of metal ions on the fish gills if the conditions are suitable for interaction of metal ions (Al, Fe) and the pH is low.

At this stage, water salinity should be considered. As the fry grow, their tolerance of salinity increases gradually as their salinity regulation system gradually develops. A small dose of seawater may therefore be added at fry stage (parr) and gradually increased towards smoltification. This can be used if the freshwater supply to the farm is limited. It can also be a tool to control smoltification (see section 7.15.5). However, it is important to disinfect incoming seawater to avoid the smolt contracting seawater borne diseases, which may transfer further when the smolt are sent out to sea cages. It is not as critical for freshwater infections because they will normally not be transferred to seawater. Fish size is also important for seawater tolerance; as the size of the fish increases, they gain more seawater tolerance. This is comparable to nature, when the smolt follow the water flow down the rivers and into the estuaries.

The saturation of dissolved gases is important for juvenile salmon. The nitrogen gas concentration in the inlet water must be below 102% supersaturation and the carbon dioxide gas concentration in the outlet water of the fish tanks must be below 10–15 mg/l (Figure 7.4). The oxygen concentration in the outlet water should be above 7 mg/l, but there seem to be growth benefits with up to 90% saturation even though the growth increase is limited once above 80% saturation. The oxygen gas concentration in the inlet water ought to be as close to 100% saturation as possible to reduce the required amount of water, therefore aeration of the water prior to entry into the fish tanks will be necessary. If adding pure oxygen to the water, oxygen concentrations above 100% saturation can be reached. However, too high a supersaturation with oxygen gas is not recommended. Pure oxygen gas in the water is actually toxic to the salmon, because pure oxygen bubbles, not dissolved in water, can oxidize the surfaces of gills (pure oxygen is an effective oxidant). However, supersaturated water when it reaches the fish tank will rapidly lose saturation as it changes from pure gas bubbles in the water to dissolved forms once the saturation in the water is below 100%. It will normally be below 100% saturation if the fish consume

Figure 7.4 Aerator used for degassing of water before the water enters the fish tank to remove any supersaturation of nitrogen gas. This is especially important after heating of the water.

oxygen in the tank. Inlet oxygen concentrations above 120–130% supersaturation are, however, not recommended.

7.2.3 Water supply and treatment

How much water must be supplied to the juvenile tanks is dependent on whether pure oxygen gas is added, and if the water is supersaturated. Tables have been developed to calculate the necessary water supply based on water temperature and fish size (see Chapter 9 on production planning). The basis for these tables is how much oxygen the water can contain at different water temperatures, assuming a saturation degree of 95–100% and that the outlet oxygen concentration at the given temperature is 70–80% saturation or 7 mg/l.

If adding pure oxygen gas to the water and supersaturating the supply to the fish tank, the volume of water supply can be reduced (Figure 7.5). How much water to add is then dictated by the need for oxygen based on fish size and water temperature. Increased water temperature increases the metabolic rate and therefore the oxygen consumption of the fish. Tables are also available for this purpose (Chapter 9). However, if adding pure oxygen and using reduced water flow together with high fish densities, problems may occur with build-up of high concentrations of carbon dioxide and eventually ammonia in the outlet of the production unit; this has to be controlled. The requirements for hydraulic self-cleaning of the tanks (see Chapter 9) may not be fulfilled if inlet water flow to the production unit is reduced when using pure oxygen; this has to be checked by calculation (for instance [3]). To have some idea, the freshwater supply to a production tank should be in the range of 10–30 l/minute/m^3, using a tank with circulating water flow and a traditional tank inlet and outlet design.

Figure 7.5 Addition of pure oxygen gas is carried out on all smolt farms to reduce the supply of new water, here illustrated with oxygen cones where oxygen gas is mixed into the water, a tower storing liquid oxygen and an evaporator transferring liquid oxygen into gas.

7.3 Production units

Different types of production units can be used including closed units with a one-way flow pattern (raceway) or with a circulating flow pattern (tanks). Cages or pens in freshwater may also be used. The production units that are commonly used are either circular tanks or square tanks with rounded corners, and a circular flow pattern (Figure 7.6). If the tank is correctly designed, including the inlet and outlet, the fish will distribute throughout the entire tank and therefore utilize the entire volume of the production unit. Typical tank sizes are 3–20 m diameter with up to a 5 m water height. Larger tanks represent a lower cost and require less maintenance per cubic metre farming volume. Owing to this advantage, fish production tanks with a diameter above 30 m have been tested. There should be a certain ratio between diameter and height in the tank to ensure good hydraulic self-cleaning with traditional inlet and outlet arrangements, which is normally 0.25 : 0.5.

Often, ongrowing of juvenile salmon is divided into a two-step process that starts with growing the fish in smaller tanks before moving to larger tanks, normally in the summer or at the end of summer, depending on hatching time and water temperature. This is also dependent on the production model used for this stage. Production of 0^+ smolt requires artificial light control and so indoor growing must be used. If producing 1^+ smolt the first stage can be indoors before the salmon are moved to an outside plant with tanks to continue ongrowing, typically at the end of summer. With the new large RAS smolt plants, the entire production process up to delivery of smolt will be indoors. A three-step model is also used which begins with start feeding, continues with fry/parr and ends with smolt.

When the fish have reached the parr size they can tolerate a much higher water current than the yolk sac fry. Therefore, the tanks (production units) can be hydraulically self-cleaned, meaning that the inlet water is used to keep the tank free from feed loss, faeces and biofouling [3]. To achieve this, it is important to have a correctly designed tank inlet and outlet, in order to achieve the correct flow pattern and water exchange. The result is tanks that both are hydraulically self-cleaned and also secure an even distribution of fish throughout the entire

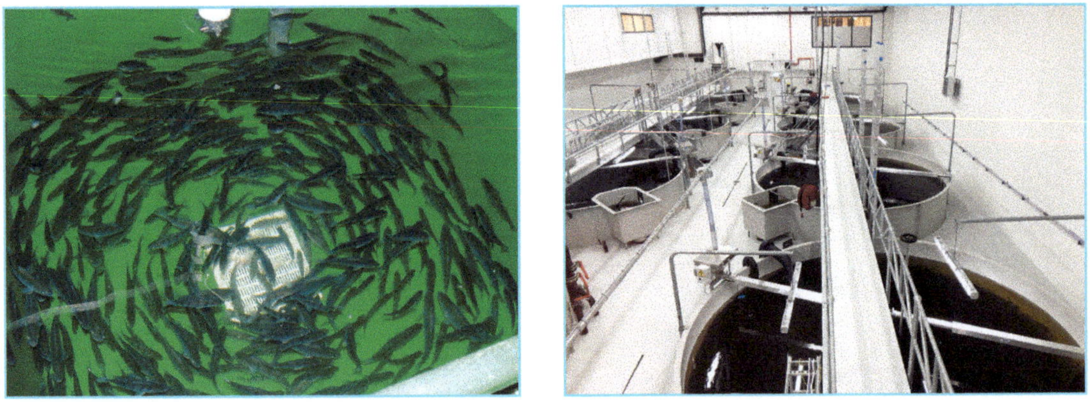

Figure 7.6 Tanks with a circular flow pattern are the most commonly used production unit for smolt production.

tank volume. The last point is important because then the entire production volume can be utilized.

The surface of the production tanks should be smooth because then removal of biofouling will be easier. Cleaning and disinfection of tanks between intakes will also be easier if the surfaces are smooth. The water flow and water velocity in the tanks is normally either based on water requirement or based on retention time (max 100 minutes using additional pure oxygen) (further explanation in Chapter 9, production planning). Through observation of fish behaviour water velocity can be controlled: the fish should remain in the same place in the current and should be distributed throughout the entire tank volume, not swimming forward or being forced backwards, or staying crowded in some parts of the tank. For salmon, an old rule of thumb states the water velocity in centimetre per second should be equal to the fish body length in centimetre. However, exercise is important for the salmon. Salmon exposed to swimming velocities of one to two body lengths per second compared to static water showed increased growth, increased FCR, a more uniform growth rate for the total fish group, fewer aggressive interactions, reduced fin and body damage, reduced relative rate of oxygen consumption, more effective swimming patterns, increased aerobic capacity and quicker recovery after stress/exhaustion [4–6].

Optimal swimming speed regarding maximum growth varies with fish size, environmental conditions and season (a higher swimming speed is seen during summer months with higher temperatures (Figure 7.7)). Over shorter time periods (30 minutes), critical swimming speed has

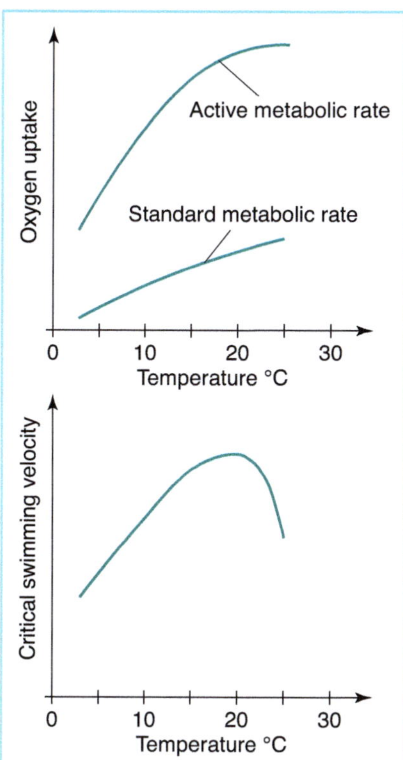

Figure 7.7 The oxygen consumption increases with water temperature and with the metabolic rate. The critical swimming velocity (maximum swimming speed) is also affected by a rise in the water temperature until a ceiling of 20°C (based on [9]).

been measured to four body lengths per second (0.8 m per second) in smolt [7]. Trials have also shown that by increasing the water velocity in the tank in order for the salmon to exercise, the percentage of fish with fin damage may decrease [8].

The flow velocity is also dictated by the water retention time in the tank: a longer retention time means reduced velocity. If velocity is suboptimal, special consideration to the design of the tank inlet and outlet, among other factors, may need to be given.

At the tank outlet, screens are installed to keep the fish within the tank. The holes in the outlet screen can gradually be increased from 2–10 mm as the fish grow. It is best to have holes as large as possible because this reduces friction and improves self-cleaning. It will also be easier for faeces particles to flow through the screen.

Cages in lakes or rivers may also be used as production units for the last part of the production cycle on the smolt farm, even though this is not very common. Here, the fry are produced inside in tanks until sizes of 5–10 g when they are transferred to the cages. The great benefit of production in cages in lakes is the low production cost. However, it also includes some major drawbacks. Light and temperature control is impossible, meaning the smolt will smoltify only in the spring as in nature, when day length and water temperatures in the lakes increase. The water quality and water temperature must also be in the range of what is acceptable for salmon, since no regulation is possible. For example, high summer temperatures may be a challenge. Another major drawback is the possibility of pollution. The production must be suited to the nutrient balance in the lake so that the lake can handle the discharge from the farm, without leading to any eutrophication. This requires a certain size of lake and also a degree of water exchange. Using slow-flowing rivers is therefore also an alternative. The inclusion of a waste handling system in the cages, either to collect all effluent using closed cages or to collect a major part of the effluent by having cages with a closed bottom but net for the sides, is a potential solution. By having a closed bottom, the major part of the particles that settle can be collected and transported to a filtration unit. With a traditional land-based smolt production farm with drainage to a lake or river particle filtration will also be important, and it should be possible to control the discharge from the farm.

7.4 Feeding routines

At this life stage, feed is delivered by automatic feeders and the amount is typically given according to tables. It may be further controlled with some additional hand feeding, or by watching for any feed particles accumulating in the outlet. It can be an advantage to slightly overfeed rather than underfeed to avoid aggression, at least earlier in the life stage. This may result in some increase in discharge but the amount is still small, and it is quite easy to remove any uneaten feed by adding a strainer in the outlet from the farm, which is required in any case. The amount of feed used is still limited so the consequences of slight overfeeding on the environment are reduced. The feeding frequency will be higher in the first part of the life stage when the fish are smaller, with the frequency reducing from once every 10 minutes to once every 30 minutes. The optimal particle size is 2.2–2.6% of the fish fork length [10], but feed suppliers have tables that give the recommended particle size for different sized fish (see also Chapter 10, feeding). The meal size can also be defined based on the number of particles per meal.

During this period, it is important to provide a feed containing enough protein and phosphorous for the fast growth of muscles and bones. The FCR in this period can be very good and it is not unusual to have an FCR below 1 on commercial smolt farms.

Automatic feeders ensure a high enough feeding frequency and enable 24 hour feeding. Different types of feeding equipment are available, including feeders installed on each tank and central feeding systems.

7.5 Fast-growing phase

When the salmon reach a length of approximately 8 cm they are reaching the fast-growing stage [11,12] (Figure 7.8). The time taken to reach this stage varies with environmental factors, genetic origin and possibly geography. The length of this stage is quite short (8–10 cm). It is important to be aware of this because the fish must be supplied with an increased amount of feed as the growth rate becomes several times higher [13]. Length increase of up to 1 mm/day has been recorded in this phase. Owing to this sudden increase in growth a bimodal size distribution can be achieved with individuals that have gone through the stage and individuals that have not, creating an upper and lower bimodal in the fish group. After the fast-growing stage, the growth rate will not be so sensitive to temperature. On smolt farms with outdoor tanks, this is the typical time to move the fish outdoors.

It is also within this stage that the fish start the process of smoltification. The fish that have gone through the fast-growing stage can now develop to become smolt ready for seawater. If the required conditions are provided, this is the trigger for production of 0^+ smolt, for delivery

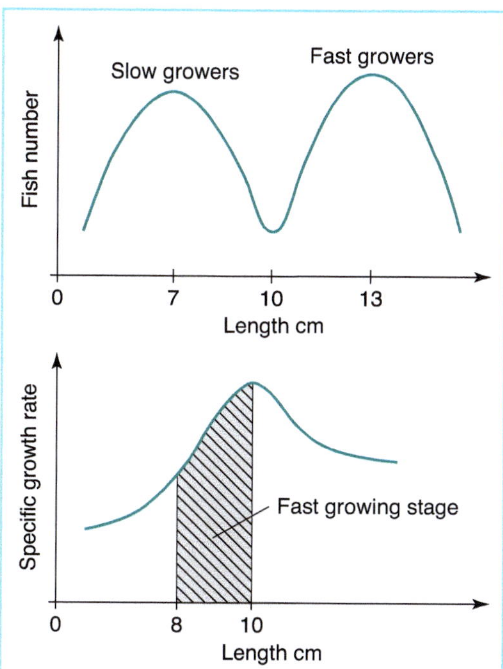

Figure 7.8 When the fish reach a length of about 8 cm the fast-growing phase starts; a bimodal size distribution will occur in the fish group (based on [17]).

in the autumn. The lower bimodal group of the population will continue to grow and develop and typically the growth rate will decrease due to the lower water temperatures in the autumn, so these fish will remain over the winter season before becoming 1^+ smolt. This again is parallel to what happens in nature, where the lower bimodal will smoltify the next spring; however, if the temperature and food supply is very good during winter and spring the lower bimodal of the group may also smoltify [12,14–16].

The growth rate during this period is highly dependent on the photoperiod. In natural light, the growth rate increases with increased day length during springtime while it decreases with reduced day length during autumn.

7.6 Light

During this period light is important. For optimal growth, 24 hour light and feed over the entire period is recommended (Figure 7.9). The intensity is less important for growth. Good growth has been achieved both with 10 lux and with 1000 lux at the water surface, but over 43 lux is recommended to achieve proper smoltification [18] (more information about light and how light is measured is described in Chapter 8, ongrowing). For the requirements of the fish and to give proper working light, 500–800 lux will suffice. However, there are no observed problems with reducing the light intensity during night-time although it is important to be aware that this may be a trigger for smoltification (section 7.15.5). Continuous light may lead to more unsynchronized smoltification if no other precautions are taken. In large deeper tanks submerged lights are used in addition to flood lights above the surface. This is especially important in RAS with more turbid water.

7.8 Stocking density

Stocking density is an important parameter as it may negatively affect the growth rate, FCR and incidence of fin damage [19]. It may also raise the stress levels which could lower immunity.

Figure 7.9 Light is used for 24 hours to optimize the growth, either above the surface or submerged in large deep fish tanks.

Stocking density varies with fish size, and in general can be increased with increasing juvenile size. When setting maximum stocking density (expressed in kilogram of fish per cubic metre of farming volume) animal welfare must be taken into consideration. National regulations or recommendations may also dictate this value, as might insurance conditions. Stocking densities from 10 kg/m^3 for fish up to 1 g, up to 50 kg/m^3 for fish above 30 g is one such national recommendation [19,20]. A number of experiments have also been done with different fish densities. In one experiment, densities up to 86 kg/m^3 on pre-smolt showed no negative effect on growth, fin condition and a number of other parameters [21], while in another experiment smolt densities of 8 kg and 30 kg/m^3 were compared and aggression was more frequent in the low-density group; however, fin biting was highest in the high-density group. The lowest growth and body condition was observed in the low-density group [22]. For Atlantic salmon parr with a mean weight of 5.8 g, a maximum density of 146 kg/m^3 was recommended as appropriate, but the preferred density was below 80 kg/m^3 [23].

When using high densities, it is more difficult to ensure access to feed for all fish and satisfactory water quality, and therefore control of these is very important when increasing the fish density. Fish density above 40–50 kg/m^3 will typically also require the addition of pure oxygen gas to the water because it is difficult to provide enough oxygen with the natural water without producing too strong a water current within the tank. Too high a retention time in a production tank with high fish densities will result in accumulation of carbon dioxide above the recommended level and so carbon dioxide aeration will then have to be included in the production unit. High fish densities also increase the necessity for fast reaction times when something fails, such as a water stoppage. The risk for disease will also increase.

A density that is too low, however, may increase hierarchy within the tank because the individual salmon become more territorial. In one experiment performed with post-smolt, better welfare and less aggression was attained at 25 kg/m^3 compared to 15 kg/m^3 [24].

Higher fish density will be more economical provided that the oxygen concentration and other water quality parameters can be kept within acceptable levels so maximum growth can be achieved. This is because more fish can be kept in the same volume, but here it is important that animal welfare is considered. It can be a problem to ensure that all fish get feed when required, the water quality for the individual fish may be a problem and a very good distribution of water around each fish is needed. Too high a density may also result in wear on fins. Based on this, densities up to 60 kg/m^3 for larger parr and up to 80 kg/m^3 just before splitting of groups or delivery of smolt may be recommended.

Instead of reporting the density in kilograms per cubic metre, percentage of fish in relation to water may be used, for example, 10% density means 100 kg/m^3.

7.9 Internal transport and size grading

7.9.1 Splitting of the fish group

In the smolt farm fish will need to be moved or transported internally during the production period for several reasons. The fish grow continuously and if the fish group is not moved or divided, the fish density in the production tank becomes too high (Figure 7.10). To avoid too

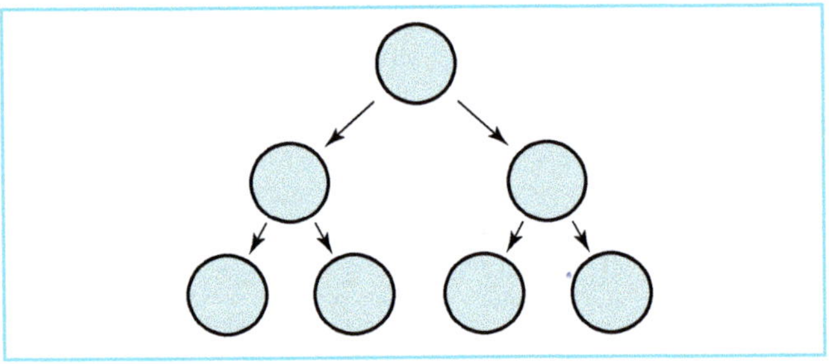

Figure 7.10 Fish growth requires splitting of the fish group to avoid too high a density in the production unit.

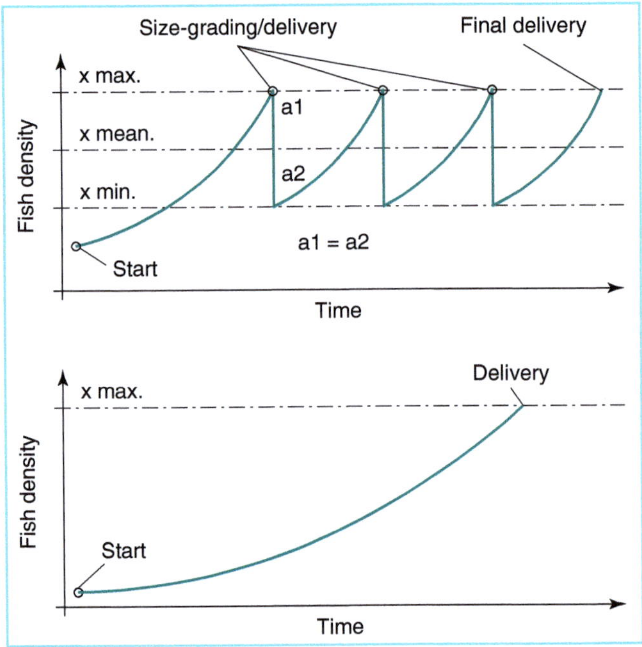

Figure 7.11 The frequency of splitting affects the mean fish density in the production unit [3].

large a variation in fish size in a production unit, the fish should be divided into different size groups, which also requires moving the fish. This size grading enables weight control in the tanks and correct vaccination of the fish. Vaccination also requires internal movement of fish on the farm.

The frequency of dividing the fish groups depends on, among other things, the desired average overall fish density and through this the best utilization of farming volume (Figure 7.11). At first, the density is low and then increases until it reaches the maximum allowed. If there is no splitting of the fish group, the intake density must be very low to avoid a high density at the end, so as not to overcrowd the tank and reduce animal welfare. The result is a low utilization of the available farming volume. The more expensive the farming volume, the higher the required average density throughout the year will be. Particularly in today's RAS plants with high investment and running costs, it is important to carry out regular splitting of the fish groups to keep a high overall fish density. Production models with high regularity of splitting also require regular intake of eggs and sending out of smolt. The smolt (number) or weight (kg) produced per unit farming volume per year is therefore an important parameter through which to evaluate the production efficiency of a farm.

There are a number of methods for internal transport of fish [3]. Pumping is most common and there are different pump principles available.

7.9.2 Size grading

Size grading on the smolt farm will lead to overall improved growth for the entire fish group (Figure 7.12). It is also very difficult to keep satisfactory production control in the fish tanks if the size variation in the production tank is large. Size grading will reduce the variation in weight (coefficient of variation) in the group. If the size variation becomes too large it will be difficult to provide the optimal feed particle size. Grading frequency varies, but at least 2–3 times during the smolt production period is recommended. This is often done in combination with splitting or vaccination.

When performing size grading there will always be a debate as to whether the smallest fish group should be destroyed or fed more. Owing to the individual variation within a fish group there will always be individuals that grow fast and others that grow slowly. An increased total growth rate of the entire fish group may be achieved by destroying the slow growers. It is an

Figure 7.12 Size grading of fish here performed manually with a cradle but typically carried out with grading machines.

advantage to remove these as early in the production process as possible, typically at the first or second grading. In addition to slow growth, these fish may also have other defects and be more susceptible to disease. To compare with salmon stocks in the wild, typically only 1–2 percent of the entire fish group will reach smoltification stage; the rest will not survive, but of course the conditions for the fry are harder.

There is some discussion about when the first grading should be performed, but it is typically when the size is around 5–8 g. Various grading programmes are then followed until smolt size is reached. One method includes grading the fish group three times in three size groups within each grading, the first grading when fish are 5–7 g; the second at 20 g; and the third at 50 g (with small-sized fish removed) [25]. Twenty per cent of the fish group is taken out, mostly at the first grading. Another possible method is to divide the fish group into 4 size groups for each of the 3 gradings: the first grading is when the fish reach 6–8 g (with destruction of the smallest fraction); the next is when they reach 25 g (destruction of the smallest fraction again); and the third grading in relation to vaccination at 45–50 g, but this is dependent on water temperature. Generally, around 20% of the population may be destroyed before smoltification and transfer to the sea cage farm. This is a very important factor for the smolt farmers to decide, because destroying high numbers will most probably deliver smolt that have a higher growth rate in the sea, but reduced numbers will be produced and the price will be higher. Included here, of course, must be a discussion of animal welfare.

7.9.3 Negative effects of fish handling

It is important to remember that all handling creates a stress response in the fish [26–29], which may affect the production results negatively. This includes reduced growth, increased mortality and reduced immunity. If the handling is too rough, the fish may suffer from physical damage. Physical damage and open wounds may in addition be an opportunity for fungal attack and the fish will also be more exposed to diseases. This represents an important aspect of animal welfare in smolt production [19].

All handling includes some kind of human work, which requires time and creates economic cost. It is therefore important to choose the best methods for handling of fish and to carry out handling as carefully as possible.

7.10 Increased focus on fish health and animal welfare

The possibility of spreading disease when smolt are sent out to the sea cage sites requires continuous efforts to optimize farm management regarding disease problems and outbreaks. For example, there are zones in the sea where fish transport with well boats using open well valves is prohibited due to the possible spread of disease. It is therefore important to obtain documents confirming the health status of the eyed eggs that are taken into the smolt farm. The smolt farm also needs to have a good health control programme overseen by a veterinarian to prevent disease outbreaks at the farm, which is also regulated by law in the main salmon producing countries. Most smolt farms also have a vaccination programme to ensure all smolt leaving the farm are vaccinated.

Public concern about fish welfare is increasing along with regulatory attention. There is scientific research [19] which suggests that fish feel some type of fear and pain [30–33]. Documentation of good fish welfare is therefore an important part of today's salmon production [5,20].

7.11 Vaccination

Today the majority of the smolt that are ongrown in the sea are vaccinated (see also Chapter 11) (Figure 7.13). Even though this increases the cost of the smolt, the overall production economy for the sea cage farm will be improved. This is also one of the major reasons for the low use of antibiotics in the salmon farming industry today. Vaccination of the fish is also positive for the reputation of the salmon industry.

Vaccines may be given by dipping or bathing the fish in a vaccine bath (see Chapter 11). However, the predominant method today is to inject the vaccine into the belly cavity of the salmon parr. This gives the best protection, but the salmon must be above a minimum size (30–35 g). The injection has traditionally been given manually by a syringe, but now it is becoming more common to use a vaccination machine.

It is important to make a good vaccination plan, including vaccine type, vaccination method, time of vaccination and which diseases the vaccine will protect against. The vaccination programme should be planned in cooperation with a veterinarian and the receiver of the smolt.

Figure 7.13 Vaccination plant.

New and improved vaccines are continually developed, and today vaccines are available to protect against a range of diseases, for example, vibriosis, cold water vibriosis, yersiniosis, IPN, winter wounds and furunculosis. The vaccines stimulate the immune response against the diseases. The vaccines given today are typically a cocktail of individual vaccines providing protection against three to seven diseases. The vaccines give protection for a certain period, reportedly up to 2 years in sea, with some differences between the vaccine types.

The water temperature at vaccination of the salmon must not be too low, as below 2°C the vaccine has a limited effect; recommended temperature is above 5–6°C.

The injection of the vaccine must be correctly performed. There will generally be some increase in acute mortality just after vaccination, but not above 0.5%; if so, something has not been correctly performed. There might be chronic effects or side effects of the vaccine, such as reduced growth, adhesion of internal organs, discolouration of the belly cavity and reduced slaughter quality. However, salmon vaccines are developing rapidly and the negative side effects are continuously reduced.

It is important to have security protocols for personnel during vaccination. When manual injection (sticking) using a syringe is the method used, it is important that the vaccination personnel do not inject themselves, which can be quite easy when thousands of fish are being vaccinated each day. Procedures detailing the required action if this occurs are therefore important. Today, as mentioned, it is becoming far more common to use vaccination machines, and capacities of up to 40,000 smolt per hour including size grading have been reported [34].

There needs to be a certain period, of at least 6 weeks, between vaccination and transfer of the fish to the sea. Therefore, vaccinated smolt will always be over a certain size, depending on growth rate in the weeks after vaccination. The vaccination programme is therefore an important part of the production plan for the farm.

7.12 Hygiene and biosecurity

Hygiene on the farm is important, both to reduce the risk of disease and to increase biosecurity. This includes daily removal of dead fish and brushing of the tanks to prevent any accumulation of faeces and feed waste (Figure 7.14). Tank outlets are often mis-designed and therefore shock draining of the outlet should be performed several times a week to avoid accumulation of waste feed and faeces in the outlet pipes. All accumulation of waste feed and faeces inside the tank will promote fungal development and bacterial growth. Brushes and dip nets should be dried between each use or be placed in a disinfection bath. It can be an advantage to have separate ones for each tank, to avoid any spreading of disease between the tanks. Floors should be cleaned regularly and different departments should be fully washed down and disinfected regularly if the production regime makes this possible. Separate shoes/boots and clothing should be used for each department.

7.13 Control and management

It is important to have good production control and inspection routines. This includes recording of environmental conditions, such as the water quality parameters. The water temperature, both

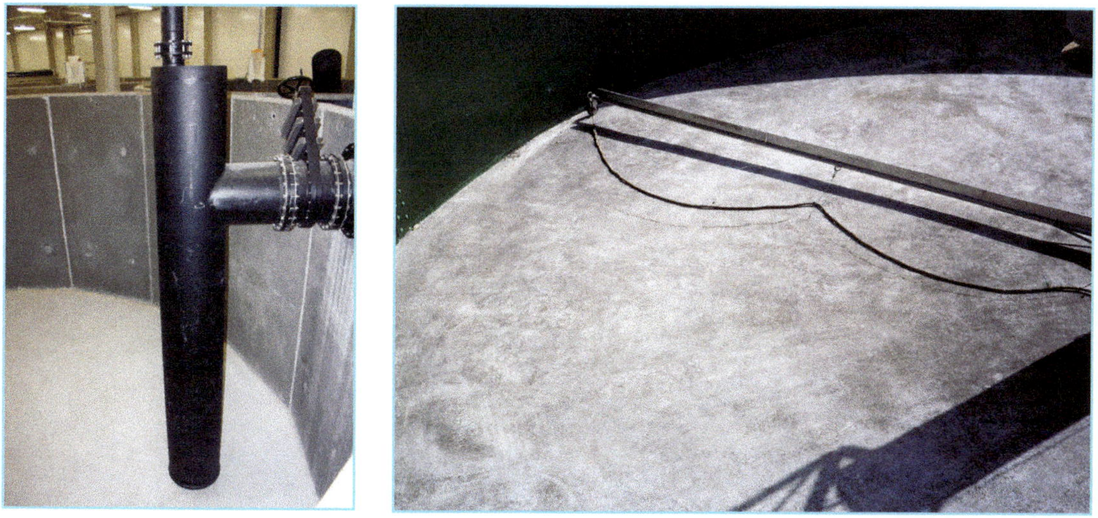

Figure 7.14 Good hygiene in the production units is important, as is the design of the inlet and outlet. If the units do not have a hydraulic self-cleaning design, manual washing or special equipment is required.

of the incoming freshwater and of the heated water in the different departments, must be monitored and recorded. The oxygen content of the inlet and outlet water, preferably of each tank, should be measured regularly. It may also be necessary to control the pH if it is low or variable, but this depends on the water quality. It is also important to monitor total gas concentration and especially the supersaturation of nitrogen, but this again depends on the site, inlet water quality and internal water treatment system. It should at least be measured when starting production. A good system may include taking water samples of 1–2 l each day for a period of 1–2 weeks, and storing them in a refrigerator throughout this period. If there are any problems, a more detailed water quality analysis can be performed on the stored water samples to find the cause. If there are no problems at the end of the period, the water samples can be discarded.

7.13.1 Growth control

Growth control is an important part of the overall control regime on the farm. This includes taking representative samples of the fish in the production units to measure their weight, and using this to estimate the average weight. By taking samples every month and keeping control of the number of dead fish, it is possible to estimate the biomass in the production units. Further, the average individual weight gain and total biomass gain for the tanks can be found and controlled. By monitoring the feed delivered over the period and knowing the increase in biomass, compensating for dead fish, the FCR can be found and the development of the FCR followed.

7.13.2 Record-keeping requirement

It is important to keep accurate records, often in the form of a journal, for the farm and in some countries this is also required by law. In the journal the intake of eyed eggs or fry must be included. Health status, both at intake and during production through the regular veterinarian health controls, should be recorded. This information will also be important for the purchasers of the smolt. Number, time, and size of delivered fry and smolt should also be included, as well as amount of feed used. Purchasers of the smolt typically also want to know of any chemicals and pharmaceuticals that have been used.

7.13.3 Weight sampling

Production control requires that fish growth is monitored during the production period. Owing to the high number of fish and total overall fish weight in each tank it is impossible to manually measure the weight of the entire biomass by counting the total number of fish on a regular basis. Therefore, a number of fish are taken out for weight sampling, although this presents the problem that the weight sample must represent the average weight of the fish in the tank. If, for instance, larger fish are not included in the weight sample, the estimated average fish weight will be too low (Figure 7.15). It also requires that the number of fish in each production unit is

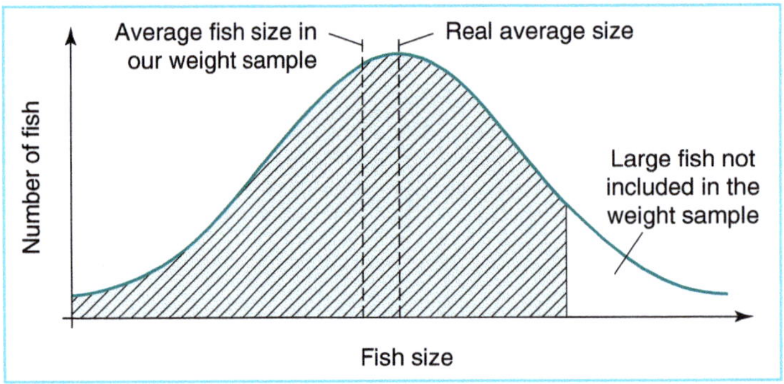

Figure 7.15 It is important that a representative sample is included to estimate correct average fish weight [3].

Table 7.1 Variation in fish size in graded and ungraded group, illustrated with three weight samples in each group.

Not graded fish groups				
Sample	First weighed	Middle	End	Avg.
1	16.1	8.7	5.2	9.99
2	15.7	12.2	4.7	10.93
3	13.6	9.6	4.5	9.31

Total avg. = 10.10, std. = 6.05, CV = 8.07%

Graded fish groups				
Sample	First weighed	Middle	End	Avg.
1	38.4	38.1	32.1	36.92
2	37.2	37.0	33.5	35.93
3	37.3	40.2	33.7	37.07

Total avg. = 36.63 g, std. = 6.8, CV = 1.69%

Example

Weight sampling of fish from a tank is achieved, showing the following results.

Sample	Number of fish	Total weight	Average weight
1	103	1102.3 g	10.5 g
2	107	1255.6 g	10.8 g
3	98	1009.4 g	10.3 g

The average weight of the three samples is 10.53 g. Assuming an almost equal number of fish in all the samples, the standard deviation (δ) between the three samples is 0.52. This gives a coefficient of variation (CV) of (0.252/10.53)100 = 2.39%

According to the procedure this is not satisfactory, and another sample has to be taken:

Sample 4 – 102 fish – total weight 10.71 g – new average weight 10.5 g.
This gives a new δ of 0.206 and a CV of 1.96%. This is satisfactory (under 2%).

However, it is not possible to achieve a CV below 2% without size grading of the fish.

known and correct, so collecting and counting the number of dead fish is of great importance. By knowing the average individual weight and the number of fish in the production unit, the total biomass can be found, which is an important parameter to follow in relation to the growth of the fish.

Several methods are used for taking weight samples, all of which are manual since no automatic methods have yet been developed. What is certain is that a number of samples of fish have to be taken out to have some certainty of estimating the average fish size. In tanks with poorly graded or non-graded fish, resulting in a group that is not homogeneous in size, taking a representative weight sample from the fish tank will be very difficult (Figure 7.16). If size grading is not performed, it will therefore limit the possibility of keeping good production control.

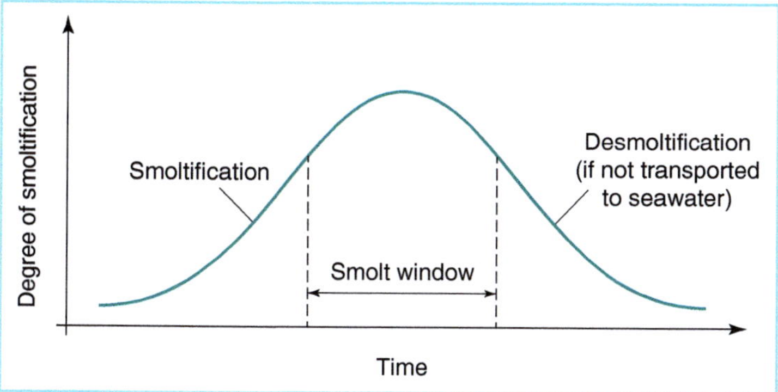

Figure 7.16 In the smolt window the salmon can be transferred to seawater without any osmotic problems occurring (partly based on [17]).

Problems with taking weight samples include that it is time consuming for the operators and stressful for the fish. Due to this, there is some discussion about how often weight samples should be taken, and there will be a great future demand for equipment that can do this automatically.

One possible way to perform weight sampling is shown below; this is adapted to smaller tanks and is quite time consuming but illustrates how it can be done. The water level is reduced so the fish are more crowded, then a dip net is drawn through the water column close to the bottom and, if possible, across the tank and to the surface. All the fish collected in the net (there should be over 50) are counted and the total weight measured. At least three samples should be taken, and if the difference between the samples is too large (coefficient of variation above 2%) it is recommended to take more samples.

7.14 Smoltification

7.14.1 The smolt window

The salmon will gradually develop from parr to smolt (the parr–smolt transformation, recommended review [11]) if the weight and environmental conditions are correct. To develop into a smolt, the fish has to be above a minimum size, there has to be an increase in daylight and the fish has to be in the upper bimodal in the fast growth period. When the smolt reach what is called the 'smolt window' [35], they are ready for transfer to seawater. The smoltification window lasts for 300–400 D°C, and if salmon that are smoltified are not transferred to seawater during this period they will desmoltify, which means they revert to not tolerating seawater (Figure 7.16). If a salmon that is not smoltified is transferred to the sea, it will die due to water loss, as will a smolt that has desmoltified.

7.14.2 Smoltification: when and how?

If using the term day degree (D°C), degree days or accumulated thermal units (ATU), with optimal management it is now possible to produce salmon to smolt size before reaching 3000 D°C after hatching. In the past it took 4000–4500 D°C, which illustrates the progress made, mainly due to optimization of feed and breeding programmes. When keeping a constant temperature of 11°C, this means that the smolt will be ready within 1 year, but this is theoretical. It is also impossible to obtain 100% smoltification in a fish group at the same time due to genetic variation between the individuals.

When the salmon changes from parr to smolt several changes occur, and this is a vulnerable period for the salmon. The environment totally changes as it goes from freshwater to seawater, the salmon are now able to swim long distances searching for food, the food composition differs, the salmon change behaviour to travel in schools and the predator pressure increases. Most importantly, the fish develop the ability to perform ionic regulation, or hypo-osmotic regulation. Going from parr stage to smolt stage includes a range of changes that can be divided into morphological, physiological and behavioural changes (see Chapter 4). Organs such as gills, kidney, gut and urinary bladder are 'changed' to be able to handle the new situations [14]. All these changes make the fish more susceptible to disease in this period.

7.14.2.1 Morphological changes

Morphologically there is a visual change when the fish develop from parr to smolt. The parr marks on the side of the salmon disappear, and their colour changes to silver on the sides and belly and black on the back. The fin bases also turn black and the fins get more transparent. The salmon become less visible to predators from above and below when they enter the sea.

The condition factor drops when going from parr to smolt. The fish change shape from fat to thin (typically from a condition factor of above 1.3 to below 1.1), which makes the smolt more adapted for swimming long distances searching for food and for survival in the sea in general (Figure 7.17). Wild salmon typically migrate for long distances when entering the sea, and their buoyancy increases so it is easier to swim in schools in the seawater.

7.14.2.2 Physiological changes

There is a range of physiological changes that occur in the salmon throughout the smoltification process. This includes hormonal changes and interactions. These can be observed by measuring, for instance, the development of growth hormone, insulin-like growth factor (IGF-I), thyroxine (T_4), prolactin (PRL) or cortisol [36, 37], and also includes the hormones that control osmoregulation in the fish. The number and activity of the special chloride cells develop during the smoltification period. These cells are located on the gills of the salmon and are responsible for removing salt from the blood of the fish to inhibit accumulation. Owing to the high number of mitochondria in these cells, they may also be called mitochondria-rich cells (MR cells). These cells have the ability to produce a lot of energy since the number of mitochondria is high. When the salt concentration in the blood becomes too high due to the higher osmotic pressure in the sea, it is removed from the blood passing over the gills by these chloride cells (see Chapter 4). The amount of the enzyme Na^+,K^+-ATPase (or ATPase), a special enzyme responsible for 'pumping' ions against a gradient, increases in the chloride cells [19], to pump salt ions out of the fish. As the amount of these enzymes increases when the fish reach smoltification, this parameter can be used for monitoring the smoltification status.

The release of growth hormones when starting the smoltification process increases both the weight and length of the smolt. Compared to parr, the ratio between weight increase and length increase also differs. The proportional length growth is increased, and the result is a reduced condition factor [38]. The growth hormones also stimulate the feed intake and the feed conversion rate of the smolt [39] (for rainbow trout [40]).

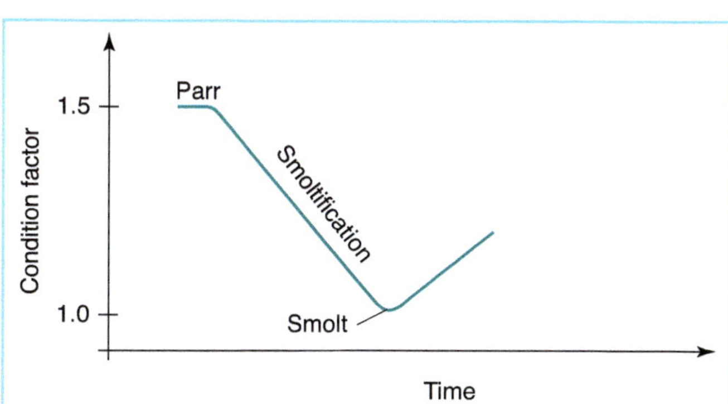

Figure 7.17 When the salmon are close to smoltification, the condition factor will drop so the salmon gain a shape more effective for swimming long distances on their migration in the ocean.

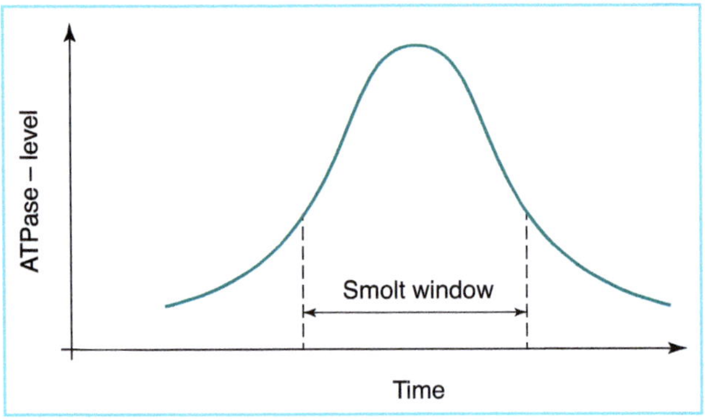

Figure 7.18 The amount of ATPase enzymes in the gills increases when salmon are close to the smolt window and drops of not sent to sea (based on [17]).

The physiological changes will also be related to the fact that the metabolism of the fish increases [11].

- There will be changes in metabolic enzymes.
- Liver glycogen will be reduced.
- Protein and lipid metabolism will increase.
- Blood glucose levels will increase.
- The density of mitochondria will increase.
- Oxygen consumption will increase.
- There will be changes in the haemoglobin composition.
- There will be a reduced amount of fat and an increased amount of water in the fish.

Energy reserves are reduced during smoltification, both the lipids and the glycogen present in liver and muscle (for steelhead trout [41]). The increased lipid metabolism and smaller reserves will also affect the condition factor [42].

Too high a carbon dioxide concentration in the water during this period may have a negative effect on the smoltification process. For instance, high carbon dioxide concentrations reduce chloride levels below normal values [43]. Plasma chloride values can be used as a general indicator of stress in fish, with values that are too high or too low indicating osmotic stress.

7.14.2.3 Behavioural changes

Fish will change from exhibiting territorial to schooling behaviour. This is what happens in nature: the parr are territorial in the river while the smolt swimming out of the river and into the sea travel in schools. Changes can also be seen in the swimming behaviour. As parr, the fish swim against the water current (counter current), but when smoltifying the fish follow the water flow (co-current), as also observed in nature when the smolt are migrating out of the rivers into the sea, so moving co-current. This therefore provides a natural grading of smolt from non-smoltified salmon. An increase in the jumping activity of smolt can also be seen in this period, and in addition the swimming frequency and endurance will be reduced.

Before releasing smolt into the sea it is important that the quality of the smolt is satisfactory. This includes reaching a certain size, which ought to be above 25 g. Today, however, a larger smolt in the range 60–100 g is preferred for ongrowing. Weights above 100 g is also very common, but this requires higher water temperatures on the smolt farm. There is now also a growing interest in producing post-smolt, which describes smolt that have gone through smoltification and are kept in seawater on land-based farms or in closed sea cages. This makes it possible to deliver larger and more robust fish, which can better handle the environmental conditions, to the traditional sea cages. It also reduces time in the sea and therefore time exposed to sea lice and disease.

When sending the smolt out to the sea cage farm, it is important that the fish are in general good health and are disease free. It is also recommended that the fish are vaccinated, and this has to be carried out at least 6 weeks before they are sent out. The fish also need to have reached a certain size before vaccination, typically above 30–35 g, so this results in a minimum size of fish, typically above 60 g, that are vaccinated and ready to transfer to sea. This, of course, depends on the water temperature following vaccination since this is the most important factor for growth.

With all the changes that are happening to the smolt during the smoltification process it is a period during which they are more sensitive to environmental factors. A suboptimal environment, such as reduced water quality, high fish density, or suboptimal feeding and water velocity, may postpone the smoltification process. Handling stress must also be avoided as much as possible and therefore there is an interest in the use of calming chemicals at this stage (section 7.15). The scales on the fish are sitting loose so handling must be as careful as possible.

7.14.3 Desmoltification

If the smolt do not go into the sea they will desmoltify. The fish will lose much of their ability to tolerate seawater, and feed intake and growth will also be reduced [44]. All the morphological changes are not reversed, however. The fish become darker again, but the parr marks do not return. If these fish are transferred to the sea after desmoltifying, growth will be low as will disease resistance [11] so this is not a recommended strategy. This shows the importance of a good smolt quality.

7.14.4 Testing of smoltification

When transferring smolt from freshwater to seawater it is important that the fish are within the smolt window and that they can tolerate seawater. Several tests are available to evaluate smoltification status, including seawater challenge tests, a chloride test and an ATPase test. Assessment of smoltification can also be based on behaviour, condition factor and morphology (smolt index) [45].

A simple and illustrative example is a seawater challenge test that can be performed in two ways. For the first, a sample of fish is taken from the production tank. These are placed in a 3.5% saltwater solution for 24 or 48 hours and then the fish are anaesthetized in order to obtain a blood sample. For example, 0.35 ml blood is withdrawn, centrifuged for 10 minutes with heparin [46] and the ionic concentration of sodium and chloride in the blood is then measured. For a seawater-ready smolt, the sodium value should be around 165–170 mM/l, while the chloride should be around 140–150 mM/l [14]. Measuring the parameters before and at the end of the challenge test should show small differences in the chloride values, with a good smolt having a variation less than 5–10 mM/l compared to reference values [47]. The test shows the effectiveness of the pump at pumping ions (salt) out of the fish. A seawater-ready smolt should be able to regulate levels back to normal blood values, but this will not be the case for a smolt that is not ready. The test is, however, not very accurate and it is difficult to use under cold water temperatures; if used alone the fish are typically sent out too early [25]. Such a test is questionable from an animal welfare perspective, because it is not known if the fish will be able to tolerate sea water or not.

The second method, to avoid taking blood samples, is to perform a 96-hour challenge test. A sample of fish, for instance 20 fish, is placed in a 4.0% saltwater solution for 96 hours and then the mortality is measured. There should be 100% survival in this test. During the test it is important to maintain the same water quality and water temperature as in the tank from which the fish for the sample were taken. To avoid degradation of the water quality the fish should be starved for 24 hours prior to the test. This test can be performed once a month before delivery, for instance starting 3 months before expected delivery date. For this seawater challenge test, the seawater is usually artificially made. It is important to use salt designed for such use, for example, aquaria salt. Ordinary consumer salt may contain anti-lumping agents that can be toxic for the fish. This test too is objectionable from an animal welfare perspective. Salmon should not be exposed to salinity above 3.5%, and if they are not fully smoltified they may die.

A better test, including from an animal welfare perspective, is to measure the ATPase enzyme activity, and this test is more and more commonly used. The amount of the enzyme Na^+,K^+-ATPase (the enzyme responsible for pumping out salt from the salmon) in the chloride cells increases when the salmon are close to becoming smolt. Typical ATPase values are below 5 (μmol Pi(mg prot.)$^{-1}$h^{-1}) for parr and should be above 10 for smolt [37]. In between, there is a transition phase. Measuring the amount of the enzyme will therefore give an indication of the smoltification status. In a further test developed for Na^+,K^+-ATPase, a real-time PCR analysis is used to measure the gene expression of various markers in the gill tissue (this is a new measuring technology using biotechnology). Typically, freshwater ATPase is measured, but seawater ATPase or a cofactor only found in the chloride cells associated with seawater tolerance can also be measured [48–50]. The freshwater ATPase that is important for parr in freshwater will gradually decrease during the smoltification, while the seawater ATPase and the cofactor will increase. Owing to individual variation, it is recommended to take samples from at least 20 fish in the population, and by following the fish over a period of time, long enough for at least three tests, a prognosis of the smoltification development status can be found. These tests will also give additional information compared to a regular ATPase test.

The ATPase and PCR analyses are performed in specialist laboratories and gill samples have to be sent for diagnostics. How they are taken and sent is described in procedures supplied by the laboratory. Here the new PCR analysis has an advantage regarding logistics because the samples can be sent through standard mail services (up to 1 week), while the general ATPase test requires that the gill samples are stored at extremely low temperatures and are required to be frozen using liquid nitrogen or dry ice.

It can also be interesting to produce some kind of smoltification index where a number of parameters related to smoltification are included (Table 7.2). This will make the prognosis more accurate. One such index can be made by scaling of visual external signs of smoltification (morphology). This could include degree of silver colour, parr marks and dark edges of the fins, and based on these an overall index can be made [51]. Each of the three parameters is graded from

Table 7.2 Use of the smolt index is a method to visually observe if the salmon are smoltified. The fish are given a total index based on the sum of points of the three factors divided by three. A higher score means more advanced smoltification [51].

	No	Weak	Visible	Silver colour
Silver colour	1	2	3	4
	Strong	Visible	Weak	No
Parr marks	1	2	3	4
	No	Weak	Visible	Dark edges
Dark edges on fins	1	2	3	4

1 to 4, where 4 represents a fully smoltified salmon. The smolt index is calculated based on the average score of the three parameters.

New methods using camera technology and measurement of spectral characteristics are also being tested. This is possible due to the morphological differences and allows automatic classifications and grading of smoltified and non-smoltified salmon to be made.

Making trend lines following the population as it nears smoltification is also a way of predicting the correct smoltification time. This will also improve the prognosis. The trend line can also represent ratios, for example, ATPase to weight or ATPase to condition factor. It can be an advantage to combine all measurements by including all signs of smoltification, including gill sampling for PCR ATPase, visual external signs, behaviour, weight and day degrees. Some fish may also be placed in seawater 1 week before transfer to document that the salmon really have smoltified.

One monitoring programme proposed for use before smoltification includes use of ATPase tests starting with a first sample when the photo manipulation period begins, or for supersmolt (section 7.15.5) [52] start at vaccination with the 0 sample, followed by 2–3 samples throughout the smoltification process. If there are unclear results, run a 4-day salinity challenge test and measure chloride concentration.

7.15.5 Triggering smoltification and seasonally independent smolt production

Today the trend is to get the salmon to smolt stage quickly and to deliver a large smolt. This is achieved by heating the water so the growth rate of the salmon increases. This applies to both flow through farms and RAS farms, with the difference being that the RAS farms can increase growth even further with stable water temperatures of 13–14°C. Some RAS farms are considering the use of even higher temperatures to increase growth further still.

Regardless of how they are produced, it is of absolute importance that the salmon are smoltified when transferred to the sea; they die otherwise. The main trigger for smoltification is that fish are of a certain size, which normally represents no problem today because a large smolt, over 75 g, is preferred. In nature, the trigger is not only the increase in water temperature, but also the increase in day length so the smolt can migrate out to the sea in the spring. At this time of the year there is much water in the rivers and the temperature in the sea is increasing, as is the available food. This illustrates how well salmon are adapted to or the product of the natural conditions in the wild.

When producing 1^+ and 2^+ smolt it is not problematic to produce a smolt ready for transfer assuming that the size is over the minimum. When using natural water temperatures in the winter before sending out and when having the smolt in tanks outside so they are exposed to natural light with increasing daylight and increasing water temperature, smoltification will occur naturally.

Today the trend is, as discussed, to get the salmon up to smolt size as fast as possible by heating the water to increase growth rate, either in flow through or RAS systems. The result is that the salmon reach a size at which they can be ready as smolt in the autumn, before 1 year old, producing 0^+ smolt. To be able to do this, however, smoltification must be stimulated in the autumn. Such production is also called seasonally independent smolt production. To achieve this, the salmon need an artificial signal to trigger smoltification. One way is to utilize light manipulation, which has been the most common method. However, smoltification programmes that use additives to the water or to the feed to stimulate smoltification have now been developed.

A major trigger for smoltification is, as discussed, an increased day length, presuming that the fish have reached a certain weight. Therefore, to produce out-of-season smolt the fish need a trigger to smoltify that is not naturally present. Water temperature can be controlled by heating equipment that is usually installed on the farm. However, in the late summer and autumn the day length is decreasing so the light has to be controlled with a specific light programme to achieve smoltification. Several different light programmes have been used for this purpose [46]. One that has been used with good results, where the aim is to produce 0^+ to be sent out in the autumn, starts with grading of the population (taking out the largest) and vaccination in late spring (May/June in the northern hemisphere). This is followed by the provision of 12 hour light for 6 weeks, followed by a 4–6 week period of 24 hour light. This is also temperature dependent, with a higher temperature resulting in earlier smoltification. Fish produced with this light regime will be ready for transfer to the sea in August–September as 0+ or ½ year old smolt. Another proposed programme is to provide the parr with 24 hour light until the fish reach 20 g, then to start providing 12 hours of light and 12 hours of darkness. This light regime is continued for a 6 week period followed by 400 D°C with 24 hour light, then the salmon will smoltify [25].

When using light manipulation, it is important that the fish can perceive the difference between day and night, so there should be at least 100 times more light (lux) in the 'day' conditions [46]. It is also recommended to gradually increase and decrease the light when turning the light on and off, to avoid stressing the fish with an abrupt change in the light conditions. Programmes that produce a gradual decrease and increase of day length may also be used. It is important to be aware that light programmes with night conditions may decrease the total growth in the smoltification period and may result in some more time being needed on the smolt farm to achieve a smolt with a given weight. A general problem with using light manipulation for seasonally independent smolt production is lack of homogeneity within the fish group, as not all will smoltify at the same time. Therefore, size grading before beginning light manipulation is important in order to achieve the most homogenous fish weights as is possible, and also to ensure the upper part of the bimodal size distribution is used.

A method to trigger smoltification without a light programme has also been developed; this is called the supersmolt method [53]. Originally it was a combined method with the addition

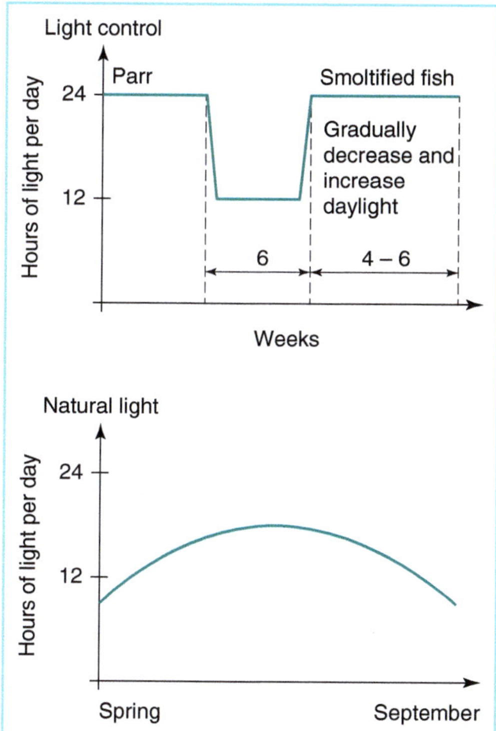

Figure 7.19 The hours of natural light will vary with latitude but will always increase in spring and decrease in autumn. A light manipulation programme can be used to smoltify salmon out of season (partly based on [46]).

of salts and other substances to both feed and water [54], but it has been further developed and today only the use of the specially composed feed is required [55]. The method is based on stimulation of the calcium sensing receptors (CaSR) that regulate intracellular activity [56] and can initiate the smoltification process. CaSR in the fish are located either internally (for example, in the gut) or externally (for example, on gills). The first method stimulates both and includes adding a dose of ions (Mg^{2+}, Ca^{2+}, Cl^-) to the production water in the morning and evening in combination with adding ions (Na^+ and Cl^-) and amino acids (tryptophan) to the feed, both in given concentrations. In the second method, it is only the feed that is specially designed, again with ions and amino acids. The chloride cells on the gills will be stimulated to develop the Na^+, K^+ATPase enzymes, and the smoltification process will begin. The advantage with this method is that a light manipulation programme is not necessary. Several studies have also observed a larger smolt using this method because it uses 24 hour light for the entire period [57,58]. When using the traditional light programme with a period of darkness this may lower the overall growth of the salmon; therefore, it may be possible to send out a larger smolt or send the smolt out earlier when using the special feed. This could be an advantage for the autumn production of 0^+ [58]. Condition factors also seem to be higher when using the special feed than for traditional smoltification, and there is the advantage of achieving a more equal smoltification in the entire group, with all smolt sent out being in the smolt window. A delay in sending out, as may occur if the sea site is not ready as planned, will not have the same consequences because the fish will not desmoltify. The fish are fed on this special feed for 3–5 weeks before full smoltification is achieved for the entire group. However, the cost of the modified feed is higher than traditional feed so this must also be taken into consideration when selecting a smoltification method.

In general, when producing smolt, either traditionally or independent of season, it is important that the water quality is optimal to control the other factors that may affect the smoltification process. High carbon dioxide can, as discussed, affect the smoltification process and the same will be the case for a drop in temperature [59]. The fish must not be suffering from any disease; for example, salmon suffering from yersiniosis show problems with smoltification [59]. The size variation in the group is also an important factor, because fish size affects smoltification with the largest smoltifying first; therefore, if the difference is too large the largest fish may desmoltify before the smallest smoltify. Therefore, the size difference in the group must be low, and the time from vaccination to being sent out should be as short as possible to avoid too much size variation but long enough to get vaccination effects (6 weeks).

An alternative approach is currently undergoing commercial testing based on experiences with wild salmon. The fact is that even if breeding has produced a salmon that grows faster, the full potential for growth may not have been reached until the ongrowing phase in sea cages. For wild salmon there have been examples where the weight of the salmon was 25 g as smolt but after 1 year in the sea has increased to 3 kg [60]. In addition, it may be possible to get a compensatory growth when the fish enter the sea with an unlimited feed supply. Therefore, 1.5^+ or 2^+ grown at normal water temperatures, similar to those used at the beginning of the salmon farming industry, are being tested [58]. The great advantage for the smolt farm with such a production regime is that it is not necessary to heat so much water on the smolt farm, therefore costs are reduced, and a simpler farm design can be used which will reduce the risk for technical failure. The advantage for the ongrowing farms is also obvious, due to the potential increase in growth. Preliminary results show that it is possible to reach a 5 kg fish after less than 1 year in sea water with this method. By producing 2^+ smolt with a size of 200 g to be sent out in April it seems possible to reach 5 kg within 9 months, but of course this depends on water temperature [61]. This can be an advantage for a site because it allows a 2 month fallow period before the next smolt transfer. In this way, the same sea cage site may be used on a yearly basis, assuming the site is fully recovered. However, there is a major drawback that will limit such production, because it increases the turnover rate (number of smolt produced per year) on the smolt farm. An increased farm size will be necessary for the same quantity of production, which will also increase the total water requirement. Typically, the freshwater supply is the limiting factor for production on the smolt sites, and the reason that many smolt farms are now changing to RAS.

7.14.6 Smolt quality

For fish that miss the smolt window, either parr or those that have desmoltified, the consequences can be fatal. Mortality will certainly occur, but the severity depends on where the development is in relation to the smolt window. Growth rate in the sea will also be reduced. The fish will be stressed and more susceptible to disease, and may then also infect other fish. Therefore, as mentioned several times, delivery of optimal smolt is of major importance.

7.14.7 Temperature of sea water at time of transfer

The temperature of the sea when the smolt are transferred is important, particularly today with seasonally independent smolt production. In nature, the migration of smolt in spring occurs with a sea water temperature of around 9°C [11], and this is also the temperature that research suggests results in the least osmotic disturbance when moving the smolt to seawater [62]. Today, however, smolt are transferred to the sea all year round, with, perhaps, a break in the coldest winter months. What is important when transferring smolt in months with low temperatures and falling temperatures is that there must be a gradual water temperature adaptation (typically lowering) on the smolt farms prior to sending out. It is not recommended to transfer smolt to the sea in the autumn when the water temperature in the sea drops below 5–6°C and is falling; it is easier to transfer at low water temperatures in the spring when the water temperature is naturally increasing.

7.15 Preparation for sending smolt for marine ongrowing

When sending smolt to the sea it represents a stressful situation for the fish, with a total change of environment including a change from fresh to seawater. The smolt must therefore be well prepared for this environmental change. Optimal feeding in the final period will of course be important, and many farms use a special smolt feed to help the fish adapt to the change. Experiments have, for instance, shown that increasing the amount of zinc and omega-3 in the feed will improve skin health [63], meaning that the fish will be less affected by problems with wounds when transferred to the sea.

It is important to carry out preparation for the transport. First of all, the smolt must be ready (smoltified) and in good shape. All handling should be finished at least 14 days prior to transport of the fish out of the smolt farm. Starving the smolt for at least 2 days before sending should be carried out to avoid accumulation of metabolic waste products in the water during transport of the fish to the sea cage farm. Great care must be taken when loading due to the presence of loose scales on the smolt, as loss of scales may result in problems with osmotic balance when the smolt enter the sea. Care should also be taken regarding water temperature. There should be minimal or no temperature difference between the rearing water, transport water and seawater where the smolt is to be transferred. As discussed, adaptation may be done by lowering the water temperature at the smolt farm prior to sending, and it is not recommended to transfer smolt to seawater when the seawater temperature is below 5–6 degree, especially when it is falling in the autumn.

Before transport, as before other handling operations, the fish can be sedated (a low degree of anaesthesia). This will calm the fish so that the stress response may be reduced. For instance, the fish will not swim so much against the water flow when they are moved by pumping, and all handling will be simpler and more effective. Owing to the likely reduction in stress response, the oxygen consumption will be reduced as will the carbon dioxide production. It may also reduce the fear and flight mechanism in the salmon and improve fish welfare. A range of chemical and natural anaesthetic agents can be used for this purpose [64]. One chemical used is based on the active substance isoeugenol, where the recommended dose for sedation is 2–5 mg/l [65].

Figure 7.20 Transport of fish out from the smolt farm with a well boat.

However, sedation may affect smoltification (osmoregulation) [66], and here more research is needed.

A well boat is most typically used for transporting the fish from the smolt farm, but trucks with water tanks may also be used even though the transport capacity will then be limited (Figure 7.20).

References

1. Ytteborg, E. et al. Molecular pathology of vertebral deformities in hyperthermic Atlantic salmon (*Salmo salar*). *BMC Physiology*. 2010. 10(1): 12.
2. Fjelldal, P. et al. Vertebral deformities in farmed Atlantic salmon (*Salmo salar* L.). *Etiology and Pathology*. 2012. 28(3): 433–440.
3. Lekang, O.-I., *Aquaculture Engineering*. 2020. John Wiley & Sons: Chichester.
4. Solstorm, F. et al. Not too slow, not too fast: water currents affect group structure, aggression and welfare in post-smolt Atlantic salmon *Salmo salar*. *Aquaculture Environment Interactions*. 2016. 8: 339–347.
5. Noble, C. et al. *Welfare Indicators for Farmed Atlantic Salmon: Tools for Assessing Fish Welfare*. 2018. Tromsø, Norway: Nofima.
6. Jørgensen, E.H. and Jobling, M. The effects of exercise on growth, food utilization and osmoregulatory capacity of juvenile Atlantic salmon, *Salmo salar*. *Aquaculture*. 1993. 116(2–3): 233–246.
7. Solsletten, V. Her tester de hvor mye laksen tåler. 2019. https://fiskeribladet.no/tekfisk/nyheter/?artikkel=60936.
8. Jobling, M. et al. The effects of prolonged exercise training on growth performance and production parameters in fish. *Aquaculture International*. 1993. 1(2): 95–111.
9. Hvas, M. et al. The effect of thermal acclimation on aerobic scope and critical swimming speed in Atlantic salmon, *Salmo salar*. *Journal of Experimental Biology*. 2017. 220(15): 2757–2764.
10. Wańkowski, J.W.J. and Thorpe, J.E. The role of food particle size in the growth of juvenile Atlantic salmon (*Salmo salar* L.). *Journal of Fish Biology*. 1979. 14(4): 351–370.
11. Stefansson, S.O. et al. *Smoltification. Fish larval physiology*. Vol. 639. 2008. Science Publishers: New Delhi.
12. Thorpe, J.E. Bimodal distribution of length of juvenile Atlantic salmon (*Salmo salar* L.) under artificial rearing conditions. *Journal of Fish Biology*. 1977. 11(2): 175–184.
13. Benfey, T.J. et al. Muscle ornithine decarboxylase activity as an indication of recent growth in pre-smolt Atlantic salmon, *Salmo salar*. *Aquaculture*. 1994. 121(1–3): 125–135.
14. Heggberget, T.G. et al. Smoltifizering hos laksefisk. *NINA forskningsrapport*. 1992. 31: 1–42.
15. Kristinsson, J.B., Saunders, R.L. and Wiggs, A.J. Growth dynamics during the development of bimodal length-frequency distribution in juvenile Atlantic salmon (*Salmo salar* L.). *Aquaculture*. 1985. 45(1): 1–20.
16. Thorpe, J.E., Talbot, C. and Villarreal, C. Bimodality of growth and smolting in Atlantic salmon, *Salmo salar* L. *Aquaculture*. 1982. 28(1): 123–132.
17. Stefansson, S.H.T. Smoltifizering hos laks, en tilapsning til sjøvann, in *Frisk fisk, om liv og død i merdene*, H.I. Wergeland, B. Aalvik and O.M. Rødseth (eds). 1996. Oslo: Universitetsforlaget.
18. Handeland, S.O. et al. Low light intensity can reduce Atlantic salmon smolt quality. *Aquaculture*. 2013. 384–387: 19–24.
19. Santurtun, E., Broom, D.M. and Phillips, C.J.C. A review of factors affecting the welfare of Atlantic salmon (*Salmo salar*). *Animal Welfare*. 2018. 27(3): 193–204.
20. RSPCA. *Welfare Standards for Farmed Atlantic Salmon*. 2018. Royal Society for the Protection and Cruelty of Animals: Horsham.
21. Hosfeld, C.D. et al. Effects of fish density on growth and smoltification in intensive production of Atlantic salmon (*Salmo salar* L.). *Aquaculture*. 2009. 294(3–4): 236–241.
22. Jones, H.A.C. et al. Social network analysis of the behavioural interactions that influence the development of fin damage in Atlantic salmon parr (*Salmo salar*) held at different stocking densities. *Applied Animal Behaviour Science*. 2011. 133(1–2): 117–126.

23. Soderberg, R.W., Meade, J.W. and Redell, L.A. Growth, survival, and food conversion of Atlantic salmon reared at four different densities with common water quality. *The Progressive Fish-Culturist.* 1993. 55(1): 29–31.
24. Adams, C.E. et al. Multiple determinants of welfare in farmed fish: stocking density, disturbance, and aggression in Atlantic salmon (*Salmo salar*). *Canadian Journal of Fisheries and Aquatic Sciences.* 2007. 64(2): 336–344.
25. Saether, P.A. *Smoltifizering, in Kultiveringsmøte.* 2017. Kultiveringsanleggenes forening, KAF.
26. Lekang, O.I. *Effects of Handling on Juvenile Fish. Studies of the Stress Response, Growth and Susceptibility to Disease in Fish Exposed to Different Handlings.* 1989. Norges Landbrukshoegskole.
27. Carey, J.B. and McCormick, S.D. Atlantic salmon smolts are more responsive to an acute handling and confinement stress than parr. *Aquaculture.* 1998. 168(1–4): 237–253.
28. McCormick, S.D. et al. Repeated acute stress reduces growth rate of Atlantic salmon parr and alters plasma levels of growth hormone, insulin-like growth factor I and cortisol. *Aquaculture.* 1998. 168(1–4): 221–235.
29. Madaro, A. et al. A comparative study of the response to repeated chasing stress in Atlantic salmon (*Salmo salar* L.) parr and post-smolts. *Comparative Biochemistry and Physiology Part A: Molecular & Integrative Physiology.* 2016. 192: 7–16.
30. Branson, E.J. *Fish Welfare.* 2008. Wiley: Chichester.
31. Huntingford, F.A. et al. Current issues in fish welfare. *Journal of Fish Biology.* 2006. 68(2): 332–372.
32. Ashley, P.J. Fish welfare: current issues in aquaculture. *Applied Animal Behaviour Science.* 2007. 104(3–4): 199–235.
33. Kristiansen, T.S. et al. *The Welfare of Dish.* 2020. Springer: Dordrecht.
34. Skalamaskon. Vaccination. 2020. https://en.skalamaskon.no/aquaculture2/vaccination.
35. McCormick, S.D. et al. Movement, migration, and smolting of Atlantic salmon (*Salmo salar*). *Canadian Journal of Fisheries and Aquatic Sciences.* 1998. 55(S1): 77–92.
36. Björnsson, B.T., Stefansson, S.O. and McCormick, S.D. Environmental endocrinology of salmon smoltification. *General and Comparative Endocrinology.* 2011. 170(2): 290–298.
37. McCormick, S.D., Smolt physiology and endocrinology, in *Fish Physiology*, S.D. McCormick, A.P. Farrell and C.J. Brauner (eds). 2012. Academic Press: Cambridge, MA, pp. 199–251.
38. Björnsson, B.T. The biology of salmon growth hormone: from daylight to dominance. *Fish Physiology and Biochemistry.* 1997. 17(1–6): 9–24.
39. Johnsson, J.I. and Björnsson, B.T.Growth hormone increases growth rate, appetite and dominance in juvenile rainbow trout, *Oncorhynchus mykiss. Animal Behaviour.* 1994. 48(1): 177–186.
40. Garber, M.J. et al. Dose-response effects of recombinant bovine somatotropin (PosilacTM) on growth performance and body composition of two-year-old rainbow trout (*Oncorhynchus mykiss*). *Journal of Animal Science.* 1995. 73(11): 3216–3222.
41. Sheridan, M.A., Allen, W.V. and Kerstetter, T.H. Seasonal variations in the lipid composition of the steelhead trout, *Salmo gairdneri Richardson*, associated with the parr-smolt transformation. *Journal of Fish Biology.* 1983. 23(2): 125–134.
42. Saunders, R.L. and Henderson, E.B. Changes in gill ATPase activity and smolt status of Atlantic salmon (*Salmo salar*). *Journal of the Fisheries Board of Canada.* 1978. 35(12): 1542–1546.
43. Fivelstad, S. and Binde, M. Effects of reduced waterflow (increased loading) in soft water on Atlantic salmon smolts (*Salmo salar* L.) while maintaining oxygen at constant level by oxygenation of the inlet water. *Aquacultural Engineering.* 1994. 13(3): 211–238.
44. Arnesen, A.M. et al. Osmoregulation, feed intake, growth and growth hormone levels in 0+ Atlantic salmon (*Salmo salar* L.) transferred to seawater at different stages of smolt development. *Aquaculture.* 2003. 222(1): 167–187.

45. Ærø, C.R., Kontinuerlig lysstimuli og sjøvannsbruk i intensiv smoltproduksjon: smoltifizeringsutvikling hos Atlantisk laks (*Salmo salar*). 2018. Masteroppgave. Nord Universitet, Bodø.
46. Stefansson, S.O., Holm, J.C. and Taranger, G.L. Oppdrett av laks og aure i Norge (In Norwegian). 2002. Universitet i Bergen: Institutt for fiskeri- og marinbiologi.
47. Rosten, T. and Fjellheim, A.J. Metode for påvisning av smoltstatus hos atlantisk laks. Tidskriftet Vann, 2010 (01) pp. 116–121.
48. McCormick, S.D. et al. Differential regulation of sodium potassium pump isoforms during smolt development and seawater exposure of Atlantic salmon. *Journal of Experimental Biology*, 2013. 216(7): 1142–1151.
49. McCormick, S.D. Smolt physiology and endocrinology, in *Fish Physiology*. 2012. Elsevier: Amsterdam, pp. 199–251.
50. Pharmaq-analytiq SmoltVision. analytiq.com/en/products/smoltification/smoltvision/.
51. Pharmaq-analytiq. Vurdering av smoltindex. 2020. https://www.pharmaq-analytiq.com/sfiles/75/1/file/sm_6.4_004-veileder-smoltindeks_v8-.pdf.
52. Stim. Supersmolt feed only. https://stim.uk/productgroups/supersmolt-feed-only/.
53. Hatchery International. Europharma's patent application gets a boost. 2018. https://www.hatchery international.com/europharmas-patent-application-gets-a-boost-3283/.
54. Harris Jr, H.W. et al. Methods for raising preadult anadromous fish. 2002. Google Patents.
55. Nordly, J.R., Lyngøy, A. and Mansilla, C.R. A fish feed and method for smoltification and prevention of desmoltification in Salmonidae, and for prophylaxis and treatment of haemorrhagic smolt syndrome (HSS) in Salmonidae. 2017. Google Patents.
56. Jury, S. et al. Aquaculture and stock enhancement technologies based on recently discovered calcium-sensing receptors in finfish. US-Japan Aquaculture Panel Symposium. 2005.
57. Lyngøy, A. and Haram, L.M. Produksjonserfaringer med supersmolt i settefisk og matfiskanlegg 2006 og 2007. 2007. https://docplayer.me/19643520-Produksjonserfaring-med-supersmolt-i-sette fisk-og-matfiskanlegg-2006-og-2007.html.
58. Kileng, Ø. Er fôrstimulert smolt like god som lysstimulert smolt?. 2019 https://tekset.no/wp-content/uploads/2019/02/PRES-3.2-TEKSET-2019-Morgendagens-smolt-I-%C3%98yvind-Kileng.pdf.
59. Reinsnes, K.O. Europharma. Medlemsmøte i FHL Nordnorsk havbrukslag, Bodø 19 oktober. https://sjomatnorge.no/wp-content/uploads/importedmedia/Europharma.pdf.
60. Farstad, O.M. Full fart fra start eller første mann til mål – Hva er den optimale produksjonsstrategien frem til 5 kg. 2018. https://tekset.no/wp-content/uploads/2018/02/4-Ole-Marius-Farstad.pdf.
61. Petteresen, M. Kjempetilvekst med gammelsmolt. 2019. https://www.kyst.no/article/kjempetil vekst-med-gammelsmolt/.
62. Handeland, S.O. et al. Seawater adaptation by out-of-season Atlantic salmon (*Salmo salar* L.) smolts at different temperatures. *Aquaculture*. 2000. 181(3): 377–396.
63. Thefishsite. Zinc and O-3 help smolt transfer to sea. 2019. https://thefishsite.com/articles/zinc-and-o-3-help-smolt-transfer-to-sea.
64. Purbosari, N. et al. Natural versus synthetic anesthetic for transport of live fish: a review. *Aquaculture and Fisheries*. 2019. 4(4): 129–133.
65. Iversen, M. and Eliassen, R.A. The effect of AQUI-S® sedation on primary, secondary, and tertiary stress responses during salmon smolt, *Salmo salar* L., transport and transfer to sea. 2009. *Journal of the World Aquaculture Society*. 40(2): 216–225.
66. Jensen, P.M. Sedation of smolts should be used sparingly, say scientists. 2019. https://www.fish farmingexpert.com/article/sedation-of-smolts-should-be-used-sparingly-say-researchers/.

CHAPTER 8

Ongrowing to harvesting

8.1 Introduction

Almost all farmed, consumer-sized salmon (4–6 kg) are produced in traditional open sea cages. The sea cage farms operate with different production models. There might be one or several smolt intakes per site per year. However, to reduce the risk of disease transfer between each group, it is normal to have one smolt intake per site; these salmon grow until harvesting, then the site is rested (left fallow) for a period before being used for the next smolt intake. Average salmon slaughter weight is around 5 kg and it takes 1–2 years to reach this size, depending on the size of the smolt and the seawater temperature. During the period in the sea it is common to divide the fish group, either by placing additional cages at the same site or by moving part of the fish biomass to another site.

Typical intake of 1^+ smolt is in spring or early summer. This enables the smolt to take advantage of the high summer seawater temperatures, resulting in good growth during the summer and autumn. 0^+ smolt (off-season or seasonally independent smolt) can also be used, if available from the supplying smolt farm, and are transferred in the autumn. This can increase the production capacity of smolt farms with restricted freshwater resources, particularly during wintertime. The midwinter months are not the preferred period for transferring smolt and can result in higher mortality, although this depends on the seawater temperature during winter. Seawater temperatures that are too low prevent smolt transfer in some locations. The production model for a site is, however, also dependent on the farming licence, which differs between salmon producing countries (reviewed in [1]).

Generally, one farming company will have several production sites. One year the smolt intake is transferred to sea at one site and the next intake, in the following year, is located at another site. When the fish are harvested from the first site it should then have a resting period or fallowing period, where no fish are at the site, to ensure the bottom fauna at the site can recover to normal levels. In Norway, for example, there is a requirement for at least 2 months fallowing depending on environmental registrations (MOM registrations). The resting period can be up to 1 year, and if so, the production can rotate on three sites with one smolt intake each year per site (Figure 8.1). Fallowing of the site will also help any sea lice problems.

Recently, more coordination between smolt farms has been seen, and it is increasingly common for smolt farms and ongrowing sea cage farms to be part of a larger, vertically integrated

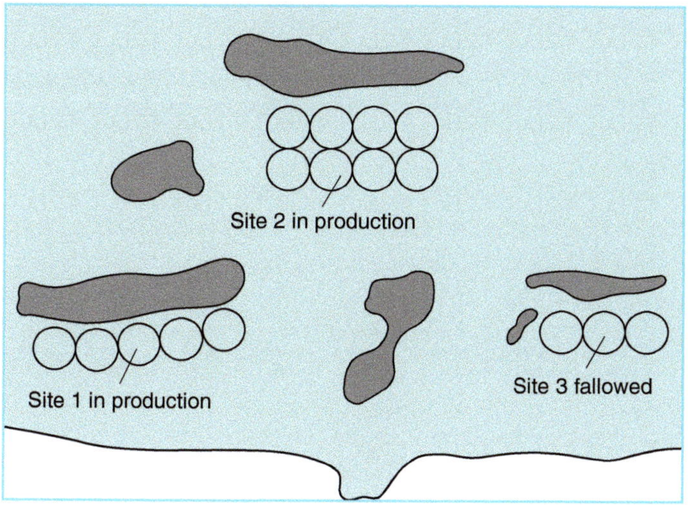

Figure 8.1 The sites should be fallowed between each group of fish and therefore several sites are used per licence [2].

company. Several intakes of differently sized smolt can then be transferred to sea to optimize production with the potential to deliver slaughter-ready salmon from the sea cage farm all year round. It is best to obtain smolt throughout as large a part of the year as possible, which also requires more transfers of 0^+ fish. There is also a trend to grade the salmon at the marine ongrowing stage. The emphasis is on maximizing site turnover, but the optimal production model to use is, as mentioned earlier, highly dependent on the farming licences.

In all countries that carry out sea cage farming, there is a requirement for a licence. How the licence functions varies between salmon producing countries (reviewed [1,3]). For example, in Norway an ongrowing licence was originally volume based but was converted to a limitation of maximum allowable standing biomass (MAB) on the site of 780 tonnes or 900 tonnes. This is termed one MAB, and a biomass larger than this on the site will result in large penalties. Several MABs can, however, be utilized on one site, but this may result in increased requirements for environmental monitoring and documentation due to the higher biomass on the site. If more than four MABs are used on a site a more detailed environmental monitoring programme is required (MOM C) (see 8.2).

Regulations require operational plans for aquaculture establishments in seawater. One such regulation dictates that an operational plan is listed for at least the next two calendar years and submitted to the authorities. Included in the plan are details of how the fish are stocked, the stocking date and number of fish. For locations where several licences are used, the plan must state which licences the stocking applies to, and any fallowing dates or potential movement of fish to other sites.

8.2 Site selection

Selection of a good site for the sea cage farm is crucial to production results and success, and will also dictate requirements for equipment. Effective salmon cage farming requires a high standing biomass on the sites, so this increases the requirement of the sites to handle waste without negative effects on the ecology or the salmon. If the water exchange is too low there may be a drop in oxygen below the level required for the salmon, with suboptimal salmon growth as a result. This can be critical in periods with high algal concentrations in the water and high water temperatures, particularly overnight when algae are using oxygen in addition to the salmon at the site (algae produce oxygen during daylight).

When selecting a site for a sea cage farm there are several factors to take into consideration. There is actually no perfect site that can fulfil all the requirements, so a multi-criteria evaluation must be done. A matrix of weighed factors can be used as a tool, and a site index may also be developed that can provide a comparison of potential sites. Plenty of literature describing more in-depth tools for selecting a good site is available (for example [4–6]). A geographical information system (GIS) is typically used as a tool when finding and selecting a suitable site for a farm [6–9], often combined with remote sensing and mapping. There are number of other tools and models that can be of help when searching for and selecting an available site [10–12]. The distance from the coast and how weather- and wave-exposed the site is will also influence the factors to include. Significant wave height will be the most important limiting factor and will set the requirements for the equipment to be used.

Many databases with information useful for a site selection process are available, but real-time measurement of important factors such as water current over a longer period should be included before the final decision is taken. However, real-time monitoring of parameters can be replaced by the use of modelling tools to estimate wave height and pattern (e.g. [13]), and a number of these models are available for different areas of the world (e.g. [14]). These include models estimating global tides, which is another parameter for selecting a site [15].

During production the effects on the environment around the farm must be controlled to avoid negative impacts, not only for the farm itself but also in relation to the environment. A fairly new proposal is to measure the sulphide levels in the seabed as an indicator, together with other traditional methods [16]. Methods and models used for selection of production sites, such as the cage aquaculture decision support tool and the MOM tool, can also be used for environmental monitoring when the production on the site is up and running [12,17,18].

Site selection factors can be categorized in a number of ways. Most categorization include the following factors: topographical, physical, chemical, biological, infrastructure, accessibility, social and legal (Figure 8.2). Some of these factors are described in more depth below.

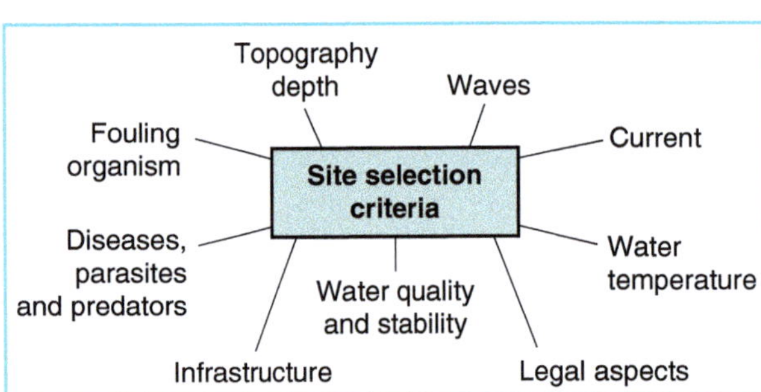

Figure 8.2 Several factors have to be taken into consideration when selecting a site for a sea cage farm.

8.2.1 Topography

Topography describes the shape of the land, the water depth and the bottom conditions, and, of specific relevance regarding site selection, how exposed the site is to weather. The degree of shelter from the open ocean is a very important factor. Areas with good sites for salmon farming are often sheltered behind islands; inside fjords may also be good sites provided that the water velocity is within a satisfactory range (Figure 8.3) and includes bottom currents that ensure distribution and dilution of the metabolic products below the cage so there is no environmental deterioration. For this reason, silled-fjords may be problematic locations, but each must be evaluated on a case by case basis, so good site evaluation prior to establishment of a farm is of great importance (Figure 8.4). Anoxic conditions may occur on the seabed if waste is accumulated,

Figure 8.3 Sheltering is an important parameter for site selection; behind islands or in fjords or bays are good, sheltered sites if water exchange is satisfactory.

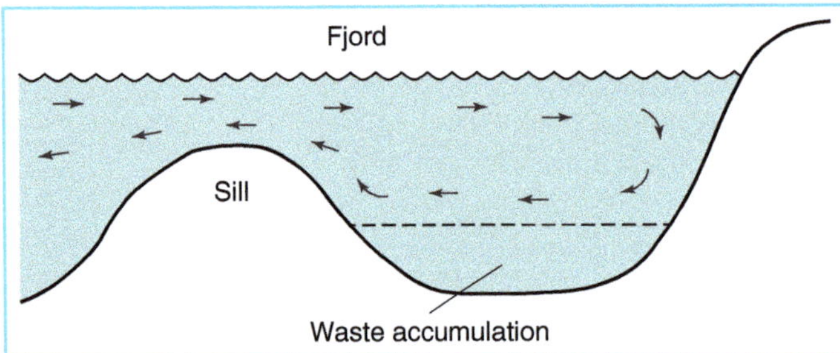

Figure 8.4 Silled-fjords may create problems when used as sites for sea cage salmon farming due to waste accumulation [19].

creating the potential for high concentrations of hydrogen sulphide (H_2S) gases in the water which are highly toxic for salmon.

The depth of the site is another important parameter for site selection. An absolute minimum depth under the cages at low tide is 10 m but this requires very good water exchange. A greater depth is recommended, but this depends on the degree of exposure at the site, particularly regarding wave height and currents. In some countries, the minimum depth required below net bags is regulated in the farming licence (for example, greater than 20 m in Norwegian regulations). However, use of deep sites increases the cost of the cage mooring system. As an indication, depths over 100 m will require mooring lines from the cages to the bottom of over 300 m (when laid at a 30° angle to ensure secure purchase) with a traditional mooring system.

The bottom conditions will have an important influence on the choice of mooring system; in rocks, bolts may be used while plough anchors are most appropriate where the seabed is unconsolidated, such as clay or sand.

8.2.2 Physical

It is important that the equipment is suited to the degree of exposure of the site. The site can be classified according to degree of exposure, mainly based on wave height and water current conditions. Equipment, cages and mooring systems certified as suitable for the classified degree of exposure are then required.

With sites that are more exposed to waves, currents and wind, heavier equipment (cages, moorings and so on) is necessary and this increases the investment cost. When significant wave heights (the average height of 1/3 of the waves) surpass 3 m, the equipment costs increase significantly.

8.2.2.1 Waves

Waves are mainly created by wind, and wave height depends on fetch length, fetch width, wind strength and wind duration, and can be estimated based on these factors [20]. Areas with too long a free fetch length from the prevailing wind direction should be avoided to limit the wave height. Another wave type is swell, which is typically also wind generated but originates some distance away. Compared to wind-created waves, swell is recognized with very long wavelengths. Direct swell is typically avoided by placing the cages in the shelter of islands, or in sheltered water in general. Placing of sites too close to regularly used shipping lanes is not advisable due to ship-created wakes and disturbance, and the risk of accidental ingress of a boat into the farm will also increase, even if the risk is low.

Wind will typically not represent an important parameter for site selection, because the area above the sea where the wind 'attacks' the farm is limited compared to the area below the water surface where the water current attacks the farm. More important is that the wind creates waves and current that are important factors regarding site selection, but this is also dependent on the fetch length.

8.2.2.2 Current

Several factors create current or water velocity. Most important is wind, but tide, local water flows such as rivers, and large global oceanic or coastal streams such as the Gulf Stream and Humboldt Stream may also create current at the site and should be included in the overall current profile of the site. Owing to the complexity of how currents are created, they are typically measured on the site for a period of time before establishing a farm, but estimation based on available models can be helpful for selecting site candidates.

The tidal current is typically limited but can be very strong in narrow fjords and narrow necks. The strongest measured maelstrom in the world was up to 30 km/h (8.33 m/s). Generally, sheltered sites will have overall current velocities in the range 0.2–0.8 m/s.

Good water exchange at the site is important, and this is created by the water current. This ensures supply of sufficient oxygen for the salmon in the cages and dilution and transport of waste products away from the farm, and should be high enough to prevent settling of particles (feed and faeces) below or close to the farm. In this way the surrounding water will handle the particulate waste products without any accumulation and negative environmental effects. However, too high a velocity, much above 1 m/s, is not recommended for traditional sea cage farms because it results in large forces on the net bags and mooring system and increases the size and costs.

Tolerance of the fish to current will also give an important input to the acceptable water current on the site, with a maximum and minimum recommended water current. Smolt have a lower tolerance of water current than adult salmon, which is a reason for locating smolt on less current-exposed sites. For example, there are recommendations for a water flow below 0.4 m/s for smolt sites. For larger salmon, a current over 1.25 m/s resulted in some fatigued salmon in one study [21]. The water temperature will also influence the tolerance of the salmon to current; tolerance drops with reduced temperature [22]. Instead of using actual velocity, velocity in relation to body length of the fish can be used. This is given in body length per second and drops with fish size. For post-smolt, critical swimming speed is reported as 2–4 body lengths per second and sustainable swimming speed 2 body lengths per second, but welfare may be negatively affected if long-term velocity is >1.5 body lengths per second [23]. One proposal recommends current velocities on production sites for fish in general, not salmon specifically, of 0.1 m/s to 1 m/s [24]. For salmon, the lower current level could be higher, at 0.2–0.3 m/s. Current velocities that are too low may increase agonistic behaviour in the salmon [25]. Problems with low oxygen may occur at tide changes on sites dependent on tide current, because when the tide turns, the current can partly stop or at least be reduced.

8.2.3 Water temperature

The seawater temperature should be well above 0°C to avoid ice problems, with a winter temperature above 2–3°C. Icing on cages may, however, also occur in freezing air temperatures combined with wind and waves, even if the sea temperature is above the freezing point. Sites exposed to floating ice should of course be avoided.

Areas with high water temperatures in the summer should also be avoided. Temperatures above 18°C may result in limited growth due to difficulties with oxygen supply, and in this case feeding should be stopped. The oxygen requirement of the salmon increases as the temperature rises, while the oxygen content in the water drops [19,23]. If the highest summer water temperature rises above 20°C it can be critical for the survival of the salmon. Water temperatures below 6°C will result in reduced growth (lower specific growth rate, SGR) and bring increased incidence of winter ulcer disease. Atlantic salmon can actually survive in a wide water-temperature window of 0–23°C [26], if survival is the only aim. However, at the highest temperatures oxygen must be supplied and distributed to ensure there is enough for all fish, which in practice is difficult to fulfil. Low and especially high temperatures will increase the risk of disease outbreak, especially when the fish are stressed due to, for example, rapid temperature changes or low oxygen content.

It is also important for there to be a gradual adaptation to temperature. Temporary fluctuation in water temperatures may have a negative impact on fish growth [27], but salmon are quite proficient in adapting to different temperatures as this will happen naturally during the year. The same is the case in deeper cages where there will be a variation in temperature between the surface and the bottom of the net bag. But even though salmon seem to have a high tolerance for water temperature variation, the optimal temperature is within a limited area [23] and in practice today appears to be around 14°C; however, with optimal supply of oxygen and water current it might be slightly higher.

8.2.4 Water quality

In general, salmon is a species that requires good water quality. Therefore, a site with suboptimal water quality is not appropriate. Sites close to potential sources for water contamination, such as industry, should therefore be avoided, for even if the water quality does not become directly affected there may be a degeneration. The same is the case with sites close to large rivers due to temperature and salinity fluctuations, which may reduce salmon growth and result in mix zone problems, with temporarily reduced water quality resulting in aluminium deposits on the gills due to pH changes if the water quality is susceptible to this. Sites close to rivers that host wild salmon migration should be avoided for the same reasons but also due to the consequences of any escape of fish and possible disease transfer.

The salinity of the site is not of major importance, because post-smolt salmon have a high tolerance of salinity; recommendations vary from 20–34 parts per thousand [23]. More brackish water (lower salinity) has some advantages regarding diseases and parasites; for example, sea lice problems will be greatly reduced. Brackish water may also be an advantage in relation to growth because the salmon do not need to perform so much 'internal work' to regulate salinity, since their internal salinity is approximately 1% (see also Chapter 4 on fish biology). However, it must be above 1%, so the salmon recognize it as seawater and do not start to desmoltify. Brackish water areas are, however, typically more exposed to water temperature fluctuations because they are affected by freshwater which has lower winter temperatures and higher summer temperatures. In a deeper net bag the salmon may adapt to this by moving up and down depending on the preferred water temperature and quality [28].

8.2.5 Biological

Fouling in net bags is a problem that must be dealt with in farming (described in 8.5.2.3). The degree of fouling seems to be site dependent, and some sites are more susceptible. Water temperature, salinity and available nutrients are important factors affecting the degree of fouling.

The disease and parasite situation will also vary with the site, and the proximity to other aquaculture production sites, especially those for salmon production, is especially important. The risk of disease outbreak increases as the proximity between farms decreases, and also depends on the water current and position of the sites in relation to the current. Therefore, distance to other farming sites is another important part of the legislation.

Exposure to predators, for example, seals and sea lions, is also site dependent, and in some locations this may present a problem. Predator nets may then be used in addition to the ordinary net bags. This involves additional investment, both in the predator nets and in a heavier mooring system. It also presents more work in general with handling and removing of fouling.

If there are rare or highly valued wild species in the area this may prevent the use of the site, as is the case in some fjords where salmon farming is prohibited in order to protect wild salmon. Before selecting a site this must of course be taken into consideration.

8.2.6 Infrastructure

Good infrastructure with a limited distance to surfaced roads is an advantage, because travelling time to the site by boat will be reduced. It is also an advantage to have access to electricity, because then a fuel based power supply from which to produce electricity on the farm can be avoided. This requires that the site is not too far away from land with easy access to electricity. Self-generation of electricity by wind power, wave power or solar panels may be a possibility, but this is still at the development stage and little information on commercial use is available, especially regarding economy. Access to a good and stable internet connection is very important today and will continue to be an extremely important site requirement in the future.

8.2.7 Legal aspects

Selection of a site may also be affected by legal aspects, such as maritime laws. It is important to ascertain whether the area is designated for fish farming, publicly designated or designated for other purposes. Special environmental issues may also prevent the use of a site, for example the establishment of a cage farm for salmonid production just outside an important salmon river due to the risk of escape. The legal requirements for access to land for an onshore base and onshore mooring are taken into account here.

Conflicts may also arise. Proposed sites may be close to the shore where public recreational areas can be important, there may be conflict with commercial fisheries on sites close to important fishing grounds, and there may be conflict with locals who claim the area for other purposes. Marine spatial planning is therefore very important and may also give local communities the potential to control a sustainable development of sea cage salmon farming.

8.3 Transfer of smolt

Before transferring smolt to a site it should be fully restored if it has been used before, with it having gone through a fallowing period. This means that on the bottom under the cages and in areas close to the cages there must not be any residual evidence of earlier aquaculture production. The bottom fauna should be undisturbed and consist of the natural and original species. If this is not the case a longer resting period is needed before the next smolt intake.

8.3.1 Number of smolt

The number of smolt at the site is given in the production plan for the farm. This also gives the necessary cage volume based on the maximum or recommended allowed biomass per cubic metre farming volume. Fish density must be decided before transfer as this will influence the planning of the moving, grading and harvesting regime, for example whether the fish are to be graded, and if so at what size, and whether the fish are to be moved before reaching slaughter size. Often the fish start in one cage but later in production are divided into two groups, either by simple separation or combined with grading. Whether there are any regulations regarding maximum allowed fish density or MAB on the site will be an important parameter for deciding the number of smolt. There may also be restrictions on the number of individual fish that can be kept in a single cage, such as 200,000 smolt in Norway for traditional cages.

The size of the smolt will also be important. If larger fish are used the number must be reduced to avoid reaching density limitation too early, but this will depend on whether any moving and grading programmes are used. Today there is a trend towards larger smolt, with up to 100 g being common, but with some interest in using even larger smolt. By using higher water temperatures at the smolt farm it is possible to produce fish up to 200 g, but these are termed post-smolt. Post-smolt are smolt that have gone through smoltification at the smolt farm and have then been kept in seawater on the smolt farm, like on the sea cage farm. In reality, this is a land-based ongrowing farm. An advantage with post-smolt is that any problems with smoltification and transfer to seawater are avoided because the salmon are far beyond the smoltification stage. Post-smolt up to 1 kg are now being tested in commercial farming. This represents, however, a three-step model: smolt production, post-smolt production (either on land or in closed cages) and ongrowing until slaughter size. The reason for the great interest in larger smolt and post-smolt is to reduce the time in the sea, due to the higher risk of disease and parasite problems. As an example, there have been large problems with sea lice, and the cost of sea lice treatment is high (described in 8.6.3). By using larger smolt to reduce the time in the sea, the number of sea lice treatments will be reduced. The same will be the case with the risk for disease outbreaks. Owing to this, larger smolt or post-smolt command a higher price.

When transferring the fish to sea cages, it is of major importance to control the number of fish in each cage. It is also important to get as much information as possible about the smolt, including the genetic history, origin, and biological and production histories, such as the time of hatching and start feeding, size grading, production time and age in total day degrees. If there have been any disease problems during the freshwater period this is of interest, as is whether the smolt are vaccinated and against which diseases. Information on the water temperature and water quality at the smolt farm will also be of interest, such as whether supersaturation with oxygen has been used and to what degree. Documentation of smoltification status should of course accompany the fish, and include which test has been performed to confirm the smoltification status.

Good-quality smolt is of major importance to production results, and therefore their history is very important. However, it is also extremely important to perform a good visual inspection when the smolt arrive to the sea cage farm. Inspection can include searching for scale loss; large loss may result in future osmotic problems. Deformities, such as of the spine or jaw or shortened gill covers, should be checked for, and there should not be any eye damage. In the first weeks, the fish have to be monitored carefully for any abnormalities. This includes monitoring appetite and behaviour. Of course, mortality must be recorded and documented on a daily basis; any later complaints about the smolt delivery could be directly linked to this. The same will be the case for growth rate.

Transference to sea cages is a challenging period for the smolt. The transportation and the totally new environment result in stress, and therefore the fish will be more susceptible to disease. The natural mortality rate is also higher just after transfer to seawater [29,30].

The smolt seem to prefer a site where a halocline exists (horizontal stratification with different salinity). Haloclines may occur on a site close to the coastline and where there are supplies of freshwater.

8.3.2 Practicalities

The smolt at a typical size of 60–100 g will normally arrive at the sea cage farm on a well boat, but transport on trucks may also be used. Either new net bags or washed and repaired nets can be used, with any antifouling coated nets cleaned well before the arrival of the smolt. To avoid problems with birds taking the smolt, bird nets covering the cage must be used at first (Figure 8.5).

The recommended mesh size of net bags varies with the fish size, but in general starts at 40–45 mm (Table 8.1) (for more information and description of nets see 8.5.2 and ISO 1107 [31]). The risk of escape due to too large a mesh size in the net bags will be higher just after a new intake and most probably occurs within the first days. The mesh size used in the net bags must therefore be of the correct size (Table 8.1) [33,34]. Studies have shown that the method for measuring mesh size is of importance; traditional measuring may lead to underestimation, especially for small-sized fish, because when the net is in the sea it is not static but behaves dynamically, affected by waves and current [35,36]. As the mesh in the net bag is not stiff, the actual mesh size will slightly increase when the net panels are semi-slack or slack, so the actual opening will be larger than for a stiff net [36] (Figure 8.6).

8.3.3 Set out time

The most common time to transfer smolt to sea cages is in the autumn (September and October in the northern hemisphere) for 0^+ smolt, and in spring to early summer (May and July in the northern hemisphere) for 1^+ smolt. There are usually no intakes of smolt in the winter months (December to March in the northern hemisphere) due to potential low sea temperatures. For all other months of the year there will be smolt transfer, but this is of course dependent on the water temperature of the site during wintertime; in some areas and countries this will not present any

Figure 8.5 Cages should be covered with a bird net when the smolt are transferred to the sea.

Table 8.1 Smallest smolt weight recommended for different mesh sizes and bar length taken (based on [34]).

Mesh size (mm)	30.7	38.5	40.7	45.7	49.2	54.4
Recommended smallest fish size in the net (g)	26	56	87	129	195	250

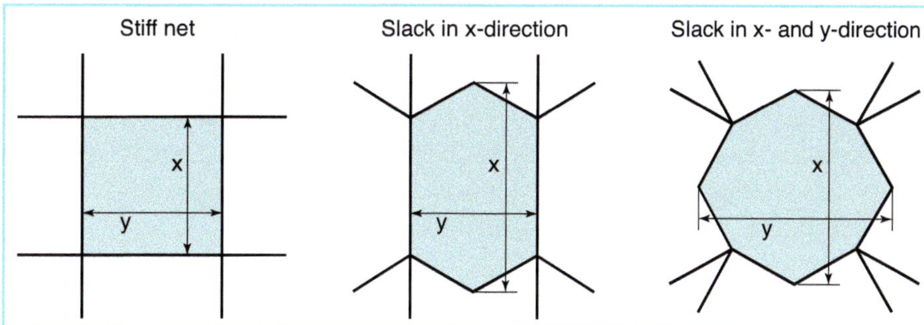

Figure 8.6 A slack net will have a larger mesh size than a stiff net (based on [36]).

problems and fish can be transferred to the sea all year round. It is recommended that the seawater temperature is above 6–8°C in the autumn and above 4–5°C in the spring for smolt transfer, but this may also be specified by national regulations. The temperature can be lower in the spring as it will increase after transfer. However, these recommendations are continuously stretched to optimize production and there are examples of fish that have been transferred in both January and March to low-temperature sites in the northern hemisphere with good results.

8.4 Daily management during the ongrowing period

During the ongrowing period, optimal daily management and control is important to achieve optimal growth, low mortality and low FCR and to reduce disease problems by early observation. This must, however, be considered in relation to the running costs (working expenses).

The important daily management operations include feeding, dead fish collection and measuring of water quality. Water quality measurement at the site typically includes water temperature and oxygen levels, together with salinity. A visual inspection of the fish and their behaviour and a simple visual check of nets and mooring systems is also an important part of the daily control routines. The last is also an important precaution to avoid fish escape.

8.4.1 Behaviour observations

The behaviour of the salmon should be observed as it is a good indicator of welfare. Typically, the salmon will swim in circles in a school inside the cage, either in a clockwise or a counter-clockwise direction. This only changes at feeding when the salmon move out of their regular orbit. During daylight the salmon prefer to swim at some depth while at night-time with lower light they usually move slightly closer to the surface. If the salmon are crowding close to the surface or do not have a distinct schooling pattern the behaviour is not normal, and something is wrong.

If the water velocity of the site gets too high, the salmon will swim against the current rather than following their normal circular swimming pattern in the cages [21]. The normal schooling swimming pattern is broken when current velocity goes above 0.3–0.35 m/s [37]. Experiments have shown that Atlantic salmon can cope with a long-term swimming velocity of two body lengths per second but growth performance is already reduced at 1.5 body lengths per second, as described earlier [25,38,39]. The behaviour of swimming in schools has an advantage for salmon, as it reduces the individual drag and therefore the energy cost of swimming, and by this increases the swimming performance [21,40,41].

However, in closed sea cages growth performance has showed an increase with increased current, although the currents used were much lower than the maximum tolerated velocity (0.63 body lengths per second [42]).

8.4.2 Feeding routines and strategies

8.4.2.1 Importance for production cost and animal welfare

Feed is the most important cost in salmon production and represents over 50% of the production cost. Achieving the highest growth rate at the given temperature on a given site is of major importance for turnover, and hence profitability, but also for effluent release from the production. Optimal feeding is very important in relation to both economy and environment. Some countries have regulations that include feeding routines and strategies; these are specifically related to good animal welfare but also cover general requirements. As an example:

> The amount of feed shall be sufficient and of a composition that promotes good welfare. The feed shall be adapted to age, development stage, weight, physiological and behavioural needs.
>
> Fish shall normally be fed daily unless this is not practical for the species or development stage. They shall be fed in such a manner that all fish have easy access to feed without being injured while feeding.
>
> Fish shall not be fed when feeding is disadvantageous out of consideration for the welfare, hygiene or quality of the fish. [43]

8.4.2.2 Practicalities

When coming to the practicalities of feeding on the site, several decisions have to be made both before starting the production and when production is up and running. This includes choice of feed type and particle size, how feed is to be supplied, how much and when to feed and how the feed is to be stored (see also Chapter 10).

The only feed in use today is dry formulated feed with a dry matter above 90%, which is typically supplied from commercial feed companies. These companies supply different feed for different fish sizes and also supply special feed, for instance, to boost growth for a period or including special ingredients to produce a high content of omega-3 oils (EPA-DHA) in the salmon. Special feed for ecological production is also available. Feed containing pigment is usually used for the entire sea production, with gradually increasing pigment content; it is possible to obtain feed which includes pigment from different sources (see also Chapter 10).

8.4.2.3 Feeding methods

Two methods can be used for feeding in sea cages: appetite feeding (ad libitum) or table feeding. These may also be called controlled feeding versus regulated feeding. With appetite feeding, the appetite of the salmon regulates the amount of feed delivered: as appetite increases, more feed is delivered. This can be performed with hand feeding where the appetite can be monitored by direct observation. However, the large amounts of feed necessary with commercial-scale sea cage farming today means that hand feeding of the entire feed amount is not possible. Even in the past when hand feeding was a common method, it was hard work for the farmers even when farm sizes were much smaller than today, and some farmers incurred shoulder injuries from handling and throwing large tonnages of feed into the sea cages.

Automatic feeding is the predominant method today, and is the only practical method on the large commercial-sized salmon sea cage farms. Feed is supplied by machines (automatic feeders, or feeding systems) and submerged video cameras can be used for monitoring appetite [19] (Figure 8.7). Operators watch the monitors and can stop feeding when the appetite of the salmon drops. Often this is combined with table feeding (see below) in order to have some control. A feeding centre is a new invention that is increasingly common; here the feeding is controlled from a centre facility that can be located anywhere as long as internet connection is available. With the introduction of such systems, several sea cage sites can be fed from the same feeding centre. It looks like a control room in a process industry where signals from cameras located on different farms are shown on monitors in the control room, and operators ensure optimal feeding in the monitored cages.

8.4.2.4 Table feeding

To perform table feeding, fish size, water temperature (the main parameter for salmon growth) and total biomass in the cage must be known. Based on these factors, tables have been developed listing expected growth rates for different fish sizes in relation to the water temperature (see also Chapter 9 on production planning). By knowing the total biomass in the cage, the correct amount of feed for each day can be estimated. However, growth seems to vary from site to site

Figure 8.7 Feeding system typically used for sea cage farming.

and also from intake to intake. Therefore, different feeding tables should be developed for each site. This can be done by starting with a general feeding table and further developing it to fit the result achieved on the actual site, including the growth history from recent years for that site. When performing table feeding, regular control weighing is very important, not only to monitor that the fish are growing as they should but also to have the opportunity to adjust the feeding if growth is suboptimal. It will also give important input to the FCR.

In practice, a combination of table and appetite (controlled and regulated) feeding is used for salmon. It is mostly appetite feeding, but there is regular control relating to the growth table and the expected feed to deliver based on the feeding table.

8.4.2.5 Feed frequency

It seems that the feeding frequency, or how many meals to feed during a day, is of minor importance for salmon beyond smolt size, with 2 or 3 meals per day seeming adequate for the fish. There is therefore no great difference in performance between separate meals or continuous feeding regimes. However, due to the amount of feed that has to be delivered to the cage each day the feeding system will need to be in operation frequently.

8.4.2.6 Feed particle size

The feed particle size should be adapted to the size of the fish. If too large, the fish will not be able to eat it while a small size will require the fish to use unnecessary energy to get enough feed particles. The feed particles used are in pellet form and sizes ranging 3–12 mm in diameter are used in sea cage farming adapted to the fish size (Table 8.2).

8.4.2.7 Number of feed particles

Based on the size of the particles the number of particles per kilograme will vary (Table 8.3). If combining this with growth tables, the amount of feed pellets that the salmon has to consume during a day for the expected growth rate can be calculated.

8.4.2.8 Feed storage

Dry feed is a fresh product and should be stored as such, and not kept for too long when the climate is warm with high humidity. It can be stored outside, but direct sunlight should be avoided and proper ventilation is important. If storing inside, the storehouse should be dark and airy. The feed can be stored in small or large sacks, or preferably as bulk. Based on the amounts of feed

Table 8.2 Recommended feed size for ongrowing in the sea from one feed supplier (based on [44]).

Fish size (g)	Pellet diameter (mm)
75–150	3
150–160	4.5
600–1200	7
>1200	9
>2500	12

Table 8.3 Examples of approximate weight of individual feed particles for different particle sizes (based on[44]).

Particle size	Weight per particle (g)
3	0.018
4	0.13
6	0.37
9	0.83
12	1.66

Example

Fish weight is 2 kg, seawater temperature is 12°C, specific growth rate is set to 1% and pellet diameter is set to 9 mm. How many feed particles does the fish need to eat during a day?

Growth rate is set to 1%, so in 1 day the fish grows 1% which is 20 g. The fish therefore needs to eat 20 g during a day. The weight of a single particle of 9 mm is estimated to be 0.83 grams, which gives a particle requirement of: 20 g/0.83 g per particle = 24 particles. The fish therefore needs to eat 24 feed particles per day.

Example

The fish in the above example is one of 100,000 in a large cage. How many kilograms of feed and number of particles must be delivered to the cage for one day?

Total feed supply: 100,000 fish × 0.02 kg/fish = 2000 kg = 2 tonnes.

Number of particles: 100,000 fish × 24 particles/fish = 2,400,000 = 2.4 million particles per day.

used, large sacks or bulk storage are the only alternatives today. It is common to store as bulk in silos either on land (where silos must be insulated) or submerged in the sea (including in feed barges). Bulk feed is typically used for supplying feed into central feeding systems, which is the most common feeding method used in intensive salmon farming today (Figure 8.8).

8.4.3 Dead fish collection

Under normal conditions there will always be some natural mortality in the cages due to the large number of individuals. Therefore, dead fish have to be collected from the cages on a regular

Figure 8.8 Feed stored in silos on a feed barge supplying feed for automatic feeding systems.

basis. How often is typically included in the government regulations and licence since it represents not only a procedure to prevent proliferation and spread of disease but also an effort against environmental pollution. Typically, the requirement is once a day but it may be necessary to perform this several times a day if mortality is high and with high water temperatures. Higher water temperatures will result in a faster decomposition rate and therefore more regular removal of dead fish is important. It could also be an advantage to euthanize and remove sick fish together with the dead fish.

A collector inside the net bag is typically used for collection and removal of dead fish. The collector is dragged to the surface and the dead fish are removed (Figure 8.9). A pump or an airlift mechanism may also be used to bring the dead fish up from the bottom of the sea cage. It can also be done entirely manually by lining up the net bag and using a dip net to collect dead

Figure 8.9 Systems for collection of dead fish from sea cages.

fish, but this is rather labour intensive and almost impossible in large cages. An automatic system (a mort collection system) that regularly pumps the dead fish from the cage directly to the treatment container and counts them has also been developed, and avoids all manual handling of dead fish.

If disease occurs the mortality may increase severalfold, as may also be the case if there are algal or jellyfish blooms. Identification of certain severe diseases may also require slaughter of all fish on the entire site, and this may also be required if there is a breakdown in a major component of the sea cage or mooring system, for instance, due to bad weather conditions. Emergency plans for how to handle mass mortality and emergency slaughter within a limited timeframe are therefore a necessity, and may also be required in the regulations. Typically, agreements with large well boats are included in such plans, because they can handle large quantities of fish within a limited time. The well boats can also have vacuum pumps to suck dead fish directly up from the cage bottom.

Dead fish are treated as animal waste, and disposal must follow any animal waste regulations (like in Europe [45]). In Europe, dead fish are generally classified as category 2 waste, so are not allowed to be used directly for human consumption. In some cases, for instance with high mortality due to lack of oxygen, they may be categorized as category 3 waste which can be used for human consumption, but no disease must be present, and this usually has to be individually approved by the food safety authority.

The system for handling dead fish includes storage in closed containers or other facilities with no drainage to the environment, in order to avoid disease development and disease transfer from the dead fish. A number of methods are permitted, but the predominant method is acid preservation. The dead fish are preserved by acid in closed containers and fish silage is produced. Naturally occurring enzymes in the viscera will result in autolytic liquification as they break down the fish protein resulting in smaller soluble units (reviewed [46–48]). The added acid will lower the pH, inhibiting bacterial growth and spoilage and speeding up the enzyme activity. Most commonly, formic acid is used which lowers the pH below 4, but mineral inorganic acids may also be used but then a lower pH (2) is required to satisfactory conserve the solution. The high lipid content will result in lipid oxidation in the silage, and an antioxidant such as ethoxyquin may be added to prevent this.

To handle normal mortality, the farm must have equipment adapted for acid preservation. Before preservation with acid, the dead fish must be ground to release/free up the enzymes and facilitate access of the acid and enzymes to all parts of the fish. The dead fish treatment system therefore contains two parts: the grinder and the storage tank for temporary storage. This could be included in the same unit in smaller farms. There may be regulations for the minimum sizes of system for handling dead fish; in Norway, for example, the requirement for aquaculture establishments with over 130 tonnes of biomass is that the capacity of the storage container must not be less than 5 m^3.

Silage is only temporarily stored at the sea cage farm before it is collected by special companies that handle such waste as a bioresource. The acid-preserved dead fish are collected by special boats but may also be collected by trucks if dead fish are taken to land-based facilities. The companies handling the dead fish will use it to manufacture products such as fat or protein concentrate. The number of industries for handling dead fish and other waste from aquaculture facilities is increasing in tandem with the growing aquaculture industry, and these are using more and more advanced handling processes resulting in improved utilization.

8.4.4 Measurement of water temperature and quality

It is important to measure water temperature regularly to aid growth predictions and most importantly for retrospective monitoring of growth, to see if the growth and feed consumption is proceeding according to budget and estimates. If the temperature is low, below 3–4°C, the feed supply ought to be very limited. If the temperature rises above 16°C, special precautions must also be taken as the available oxygen in the water will be reduced and must be frequently measured. If it becomes too low, around 60–70% oxygen saturation, the feeding should be stopped. The lower percentage level of oxygen saturation in the water to achieve maximum feed intake is highly dependent on water temperature [49]. When temperatures are close to 20°C mortality may occur and feeding must be stopped to reduce the oxygen consumption of the fish. If the salmon start to stress due to low oxygen content, this will increase the oxygen consumption further.

In addition to measuring the seawater temperature, the levels of oxygen should be measured continuously on the site, either with more than one sensor per site or preferably with individual oxygen sensors in each cage (Figure 8.10). Critical periods where lack of oxygen may occur are summer and autumn when the water temperature is high and growth rate is fast, with high oxygen consumption. High algal activity during this period may result in a critical drop during night-time, especially close to sunrise. Some farmers have started to use additional

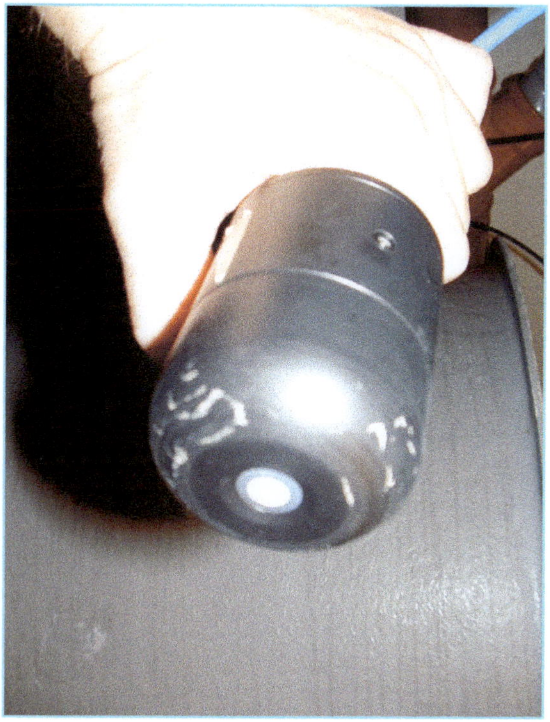

Figure 8.10 Water temperature and dissolved oxygen concentration ought to be measured on a regular basis in sea cages and can be done by a combined sensor.

oxygen (liquid or produced on-site) to avoid oxygen drop and the subsequent reduced growth in these periods. It is also possible to obtain increased growth by using oxygen when water temperatures are high.

8.5 Routine work: non-scheduled and scheduled

During the ongrowing period in the sea cages a number of working operations must be performed, which includes splitting and/or size grading of fish groups, changing net bags to increase mesh size, removal of fouling, any required disease treatment, sea lice treatment, predator control, light installation and light control, environmental control, escape control, control of allowed biomass and production control (weight, fish number, total biomass). The intervals for performing these are dependent on how the site is managed.

8.5.1 Splitting of the fish group

Owing to fish growth, density in the cage may increase to levels where it is necessary to divide the group, either because the density is so high that the water quality is degraded (which has negative effects on welfare) or because it reaches the limitation given in the farming licence (which varies between countries: from 5–10 kg/m^3 up to 25 kg/m^3). Several studies have been performed on fish densities but for a satisfactory welfare score it should generally be in the range of 20–25 kg/m^3. Below 22 kg/m^3 has been reported in some studies (reviewed in [23,50,51]), and higher densities resulted in increased fin damage. In one study the welfare score was lower at 15 kg/m^3 and 35 kg/m^3 when compared to 25 kg/m^3, and the highest aggression was observed at 15 kg/m^3 [52]. Another study found that 7–11 kg/m^3 was better than 18–27 kg/m^3 because the fish did not have the space to move to a depth with the preferred temperature [53]. With too low a density the salmon may change behaviour from a schooling to an individual behaviour, which increases aggression against other individuals.

There are several ways to split a fish group in a sea cage. Two cages can be connected by sewing together the two net bags. With the use of a bullet line the fish can be crowed together in the original net bag, and some of the fish will then swim voluntarily over to the new cage (Figure 8.11). With this method, however, it is difficult to keep control over the biomass in each of the cages after the operation. A more common way today is to use a well boat (Figure 8.12). The fish are sucked out of the original cage and into the well boat, and are then pumped into two new cages. By using pumping, the salmon can be sent through a counting and weighing unit which allows the number of fish, weight and total biomass in the respective cages to be determined, and this may be combined with grading, so that different sizes of fish are directed to different cages. This method also allows the net bags to be changed so that a larger mesh size and nets coated with new antifouling agent (see 8.5.2.3) can be used.

Figure 8.11 A bullet line may be used for crowding the fish when splitting a fish group.

Figure 8.12 Use of well boats is more common when splitting and grading fish groups. The boats are equipped with fish counters, so it is possible to have control of the number of fish.

One production regime used is to transfer smolt to one site, then after a period of growth the fish group is divided in two and both the original site and a new production site are used. The second site can be more weather exposed, because the larger fish can tolerate faster currents and higher waves.

It is possible to move the individual cages around within the site or to new sites. Cages with fish ready for harvesting, for example, can be dragged to the slaughterhouse if this is located near the sea cages. Both cages with net bags and cages without net bags (with only a collar) may in general be moved (Figure 8.13). Recommended dragging velocity is up to 0.5 m/s for a cage collar with a net bag and 1.5 m/s for a cage collar without a net bag [54]. Towing of cages is not recommended with a wave height above 0.25 m and wind velocity above 12 m/s [54]. It is advantageous to add a diagonal rope within the cage to stabilize the cage collar when towing and the towing line should be divided into several lines before being attached to the cage (looking like a hen's foot) to distribute the forces on the cage collar. Cages can also be towed with salmon inside the net bag but then the towing velocity must be low, below 1 knot (0.5 m/s), depending on fish size. However, the net bag will be deformed when towing, and the cage collar and net bag will be exposed to large forces. Animal welfare must also be taken into consideration when towing sea cages [55].

Figure 8.13 Towing of sea cages.

8.5.2 Working with the net

8.5.2.1 Some definitions

An international standard regarding fishing nets is used to define the terms and expressions in a universal way [56]. This standard gives the principal terms related to fishing nets, which can also be used for net bags in aquaculture. Typically, nets with a square mesh opening are used but mesh openings can also be hexagonal. The bar length, also described as length of the mesh, is the distance between two knots on one thread in a fully stretched net in x and y directions for square mesh. The mesh width or opening is the size of the opening in a fully stretched net in x direction, and is also called mesh size (Figure 8.14). This will vary with the type and size of knot used. Today both knotted and knotless nets are used in net bags.

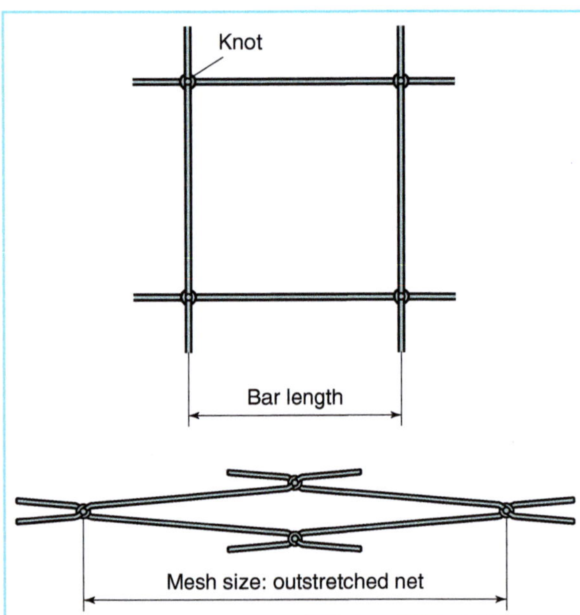

Figure 8.14 Definition of sizes of the net bags in sea cages.

8.5.2.2 Changing nets

In addition to splitting of the fish group in a cage due to high density, the net bag should be changed to one with an optimal mesh size. To reduce problems with biofouling, as large a mesh size as possible should be used, which also ensures good water exchange in the cages. However, if the mesh size is too large the fish may escape. For small fish, up to 1–2 kg, a ½ mesh size (bar length) of 12–16 mm is recommended, and for fish larger than this a 19–29 mm 1/2 mesh size (bar length) can be used. Net bags may also be changed without changing the mesh size depending on the degree of fouling (see 8.5.2.3).

Working with nets is often performed by sub-contractors or by internal divisions of the company, with a working team of four to six people. With the increasing cage size (typically 160 m circumference and with a net bag depth of 20 m to 50 m) this is hard work that requires quite large vessels suited to the task, usually equipped with cranes with sufficient lifting capacity (Figure 8.15).

Net bags have to be monitored regularly to avoid breakage that can lead to fish escape. It is necessary to document information about the net, such as the strength and any work or maintenance that has been performed. This can be stipulated through regulations, for example the Nytek Regulations [57], which include certification of the individual net bags for strength among other things. A manual for use and maintenance (a user handbook) should accompany the net bag when delivered from the supplier, and there should be a net card index for all net bags on the farm. Every instance of net handling, including washing, coating, maintenance and so on, should be documented in a journal that accompanies the net bag. The mesh in the net bag should be strength tested at given intervals to ensure that it retains the certified strength. The net bag has a limited lifetime which can also be defined in the regulations.

8.5.2.3 Control of fouling

Fouling of nets, ropes and cage frames occurs in varying degrees in aquaculture [58–60] (Figure 8.16). The degree of fouling increases at warmer temperatures and the main problems

Figure 8.15 Working with cages and nets on-site requires boats with cranes with sufficient lifting capacity.

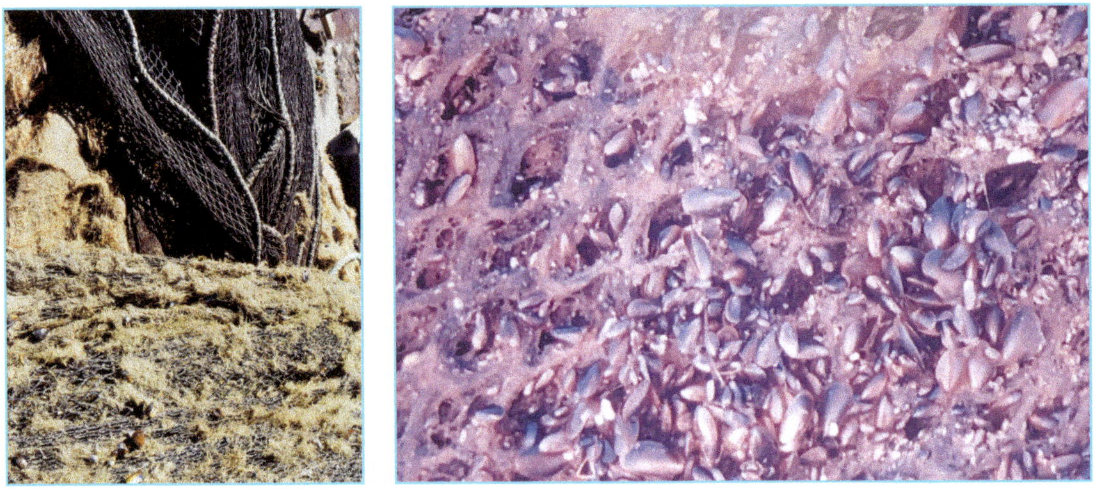

Figure 8.16 Net bags are exposed to fouling.

therefore occur in the summer and autumn. The degree of fouling varies from site to site and can become a serious problem, typically in nutrient-rich areas. Fouling of nets depends on the material used in the net bag, the mesh size and whether the mesh is knotted or knotless, which affect the areas available for organisms to settle on and grow.

When discussing fouling in aquaculture, biofouling is the cause (overview [60–62]). Biological organisms use the net as a base for growth. The fouling starts with a colonization of smaller organisms, such as bacteria, which form what is called a primary biofilm. This film provides the basis for new biological growth. The diversity and development pattern of the fouling organisms varies with the environmental conditions at the site, while the material used is important for the initiation of the fouling and how it is attached. The degree of fouling will, as mentioned, be controlled by the available nutrients in the water and the water temperature: more nutrients and higher temperatures means more rapid growth. The fouling organisms are mainly photosynthetic, meaning that their growth is dependent on light, so a higher light intensity increases the degree of fouling. The water current on the site will also be important to a degree. A slow current will result in fast growing of fouling organisms, whereas higher velocities will result in shear forces on the fouling that inhibit growth. Salt seems to promote the degree of fouling, so cages in brackish water are not so susceptible. The degree of fouling on net bags is highest closest to the surface due to the higher intensity of light. Various different species cause fouling and organisms vary from site to site. Some major groups include mussels, algae, hydroids, tubeworms, barnacles and ascidians.

The main negative effect of fouling is that water flow through the net is reduced [63–67]. As fouling increases, the effective mesh size of the net bags drops and what is termed net solidity increases. This will reduce water flow into the cages thereby reducing both the level of oxygen and the dilution of metabolic waste products from the salmon in the cage. The reduction in available oxygen within the cage may result in reduced growth of the salmon and there will be accumulation of waste products within the net bag. If nothing is done the fouling will continue to grow and may clog the net entirely, resulting in up to 100% reduction in water flow through the net in the worst cases; therefore, it is vital to control fouling.

The prevention of current flow through the net bags by fouling may result in another severe problem, as the current forces on the construction will increase due to the increased resistance to water flow; more than a 10-fold increase in current forces has been observed [59,68]. This may result in increased deformation of the net bags and an increase in the total forces on the cage collar and mooring systems, with the possibility of the forces rising above the maximum values calculated for the system. Construction and mooring system collapse may therefore be the result. This is very critical, for instance, in relation to escape of fish. A number of collapses of entire or parts of sea cage farms have been caused by fouling. A high degree of fouling will also increase the weight of the construction and all handling of nets will be more problematic. It is estimated that fouling increases the labour cost by up to 20%, and can result in a 200-fold increase in weight [59,69].

Another negative impact of fouling is that it may harbour unwanted microorganisms, such as bacteria and parasites. Free-living stages of sea lice may also be hosted in heavily fouled nets [70], and the marine amoeba which causes amoebic gill disease (AGD) has been found in the fouling layer on the net [71].

8.5.2.3.1 Fouling prevention

If the fouling at a site is so severe that action must be taken there are a number of options. Which strategy to use depends on the management of the farm. There are a range of possibilities for handling fouling and the methods are categorized in different ways (Figure 8.17). One way to divide the approach to control biofouling is (after [72]): (1) mechanical detachment; (2) killing or inactivation; and (3) surface modification of material to promote low-fouling/non-adhesion.

The first step to prevent biofouling is done before the net bags are placed in the sea, because the choice of material of the net bag is important regarding how easily the fouling organisms can attach. Different materials prevent fouling to different degrees and therefore more fouling-resistant materials can be used for the net bag, mooring lines and cage collar. Fouling problems are reduced by using polyethene net (PE) compared to polyamide net (PA), which is most commonly used in net bags. Copper wires may be braided into the plastic netting to prevent fouling, but this is significantly more expensive. Rigid net bags made of a metal may also be used and then the material can be made more fouling resistant, such as wire with copper, a copper-nickel compound, or galvanized steel, but the cost for this solution is much higher. Also, the choice of cage design can be important; cages with rotating designs can be effective for controlling fouling because part of the net is above the sea at all times. This allows the net bag to be dried by the sun

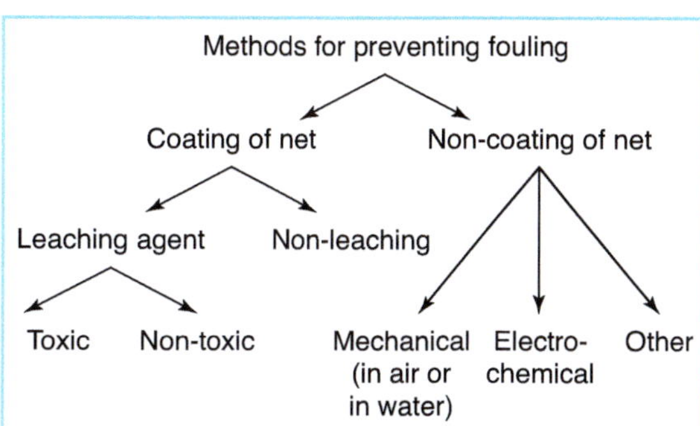

Figure 8.17 Methods for preventing fouling (based on [73]).

and this reduces cleaning to a minimum. A dual net system on each cage has also been tried. One net is in the water and one is out of the water, and periodically the nets are alternated so that the fouled net is air-dried on the cage for a period. The organisms dehydrate and die, a process that typically takes a week or so, and can then be brushed off. Care must, however, be taken in areas where levels of UV light are high because UV light can accelerate the net material degradation.

Instead of using a material in the net bag that is less prone to fouling, a coating layer with similar properties can be added to the net (Figure 8.18) [74]. Such coatings are either chemically or biologically based and can either leach active ingredients that prevent fouling organisms from coming into contact with the nets, or inhibit the fouling organism from settling. Another way to categorize coatings may therefore be as leaching or non-leaching coatings.

Most commonly used is a leaching chemical agent in the coating, also called an antifouling agent. This includes a slow-leaching biocide that permanently surrounds the net surface with a thin layer of material that is toxic for fouling organisms. This agent will leach out over a given period of time, and when there is no antifouling agent left the protection drops rapidly and a new treatment must be performed. The length of this period varies with substance but is typically around 6 months. Today the coating, or painting as it is also called, is usually water based, and two major principles are used to build up the coating, either a soluble matrix or contact leaching.

The most commonly used biocide in the coating is copper oxide which can be purchased in a variety of forms and brands. Copper is a toxic substance, particularly to algae and molluscs. Coating of nets is typically performed in net treatment stations. Here the nets are dipped for

Figure 8.18 Coating of nets, for instance copper-based coatings, are typically used to avoid fouling.

3–5 minutes in the copper solution before it is dried, so adding a layer of coating to the net. When first placing net bags that are coated with copper into the sea it is normal to avoid placing fish into the nets for 24 hours due to possible high release of copper. Organic biocides may also be used, and instead of biocides other active substances, such as enzymes, natural substances or living microorganisms, which release toxic substances, can be used.

Non-leaching coatings may also be used, with the aim of making a surface that is so smooth that the fouling organisms are unable to settle; silicone, or metal can be included in the net material, or copper wires in the net threads.

By adding a coating or antifouling agent before the nets are placed in the sea, the intervals between cleaning can be extended. On most sites, handling of fouling will be an important working operation and it is of great interest to find solutions that could prolong the cleaning intervals.

Frequent changing of nets or frequent on-site cleaning are strategies that must be used more intensively if no antifouling agents are used. The frequency of net changing varies from more than once a week to once a year, depending upon site location, net material used, season, management and design of net bag. In general, fouled nets are removed and exchanged with clean nets more frequently in summertime than in winter, due to the more rapid biofouling process.

Net bag cleaning can be done in several ways, which are split in two distinct groups: the nets can be either cleaned on the site or removed from the water and cleaned on land. In the past it was common to clean net bags on the land base of the individual farms, which were equipped with simple net-cleaning machines, but today this is done mostly in connection with other operations, such as net changes to increase mesh size or recoating of nets, and is performed by special net handlers that work for a number of cage farms. They also have special equipment to carry out such operations more effectively.

Large washing machines specifically designed for this purpose are used when cleaning nets on land (Figure 8.19). Today this is usually performed at special stations that focus entirely on net

Figure 8.19 Large washing machines can be used for removal of fouling from the net bags.

handling operations, such as washing, maintenance, repair and coating. The cleaning stations also have equipment to treat effluent water from the washing procedure. There are various other methods to remove fouling when the nets are on land, but these are infrequently used as they are very labour intensive. These methods include covering the fouled nets with black plastic sheeting and leaving the organisms to decay prior to cleaning, or drying the net in air for some days. High-pressure washers may also be used, but this involves much work and water.

Removal of fouling when the nets are in the sea is also a commonly used method. This can be performed by divers or by specially designed net cleaners. In the past, net washers operated from the surface of the cage, using the walkway, but such washers only clean the upper part of the net bag and are labour intensive to use; they are not suited to today's large cages, where the use of specially designed net cleaners operated by remotely operated vehicles (ROVs) seems to be increasing. The cleaners use several rotating discs spraying high-pressure water on the nets to remove the fouling (Figure 8.20). A newly introduced method is to continuously remove the fouling, because then the forces required for removal will be significantly reduced. This can be performed by utilizing a smaller washing robot that is continuously moving around washing inside the net bag, thereby preventing the fouling organisms from growing significantly before they are removed. In principle it functions like a robot vacuum cleaner. A challenge with on-site or in situ cleaning of the net is that the removed fouling organisms remain in the water and reduce the water quality.

Cleaner fish can be used to a limited degree to keep the net free of fouling. The cleaner fish will feed on the fouling and keep the growth low, and are stocked inside the net bag together with the salmon. The same cleaner fish may also be used to eat sea lice (biological delousing, described 8.6.3.4), but are not commonly used for keeping the net free of fouling.

Biofouling presents problems for many industries, such as shipping, and research is ongoing and several novel methods are being tested, for example, electrical pulses, electrochemical mitigation (for instance generating oxidants), vibration, magnets, deterrents to inhibit fouling organisms from entering the cage and mechanisms such as an air bubble curtain to block incoming organisms.

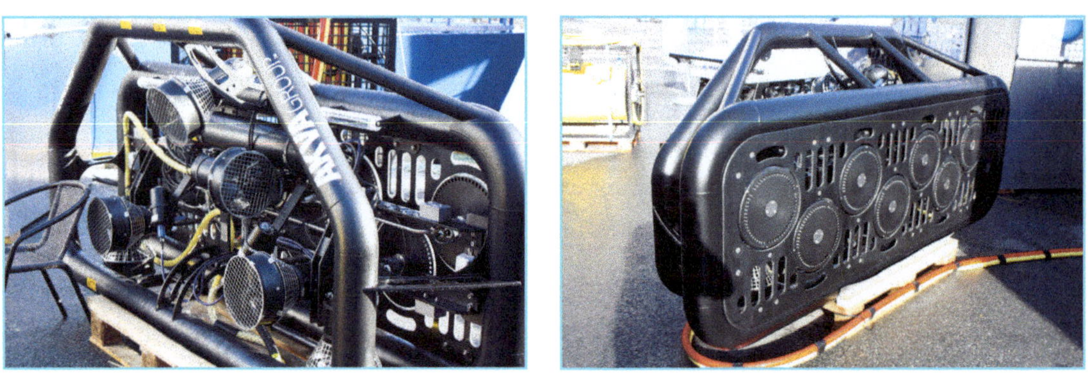

Figure 8.20 ROV net cleaners can be used to carry out in situ cleaning.

8.6 Control of diseases and parasites (lice)

Fish in sea cages are exposed to diseases and parasites and since the net bags are 'open' microbes can enter the cages with the water. There is no possibility of controlling this without changing to closed cages where cleaning and disinfection of the incoming water can be carried out. A skirt hanging around the net bag will protect to a degree because this prevents surface water, where the concentration of microbes is higher, from flowing into the cage. A disease outbreak in a cage on a sea cage farm will easily spread to neighbouring cages, and therefore it is of major importance to prevent disease from entering the farm.

Detection and treatment of diseases is a job for veterinarians, and so each individual farm must employ or have an agreement with a veterinarian. Therefore, this will not be described in depth in this textbook, but a summary is provided below and in an overview in Chapter 11. However, the work done on the farm to prevent outbreak of disease is of major importance for the farmer, and this includes biosecurity plans.

8.6.1 General disease control

The major factors in controlling an outbreak of disease is to have as optimal an environment as possible and not to stress the fish unnecessarily. Both factors are important to ensure an optimal immune defence in the salmon and require good husbandry and management of the farm. Observation of salmon behaviour and monitoring of the salmon's appetite is important. If odd behaviour or low appetite are observed, it may be an early warning of a disease outbreak.

It is also of major importance to have regular veterinary control, which will allow early warning or detection of any disease. However, some type of veterinary control may also be obligatory according to national regulations. In general, it is also an advantage to use vaccinated fish, who will have immunity against some of the potential diseases. Control of sea lice, which during recent years have become a large problem, is also very important. A high number of sea lice will represent a large animal welfare problem.

In government regulations, precautions to avoid an outbreak of disease are typically listed. The regulations are specific to each country, but much of the basics will be similar. As an example, below are some operational regulations for aquaculture facilities in Norway (direct quotation [43]).

- Risk-based health checks shall be made of aquaculture animals to prevent and treat disease and injury.
- In the event of increased mortality or suspicion of contagious or non-contagious disease in one or more production units, a veterinarian or fish health biologist shall be notified immediately, and a health check shall be carried out to determine the cause.
- In the event of persistent increased mortality, a new health check shall be carried out within 14 days, unless the cause is unambiguous and determined.
- When aquaculture animals are taken into an aquaculture establishment, at least one health check shall be carried out before aquaculture animals are taken out of the aquaculture establishment.

- The health check shall be performed by veterinarians or fish health biologists. With the permission of the Norwegian Food Safety Authority, other personnel with equivalent competence may be used.
- The operating journal shall be reviewed during each health check. Based on a risk evaluation, a representative sample of the production units shall be inspected. A representative sample of newly dead animals or animals exhibiting abnormal behaviour shall be examined and relevant tests shall be performed.
- In addition, all production units shall be inspected in the event of increased mortality or suspicion of contagious or non-contagious disease. Specimens shall be taken, and tests performed to establish the cause.
- Ongrowing fish in aquaculture establishments with more than 3000 and fewer than 50,000 ongrowing fish shall have at least four routine health checks per year. These shall be scheduled so that the maximum time between each health check is 4 months.
- Ongrowing fish in aquaculture establishments with more than 50,000 ongrowing fish shall have at least six routine health checks per year. These shall be scheduled so that the maximum time between each health check is 3 months.

The health status of the salmon industry varies from nation to nation, but generally the health conditions of ongrowing salmon have improved. A major contributor to this is the development of vaccines to provide immunity against a range of diseases, and new vaccines are continually being developed. In the past, disease was treated after an outbreak, but today, vaccines provide preventative treatment. Selective breeding has also been an important contributor resulting in today's salmon being more robust against disease. However, the industry is struggling with some diseases. In general, virus-related diseases (such as infectious salmon anaemia [ISA], infectious pancreatic necrosis [IPN] and pancreas disease [PD]), some wounds (such as bacterial-related winter ulcer) and salmon rickettsial septicaemia (SRS) have created large problems (overview of actual diseases in Chapter 11). Broken heart syndrome, heart crack, is a disease becoming more significant recently. It is therefore important to have a constant focus on prevention of disease outbreak in the daily routines.

8.6.2 Use of medicine/pharmaceuticals

Compared to other animal production industries, the ongrowing salmon industry uses less pharmaceuticals per kilogram of meat produced. There has been a general decrease in use, especially in the amount of antibiotics, even if the reduction varies slightly between the salmon producing countries. Taking Norway as an example, the use of antibiotics was approximately 50 tonnes in 1980 when the production was 80,000 tonnes of salmon, while today the production is over 1.2 million tonnes and the total use of antibiotic is less than 1 tonne.

Typically, there will also be national regulations regarding the use of pharmaceuticals and chemicals and below are listed some from Norway [43].

- When using pharmaceuticals and chemicals, great care shall be taken to avoid their release into the surrounding environment.
- If aquaculture animals are being given pharmaceuticals that entail a duty to retain the animals (retention period), due notification shall be given on a sign placed beside the locality/

site sign. The sign must be clearly visible from the sea and other natural approaching ways. The duty of notification applies from the commencement of treatment until the retention period has expired for the medication that was used.

Even though the overall use of pharmaceuticals and chemicals in the salmon industry has generally reduced during recent years, there has been an increase in use of anaesthetics due to more weight sampling and especially sea lice counting.

To prevent disease outbreaks it is, as discussed, of major importance to observe salmon behaviour for as early a warning as possible. This includes swimming behaviour and appetite, together with any increased mortality. Regular removal of dead fish as described earlier is also of major importance to prevent and control disease outbreak. Dead fish are an excellent substrate for microbial development, which may include pathogenic species, resulting in easy transfer of bacteria and viruses to the other fish. For this reason, it is especially important to remove dead fish frequently when fish mortality is caused by disease. If fish have died due to disease, proper handling of the fish is important, including grinding into smaller pieces, using acid and delivering to an approved receiver. It is important to be aware that delivery of dead diseased fish is much more restricted than delivery of ordinary dead fish.

8.6.3 Sea lice

Among all diseases and parasites, sea lice present the largest problem overall in the world salmon farming industry today. It is an area where the daily work of an ongrowing sea cage farm is heavily affected, and therefore it is described in this chapter in detail, but it is also included in the Chapter 11 on disease, where there is a more basic description of sea lice biology. To illustrate the importance of the effects of sea lice on the salmon industry, the overall sea lice cost for the Norwegian salmon industry was estimated to be over NOK 4/kg ($0.40/kg) salmon produced in 2018 [75], but the salmon industry in Scotland, Faroe Islands, Canada and Chile also face large costs due to similar challenges with sea lice. It is not only the treatment that results in cost, but also the increased mortality, the work with counting sea lice, the reduced FCR, the starvation before lice treatment and the cost of the increased smolt size to reduce the time in the sea and therefore the exposure to sea lice.

There are three different sea lice species that create problems for farmers (described more in Chapter 11). The salmon louse *Lepeophtheirus salmonis* has been causing most trouble in the northern hemisphere, but problems are also occurring with the smaller *Caligus* genus. In Chile, *Caligus rogercresseyi* is the main problem.

8.6.3.1 Occurrence of sea lice

Sea lice naturally occur in seawater, but occurrence is site dependent; on some sites there are huge problems while others have few problems. Sites far from shore or more weather-exposed sites seem to have fewer initial problems but when sea lice become establish the number may rapidly increase, although little documentation is yet available. Closed cage systems with water from deeper layers seem to partly solve the lice problem, but here lice too have also been observed [76]. When using closed production units to avoid sea lice problems it is important that inlet water is taken from a depth that is lice free. How deep to go to get lice-free water varies from site to site depending on vertical currents, since lice drift freely with water currents, but is typically below 25 m. Results from one experiment using closed cages showed that even if lice came into the cage no reproduction of lice was observed, probably because of the fast water exchange rate [76].

Inside fjord systems or on protected sites there may be serious problems with sea lice. Fjords with reduced salinity have fewer or no problems due to the fact that lice do not tolerate low salinity (this may therefore also be used in an attempt to get rid of lice on fish by freshwater treatment) [77,78]. Cold water also reduces the problems and below 4°C the sea lice will not successfully reproduce. In one experiment where the water temperature was 3°C, lice did not reach the infectious stage ([78–80]).

Regarding production planning, it will be an advantage to transfer smolt to the sea in the autumn when the temperature is dropping. The smolt will grow over the winter and will be larger and more robust in the following summer when seawater temperatures rise again and the sea lice problem increases. It may therefore be possible to operate until the autumn before taking real action regarding sea lice treatment, and if a good growth rate is achieved and large smolt are transferred it may be possible to avoid treatment altogether, but this will of course depend on the infection pressure of the site.

8.6.3.2 Treatment against salmon lice included in regulations

The salmon industry invests significant resources in preventing and combatting sea lice. Local regulations have been introduced in several salmon producing countries with the aim of combatting salmon lice. The major concerns are the possible transfer of sea lice to wild salmon and development of resistance against the method used for sea lice removal.

8.6.3.3 Counting and reporting

The regulations regarding sea lice typically include a standardized programme for counting, registration and reporting of sea lice to the government, and varies somewhat between the salmon producing countries. Included in the Norwegian regulations are the following [81].

- The seawater temperature must be measured at 3 m depth at least every week.
- Number of lice must be counted in a given way every week at seawater temperatures equal to or more than 4°C and every second week at seawater temperatures below 4°C.
- Number of adult female lice, number of lice in moving stages, number of treatments, seawater temperature, use of wrasse or lump fish shall be registered and reported to the government every week
- The report to the government shall contain information about sea temperature, treatments done, chemicals and amount of chemicals used for treatment, result of any sensitivity investigations, suspicion of resistance development, number of adult females registered in the counting, number and species of wrasse stocked in the farm since last report.

In Scotland the regularity of sea lice counting is monthly, so not so often, but it must also be reported to the government and results can be seen in fish health management reporting areas.

When sampling for lice counting it is important to take a representative sample since only a very limited number of fish in relation to the overall biomass should be taken out. One method used is a cast net or a dip net dragged through the complete water column to remove fish. It is important to report the number of fish that are counted in total, as described in the regulations. If there are three or fewer cages in the farm then all cages should be checked every time, and if there are more than three cages then fish from 50% of the cages should be counted, so that after two countings all cages in the farm will have been checked.

The following method can be used for counting (example from Norwegian regulations). However, first the fish need to be anaesthetized under direction from a veterinarian. When counting, the following should be registered.

- All adult female lice with or without ova. They are often located on the gill cover, behind fat and anal fin and on the tail fin.
- Moving lice, male and preadult stages. These are not fixed and can fall off when touched, so care should be taken when handling.
- The larval stages (chalimus stages), which look like short pencil lines and are smaller than preadult and adult stages. Often these stages are on the side of the belly, below fins or at the fin base.

It is important to differentiate the species of sea lice. When emptying the anaesthetic tank any lice that have fallen off during the operation should also be counted. A standard form should be used for reporting results to the government.

The following values are given for regulatory removal of salmon lice from fish, taking Norway as an example. If showing an average of 0.5 or more adult female lice with or without ova per fish counted, treatment must be performed on the entire farm. Treatment of single cages can be neglected if there is less than 14 days to slaughter.

Equipment for automatic counting of sea lice when the fish are in the cage is also being developed, which will not require that the fish are taken on land for counting of the lice.

8.6.3.4 Removing salmon lice

A number of methods are available today for removal of sea lice [82–85]. New methods are continually developed because none of the currently available methods are optimal regarding effectiveness, treatment cost or animal welfare. The methods may be grouped as follows:

- physical (skirt, freshwater, hot water, laser, washer and so on)
- biological (wrasse, cleaner fish)
- medical (feed or bath)
- chemical (for example H_2O_2).

It is important to note that sea lice have developed resistance to some of the medications and chemicals used [86,87], which has occurred due to the regular use of one substance. It is therefore an advantage or even a necessity to vary the medications used. Today a toolbox for sea lice treatment is available, even if none of the tools are optimal. All involve cost and require the fish to be handled several times, so can raise questions about fish welfare. As a result of this the interest in development of new methods for sea lice treatment is high. As mentioned, it is an advantage to use a range of methods from the toolbox rather than a single method [88].

8.6.3.4.1 Physical methods

Several physical methods are available for handling sea lice problems. They can be divided into methods that keep the lice away from the fish or methods that actually remove the lice.

A dense skirt or partly dense skirt/plankton net can be used around the net in the cage, creating a barrier between the salmon and the sea lice in the surrounding water (Figure 8.21). This may prevent the flow of sea lice into the cage [89,90]. The skirt is typically from 5 m to 10 m deep and since the occurrence of lice is highest in the first few metres it is partly effective, though lice can be drawn deeper due to vertical currents and therefore still enter the cage. Both tarpaulin, which is totally dense, and a fine mesh net is used in the skirt. The mesh size of the net is smaller than the sea lice stages. A problem with the skirt system is that the oxygen level in the cage beyond the skirt is reduced, so an additional oxygen supply might be needed [91]. Another problem is that the fish do not remain near the skirt but stay deeper in the cage, and so do not utilize the entire cage volume. Additional forces will also be added to the construction and mooring system because the skirts result in resistance to the water currents, so the skirt and its depth must be taken into account when considering the dimensions of the mooring system of the sea cages. Even if the preference is to use a deeper net, the increased forces that are added on the construction and the mooring of the cage may prevent this.

A specially designed cage called a snorkel cage which uses some of the same basic principles has been developed and is in use [92,93]. This cage has a limited surface area and a bottle-like shape where the neck is made of a dense material, preventing entry of sea lice, but then expands and the material changes to regular net. The salmon have access to the surface so they can snap air. If sea lice levels are low, experiments have shown that their effect can be limited [94].

Completely closed sea cages may, of course, also be used and will effectively prevent sea lice from entering if water is taken from some depth. Trials have also been done on pumping water from more shallow depths but with a water inlet equipped with a fine mesh filter to prevent lice entering the cage.

A laser system has been developed for removal of lice and is used in commercial sea cage farming (Figure 8.22). The laser unit is lowered into the sea cage and an integrated camera in the unit monitors the fish. If a louse is discovered on the salmon, a laser beam is sent automatically from the laser unit to destroy the louse on the salmon.

Purely mechanical equipment has also been developed and is in use (Figure 8.23). In one system the salmon are pumped through a unit equipped with brushes and high-pressure water

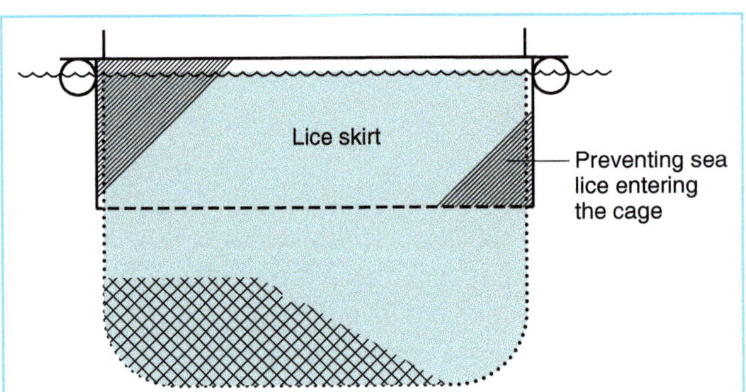

Figure 8.21 A dense skirt can prevent sea lice entering the sea cage to some degree.

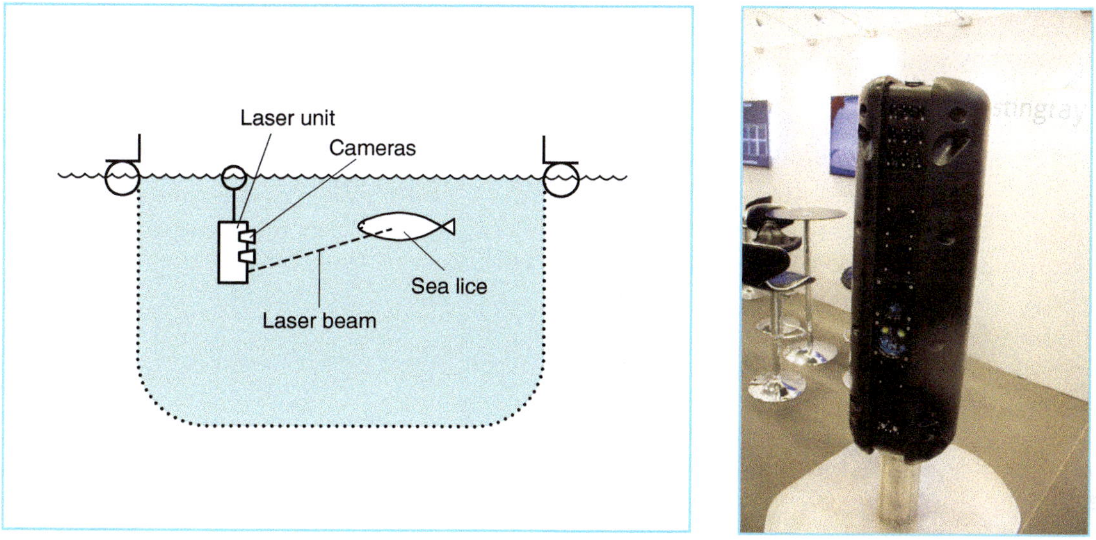

Figure 8.22 A laser system for removal of sea lice.

Figure 8.23 Sea lice can be removed mechanically by brushes and water spraying.

nozzles which flush and brush off the sea lice. In another, equipment with a similar principle uses an ejector to flush the sea lice off the salmon. When using mechanical delousing it is important to use it correctly to ensure animal welfare.

Thermal delousing with either hot or cold water may also be used to remove sea lice. Hot water has been used commercially for a number of years and in one system the salmon are pumped through a chamber with hot water which causes the sea lice to drop off the salmon. The temperature of the water is in the range of 28°C to 34°C and the exposure time is 30 seconds [95]. Using hot water on salmon, however, raises questions about animal welfare. Increased mortality after treatment, pain and tissue injuries have been observed during such treatment [96,97].

An optimized method with a lower temperature is therefore used today. Cold water may also be used, as exposing sea lice to a cold shock in water at −1°C for 10 minutes or 1°C for 240 min reduces the mobility of the lice [79]. However, again an animal welfare issue is raised as the tested temperatures have a significant effect on the salmon [79].

Keeping the salmon in freshwater for a period of time will also result in sea lice dropping off, and the earlier lice stages are also sensitive to freshwater [98]. This method is also used commercially today, either by using a freshwater bath within the cage, created by dragging a tarpaulin around the net bag, or by using a well boat where the salmon are pumped into freshwater-filled wells and remain there for a period of time before being pumped back to the cage again [98].

8.6.3.4.2 Biological

Biological lice removal includes the use of a fish species, such as wrasse or lumpfish, that feed on the lice present on the salmon; such fish are also named cleaner fish, and these may also eat some fouling on nets [99–101] (Figure 8.24). The wrasse can be caught through regular fishing of wild fish stocks, but this raises questions about not only the knowledge of the size of the natural stocks to ensure sustainable fishing but also the possibility of disease transfer from the wild fish. Today there is an increasing interest in aquaculture production of both wrasse and lumpfish. They are produced in separate hatcheries, often using RAS technology. Breeding programmes for cleaner fish to select the best strains for eating lice and tolerating low temperatures have also been introduced. Use of cleaner fish is normally the first choice for keeping sea lice at low levels, and the other methods are then used when sea lice concentrations rise above acceptable levels. When using cleaner fish, it is important to have the correct ratio of salmon to cleaner fish. The cleaner fish need sufficient food in the form of sea lice, or else they may injure the salmon. Stocking density varies with cleaner fish species, but stocking ratios between 5–10% are used, and are highest for lumpfish [102].

8.6.3.4.3 Oral treatment

Feed to which a pharmaceutical/chemical has been added that prevents sea lice infection is also a possibility. Different substances may be used which are typically included in the chemical groups

Figure 8.24 Cleaner fish feeding on sea lice can be used to keep sea lice levels down. Cleaner fish can be produced in large RAS plants.

avermectins (emamectin benzoate) and benzoylureas (diflubenzuron and teflubenzuron). The first group affects the nervous system of the lice while the second affects exoskeleton synthesis. Most commonly used is emamectin benzoate, for instance, the brand Slice. Use of feed additives such as emamectin benzoate also have some preventative effects which generally last for 2–3 months following treatment. After using these feed additives, the fish must be held back from the consumer market for a certain length of time, normally 2–4 months, but this depends on the active ingredient used. The use of lice feed has been reduced over recent years due to development of resistance [86,103,104]. Special additives may also be included in the feed to make the fish more robust or to increase the mucus layer, thereby increasing the tolerance of the fish to sea lice [105].

8.6.3.4.4 Bathing in chemicals or pharmaceuticals

The salmon may also be bathed in a solution of chemicals or pharmaceutical agents effective against the salmon lice [106,107]. A range of chemicals can be used for this, but there is a time window between when the lice are killed and when the salmon start suffering and, in the worst case, die. This time window is dependent on the concentration of the chemical, and typically a higher concentration will result in a reduced time window. Good control of both concentration and time is therefore an absolute requirement when using chemicals for treatment of sea lice. During the treatment the sea lice drop off the salmon but they may still survive, so reinfection can result; this also highlights the importance of using the correct dose and time. Recommendations from the supplier of the chemicals regarding dose and time must be followed. A retention time before the salmon can be sent to the consumer market will also be an issue. The time varies from chemical to chemical and is listed by the supplier.

When performing a bath treatment, the fish in the cage are crowded together and a tarpaulin is placed under the net bag creating a closed volume. An alternative is to use a tarpaulin skirt around the bag, but this is not so effective as it only creates a partly closed volume and more chemicals are required. Instead of treating the salmon in the net bag, a well boat may be hired and the fish pumped on board to perform the delousing inside the well boat. This is a method that is used more and more because, among other reasons, it is easier to control the number of fish and the volume of the bath. In order to calculate the correct dose, control of the volume is required.

A chemical toxic to the lice is then added to the closed volume (the bath), and the fish are kept in this bath for the time period required depending on the chemical used. It is important to remember to add pure oxygen during this period to ensure that the fish are not suffering because of low oxygen levels, as the fish are already stressed and have a higher oxygen consumption than under normal conditions.

The amount of chemicals used has been reduced over recent years, mainly due to more effective and better-controlled treatment but also because of parasite immunity to the chemicals. Increasingly other methods are used instead, or chemicals are used in combination with other methods.

When crowding the fish together in the cage with a tarpaulin, care must be taken. More crowding will result in a lower volume and then less pharmaceuticals/chemicals can be used. This may be economically preferable, but problems can rapidly occur. If there are problems with water quality (oxygen, carbon dioxide or ammonia) the result may be a requirement for premature release from the bath/tarpaulin. This is not good because the lice with a low tolerance of the treatment agent may have been killed but the more resistant ones will be left. This essentially creates a breeding programme for the most drug-tolerant sea lice and then future lice problems will escalate. Higher doses must then be used, further driving future sea lice immunity.

There are a number of chemicals (pesticides) that can be used for sea lice treatment. Although the active substance may be the same, the sales name will vary between different suppliers. Some chemicals used in the past have been prohibited either due to environmental concerns or because resistance has developed; an example is the organophosphate dichlorvos [108]. Today active substances used for bath treatment include substances from the chemical group organophosphates, such as azamethiphos, and from the pyrethroid group, including cypermethrin and deltamethrin. All act on the nervous system of the sea lice [109].

It is, as mentioned, recommended to use different methods and chemicals in the production cycle to avoid the lice developing resistance. Some resistance has already developed for the chemical groups organophosphates and pyrethroids [110,111].

The chemical hydrogen peroxide (H_2O_2) is an effective agent for handling of parasites in general and is also suitable as a delousing agent [112–114] (Figure 8.25). Simply explained, the parasites are oxidized when exposed to hydrogen peroxide. The treatment is carried out either in a bath created by tarpaulins or by using a well boat. It is important to use the correct dose and time for treatment. Again, there is a time window in which lice will be killed without harm to the

Figure 8.25 Use of hydrogen peroxide for sea lice treatment in a well boat.

salmon, but beyond this window the fish will suffer and mortality may occur. There have been several episodes where fish have died during such treatment [115]. During recent years, there has been some discussion as to whether the thickness of the 'shell' of the lice is increasing, resulting in higher concentrations of hydrogen peroxide being required and therefore reducing the window between killing the sea lice and killing the salmon. This shows how adaptive parasites can be to new conditions when it comes to survival. The most robust parasites tolerate the concentrations, and they survive for further breeding so that the next generation also have more tolerance. The shorter the parasite generation interval is, the more rapid this naturally occurring selective breeding will be. It is the same process that can occur with bacteria and viruses, presenting problems with disinfection (also described in Chapter 11), and is the reason for the occurrence of immunity of sea lice to different chemicals as mentioned earlier. This reinforces why it is recommended to use a spectrum of treatment against lice, and in this way delay immunity.

Even though hydrogen peroxide will decompose into water and oxygen, and in this respect seems fairly unharmful, in practice this process is not as rapid as needed. The result is that the concentration will drop very slowly and even when diluted into the surrounding sea, the remaining concentration of hydrogen peroxide can be at a level that can have a negative effect on the environment and fauna in the surrounding area [116–118].

In general, the use of chemicals or pharmaceuticals for treatment of sea lice has been shown to have a negative impact on non-target organisms in the area when the bath is released into the sea [107,119]. Care should be taken when treating and deciding where to release the bath, and it is best if detoxification of remaining chemicals is carried out before releasing the bath. By using well boats, the treatment water can be filtrated and this will also ensure no surviving sea lice are released to the environment. It will also be easier to add detoxification agents before releasing the water.

8.6.3.4.5 Vaccine

For many years vaccines against sea lice have been under development, but it seems it is not easy and no vaccine is yet commercially available on the market [120,121]. In general, vaccines against ectoparasites are difficult and time consuming to develop and therefore their use is not common in other animals or humans. However, a number of experiments are ongoing with vaccines against sea lice on salmon both in laboratories and as field trials, including some with quite promising results [122,123]. A number of large R&D projects are also ongoing including a search for development of oral vaccines [124,125]. The large amount of activity in the field can be seen by the number of patent applications within the area. The hope is that it will be possible to stimulate the fish to develop antibodies which kill the lice or prevent the lice attaching to the salmon, and both tracks are followed in research projects.

8.6.3.4.6 Breeding

Results show that selective breeding can also be a tool against sea lice [126–128]. It has been documented that the attachment of sea lice varies between different salmon families [129,130]. Some families seem to be 'lice collectors', where the number of lice is much higher. The time frame for breeding, however, is long, so breeding will be a method that will result in improvement in the long term.

8.6.3.5 Delousing strategies

It is important to have an overall strategy for lice treatment. This will depend on the water temperature of the site. In the northern hemisphere it is recommended to delouse in December to avoid the presence of mature lice when the smolt are transferred to the sea in the spring. Wild salmon and sea trout are also migrating out of the rivers from April.

It is far better to perform delousing across a region (synchronized delousing) than to perform individual delousing at different times for the different sites in an area. In this way an entire area can be deloused and drifting of young stages of lice in the sea from farm to farm on water currents can be avoided. How this should be done is, of course, highly dependent on the currents in the area. This is recommended in a code of good practice for finfish aquaculture in Scotland [131].

Even if the number of sea lice is low when sea lice are counted, it can be an advantage to implement actions against sea lice early to avoid the levels increasing to a number that is difficult to handle and requires more costly treatments [132]. This includes installations of barriers such as skirts and introducing some cleaner fish in the cages. These are the usual treatment methods to begin with.

Fallowing of sites may help to cut the production cycle of sea lice and coordinated area fallowing may help, but divergent results have been observed [133].

Today sea lice treatment is performed more and more by specialist companies that have the necessary equipment. Often a well boat is included, or the equipment used requires vessels larger than the typical vessels handling the sea cage farm. High capital-cost equipment such as this must be continuously used and it is not economically viable for a farm to purchase its own.

8.6.4 Other harmful organisms

Harmful organisms may bloom in the sea creating problems for the salmon farms, for example algae, harmful algal blooms and jellyfish in high densities [134–137]. Algae may be directly toxic and therefore lethal [138]. However, algae and jellyfish may not be directly toxic for the salmon but, as they use oxygen, the oxygen level can be reduced to a critical level or the number of algae or jellyfish can be so high that they clog or damage [139] the gills of the salmon and thereby inhibit oxygen transfer [140,141]. Therefore, sites where frequent algal blooming is a known problem should be avoided. Large mortality may be the result and therefore early warning is very important. Instant slaughter may be the only possibility, but then very high slaughtering capacity is needed. Alternatively, the salmon can be moved away from the exposed site to clean sites, either by loading into well boats and then being transported or by towing the cages. Lifting of water from lower depths has also been tried, as has large propellers to increase the water flow through the cages. The last is a solution only to low oxygen levels in the cage, not to gill clogging by algae clog the gills or toxicity.

8.7 Use of light and light control

8.7.1 Definition of some terms related to light and measurement of light

Light is electromagnetic waves that travel through fluids and gases: in this case water and air. The visible light spectrum has wave lengths between 380 and 760 nanometres (10^{-9} m); at the lower end of the scale is violet and blue, while the longest wavelength is red. Below the visible light spectrum, with a shorter wavelength, is ultraviolet light while above the visible spectrum, with a longer wavelength, is infrared light. The colour of the visible light may also be given in degrees kelvin (K), as the temperature that a blackbody (a special object in physics) will have to represent the same colour. As an example, a candle has a colour temperature of around 2000 K while an electronic flash has around 5500 K. A low temperature in kelvin, 1000 K, is orange to red while a high temperature, more than 7000 K, is bluer. Normal daylight has a light temperature of 5000–6000 K. Today higher light temperatures are used, up to 20,000 K.

Electric current is supplied to a light source and the amount of energy supplied per second, the effect, can be defined as a watt (J/s). The important aspect of lighting sea cages is, however, not the energy supplied to the light source, but the amount of energy transformed into light in the visible spectrum that leaves the light source and reaches and is perceived by the salmon.

The energy leaving the light source can also be described in watts, also named radiant flux. It is only energy with wavelengths in the visible spectrum that is of interest, and this can be called luminous flux. Depending on the design of the light source the intensity can be equal in all directions, or equal in all solid angles as it is more correctly called. Light radiation, or luminous flux, can be described in terms of radiation intensity in the visible spectrum per solid angle (W/Sr). Solid angles with unit steradians (Sr) are used for describing angles on a sphere such as radians in a circle.

When talking about light, candela is the basic unit of description, and is a similar unit to a watt. More exactly it describes the light from one candle, and is defined as a source that emits 1/683 watt per steradian with monochromatic light with a wavelength of 556 nanometre (green light). The term lumen represents the total light sent out from the source, or the total amount of candelas, and is presented in candelas per steradian; one candle is one candela and two candles is two candela, and so on.

When talking about light in a salmon farm, the important aspect is the illuminance at different points at given distances from the light source, or the irradiance. This is measured in lux and differs from the term lumens, which is the measure for the intensity leaving the light source. One lux (lx) represents one lumen per square metre, which represents the illumination of one lumen distributed evenly over a 1 m^2 area. The lux value can be converted to illumination value in W/m^2, but the conversion depends on the light source. Different light sources have different luminous efficacy (lumens per watt). LED and metal halide lamps (described more later) are typically in the area of 75–100 lumens per watt, while halogen lamps have 16–24 lumens per watt [142]. Other units are also used to measure irradiance based on the number of photons per square metre per second, or the photon flux density in the terms of micro Einstein per square metre per second ($\mu E\ m^{-2}s^{-1}$) or the term micromole per square metre per second ($\mu mol\ m^{-2}s^{-1}$) [143]. Integration of a number of photons over the actual wavelengths can also be a method. Conversion between the units is not straightforward and depends on the light source and wavelength, and specific conversion ratios must be used for the different light sources [144,145].

What can be confusing is that the human eye does not see how much light is hitting a surface but what is reflected from the surface; to the human eye, this is named luminance and is also measured in lux. Even if the same amount of light hits a surface, a white glossy surface will reflect more light than a black matt surface. The further away from the light source, the lower the illuminance (lux/m^2 – luminous flux incident on a surface), and it will drop by a second power related to distance.

When light is sent out from a light source and through a medium it can either be absorbed, reflected or transmitted. In our case, it is the light transmission or light penetration of the water that is important. The light transmission will vary on whether it travels through air or water, and if it goes through a water surface it will also be affected by a scattering effect. Almost all light is transmitted through air and it is only the distance that decreases the lux value. However, in water light will be scattered and absorbed, and the transmission depends on the water quality as particles can result in scattering or absorption. The density of water is 784 times higher than air at the water surface which of course will have an effect.

The light penetration in freshwater will vary a lot due to the high variation in water quality; in crystal clear water only a few per cent of the light present at the surface will penetrate to a depth of 40 m [146]. Seawater is typically more uniform in quality than freshwater, but most of the visible light will be absorbed within the first 10 m [147] and almost no light from the surface reaches below 150–200 m. However, in seawater the absorption may also vary quite a lot based on the quality, especially close to the surface, and therefore classification of water versus spectral transmission may be performed [148,149].

The light absorption of the water may also vary with the wavelength within the visible light spectrum [150,151]. The longer wavelengths, such as red, are absorbed much more than the green and blue light. With the short wavelength of the visible light spectrum, purple, again high absorption occurs and the depth of penetration is reduced. This means that the blue and green parts of the spectrum will penetrate much lower than the red and purple, so the light here will be blue-green.

The light absorption in oceanic water can be described by Beer–Lambert's law [152,153]:

$$E = E_0 e^{-kz}$$

E = actual irradiance
E_0 = surface or initial irradiance
k = diffuse attenuation coefficient for the actual water
z = depth or distance.

As can be seen from the equation the irradiance of an object in the ocean decreases in power with increased distance from the light source. The diffuse attenuation coefficient differs in relation to the actual water quality. A typical value for white light in shelf seas is $0.14\ m^{-1}$, also named PAR (photosynthetically active radiation), which represents the spectral range of the solar radiation where photosynthesis will occur, typically from 400 nm to 700 nm, the visible part of light. In clear ocean water it will be lower and in turbid water, higher [153–155]. The highest penetration is achieved by a wavelength of 490 nm, which is also where k-values are typically measured when finding the diffuse attenuation coefficient [156]. But the light penetration will vary with wavelength. In clearest ocean water, the size of the diffuse attenuation coefficient was found to be at the following values for different wavelengths: 200 nm – 3.14, 300 nm – 0.154, 400 nm – 0.021, 500 nm – 0.027, 600 nm – 0.245, 700 nm – 0.650, 800 nm – 2.07, and the lowest coefficient was found at a wavelength of 450 nm with 0.0168 and the highest at a wavelength of below 220 nm and above 720 nm [157]. This is related to the optical quality of the water and may also be named apparent optical properties (AOP).

However, in a sea cage the salmon will decrease the light penetration so the above equation is not appropriate. In one experiment, the attenuation of light irradiance by the fish approximated to 50% [158]. In this experiment the irradiance was measured to be down to 1% at 4 m compared to the measurement at 0.5 m. However, it is not known if shadowing from light for short periods will affect the fish.

8.7.2 Different light sources

At first, light sources in salmon cages were located above the water surface and tungsten halogen lamps or sodium vapour lamps were used. Tungsten halogen lamps produce a white light while the sodium lamps produce a yellow to orange coloured light. When the light source changed to submerged lamps, as they are today, the light source changed initially to metal halide lamps and today LEDs are increasingly taking over (Figure 8.26). Metal halide lamps produce a white-blue light and, as for tungsten halogen and sodium vapour lamps, the colour of light cannot be changed because they send out a fixed spectrum of wavelengths. With the introduction of LEDs, the colour of the light can be varied with the selection of the diodes. LEDs are also much more energy effective, and the same light intensity is achieved with much less electrical energy (around one-quarter of the energy). Since diodes are chosen for a given range of wavelengths, wavelengths that are not useful for salmon nor highly absorbed in the water such as red and yellow can be avoided and this makes the light source more effective. In addition to being more energy efficient an LED lamp also has a much longer lifetime (over 10 times), even though the investment cost is a bit higher. Improved growth and lower levels of early maturation have also been achieved with the use of LEDs compared to metal halide lamps because the light spectrum can be controlled

Figure 8.26 Different light sources can be used.

and optimized by using blue LEDs [159]. By optimizing the spectrum, the transmission in the sea and therefore the light intensity at a greater depth will increase. This can also ensure that all salmon throughout the entire cage get enough light.

8.7.3 Location of light sources

In the past, as discussed, lights above the water surface were used. The lights were quite strong because the light had to pass through the water surface and down through the water mass, and these lights had a negative effect on the surroundings, for instance, there were complaints from boats. In addition, lamps above the surface result in crowding of the fish close to the surface at night (see below). This may result in high densities, reduced water quality, stress and increased susceptibility to microorganisms and sea lice. On a cage of 15 m × 15 m, for example, four 500 W floodlights were used. The lamps used were tungsten halogen lamps or sodium lamps.

Today, light sources lowered below the surface are used, and they are submerged in the water inside the net bag (Figure 8.27). Submerged lights are placed at a typical depth of between 3 m and 7 m.

8.7.4 Reasons for using light

The use of light on Atlantic salmon affects their physiological response, which can be beneficial. The physiological response depends on photoperiod (periodicity), light intensity (quantity) and spectral composition (quality) [160,161].

During the winter season, it is an advantage to use artificial light in the sea cages as it will delay and reduce the degree of early maturation [162–164] and can therefore also be called anti-maturation light [165]. It has also been shown that use of artificial light results in increased and more uniform growth [160,163,166,167], easier control of growth [28] and improved slaughter quality.

During the day salmon move deeper in the water column, while at night they move higher in the water column, and these changes in swimming pattern can be controlled by using submerged lights [28,168,169]. Submerged lights will also help to keep the fish away from the surface layers where

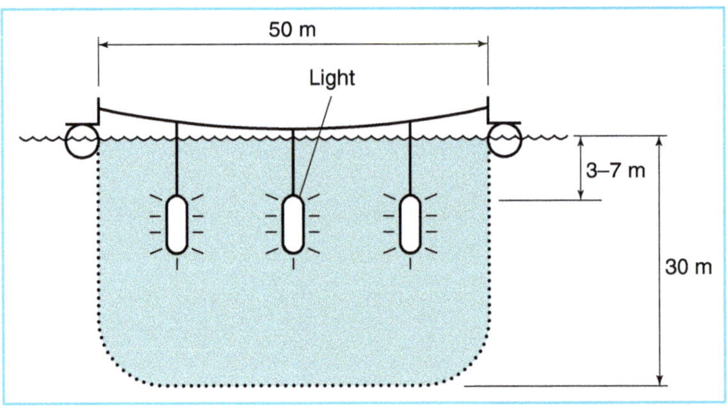

Figure 8.27 Light installation in sea cage below the surface.

the levels of microorganisms and sea lice are higher [88,170,171]. Therefore, deeply placed lights have been proposed as a tool for sea lice control, especially when used in combination with supply of feed at a depth, rather than at the water surface. However, there can be problems using high light intensity in the autumn because it can trigger early maturation. It has been shown that the vertical movements of salmon in the cage can be controlled by quite low intensity lights (0.1 μE) [172].

Light will affect the salmon because seasonal changes control the internal rhythm of the fish [173,174]. As discussed earlier, the process of maturation is influenced by seasonal changes. The growth pattern is also controlled by the season to some degree. In the winter, there is less natural prey in the sea and salmon growth is naturally reduced due to its genetic inheritance from wild salmon.

Since intensity is measured with different units it is difficult to compare, but if the intensity in the cage drops below 5 lux at any place in the cage, maturation may be triggered [175]. Based on this, sheltered and dark areas in the cage should be avoided, but since the salmon swim continuously in a circular pattern in the cage they will always be exposed to some degree of light, and practical experience shows that this is enough to avoid maturation.

The plasma melatonin level changes from day to night; it is high during the night since the salmon produce melatonin when it is dark, and low during the day; this can be the trigger for early maturation and can be avoided by use of artificial light [164,176]. The minimum light intensity to inhibit rhythmic melatonin production is reported to be 0.016 W/m^2 [174]. Other experiments reported light intensities at the cage side and bottom in the area of 8–16 lux [164,177], but this was not with the use of LEDs.

The period for use of addition light can be for 4 months from mid-winter, January in the northern hemisphere [178,179]. High light intensity in the autumn may, as mentioned, induce early maturation [180]. As light can be used to keep salmon away from microbe-rich surface water and may have a positive effect regarding reducing the numbers of attaching sea lice, it can be used as a tool for controlling a sea lice problem [170]. However, this method cannot be used in decreasing daylight because it may induce early maturation. Trials with lower intensity lights show that when the intensity is decreased the number of fish staying at some depth is also reduced, but intensities of 0.1 μE (measured 1 m from the light) still have an effect on keeping the fish below the major lice belt [165]. The minimum threshold for light to influence maturation in sea cages has been measured at 0.06 μE (0.012 W/m^2) [179], but this also depends on wavelength [174,181]. Blue and green light is much more effective at suppressing melatonin compared to red light [181].

The salmon have light-sensitive organs in the eyes and in the pineal gland (on the dorsal surface of the brain). It is the pineal gland that is responsible for the release of the hormone melatonin [182], also called the darkness hormone. To achieve release of melatonin the light reaching the pineal gland must be below a certain level.

Light intensities that are too high may result in stress in fish, but this is temporary and the response is reduced after a period [183], although this also depends on wavelength. Too high an intensity may also cause retinal damage, but again depends on wavelength [183,184]. Practical observations indicate that salmon will not go close to high intensity lights if they are able to move some distance away. Salmon have eyes that are sensitive to light, and can see colours and shapes. Light is absorbed by photopigments in the retina in the eye [185]. The photopigments are in the two types of photoreceptor cells in the eye, the cones and the rods [186]. The cones are responsible for colour vision in bright light (photopic vision) and rods see shapes and contrast with reduced light intensity (scotopic vision). The eye has three types of colour-sensitive cones in the retina, roughly separated into blue, green and red. Salmon also have colour-sensitive cones in the UV area, at least as fry [187,188]. The main light absorption in fry within the four colour-sensitive cones has been reported to be at a peak at 361 nm (UV), 435 nm (blue), 518 nm (green) and 578 nm (red) with a peak absorbance of 0.02 to 0.03 [187]. The light sensitivity of the retina of the fish is assumed to be adapted to the exposed environment [185,187].

In total, therefore, the benefits of using light in sea cage ongrowing of Atlantic salmon is based on an optimal combination of photoperiod, time when light is used, intensity and wavelength. The intensity of the light has to be above a certain level taking spectral composition into consideration. A wavelength that is both effective on the fish and not overly absorbed by the surrounding water should be used, and here blue and green light seems beneficial. Therefore, a decision should be taken whether the full light spectrum or only parts of the spectrum should be used. What is known is that water absorbs light, and the absorbance is dependent on the light spectrum. Long wavelength light, such as red, is absorbed more than short wavelengths, such as blue and violet. This means that if sending a full spectrum of light down into the sea only the blue and violet will remain at some depth, and due to this it is best to use blue or green light.

8.7.5 Suggested lighting setup

Typically, four or six light sources (LEDs) are distributed in a cage with a circumference of 160 m, and the submergible lights are lowered to 3–7 m below the water surface. The usual effect is around 500 W for each light source, and the output intensity is up to 150 lumens per watt for highly effective LEDs. The installed efficiency is in the area of 1–2 W/m^2 cage surface area. The lights may also be directed downwards but this depends on the depth where they are placed.

8.7.6 Planning light installation

Good planning of the light installation on the sea cage farm is important. This includes considering the type of light source, the number of lights, when to use the lights and for how much of the day, the energy source and the means of supplying the energy to the cage, for example, whether electrical cables are to be laid from land or whether generators are to be used.

8.8 Escape

The possibility of escape of farmed salmon from sea cages presents a problem that must be taken seriously, and all possible preventative efforts must be taken. Escaped salmon can present a number of threats to wild salmon stocks including disruption of habitat, interbreeding (introgression), which results in dilution of the original genetic pool of the wild salmon, and transfer of diseases and parasites [189–193]. Escape of salmon from sea cages can occur for several reasons and happens in all salmon producing countries. The wild local strains have developed to adapt to the conditions in the natural habitat for generations, while the farmed strains are bred for high performance under farming conditions. Even though they have the same origin, breeding has resulted in distancing them from their origin. When interbreeding occurs, genes are transferred (gene flow) from domesticated salmon to wild populations and this can result in genetic change within the wild populations [194]. Escape of farmed fish is especially critical when the local wild strains are low in numbers or have declining stocks, as is the case for wild Atlantic salmon [195].

The negative effect of escapes will of course be highest in areas where Atlantic salmon is a native species (areas in the northern hemisphere with a coastline on the Atlantic Ocean), because here interbreeding may occur. In the Norwegian national monitoring programme identifying the genetic effects of escaped salmon on wild salmon stocks, almost 70% of the monitored wild salmon stock was affected; introgression was registered in 2019 and 30% of wild stocks have large genetic changes [196]. Another study showed introgression in half of the adults sampled in rivers [193].

In areas where Atlantic salmon is a non-native species, such as the west coast of Canada (British Columbia) and Chile, escaped Atlantic salmon may theoretically establish a population in the wild, but so far there is no evidence for such claims. Experiences from Chile show that escaped Atlantic salmon do not thrive well in the local wild conditions [197,198]. However, there may be other negative effects of escape, particularly transfer of diseases and parasites to local species, and this can be linked to possible transfer of sea lice to wild fish populations.

8.8.1 Recording and monitoring escapes

Salmon farmers need to report escape of salmon to the local government which allows the overall number of escapes in each individual country to be monitored. However, it is not easy to identify small escapes of a few fish, because control over the exact number of fish in large sea cages is difficult due to the high numbers of individuals. In Norway, for example, the reported number of escapes in 2019 was 284,428, while 112,743 were reported as re-caught. The reported numbers have varied a lot during the last 5 years (from 17,187 to 286,821 from 2014 to 2019) [199]. In Scotland the reported number in 2019 was almost 33,000 [200]. In a project searching for reasons for escape of salmon in the period 2010 to 2018, it was found that on average 0.05% of the fish in Norwegian cages escaped [201].

Monitoring of the number of escaped salmon migrating up salmon rivers will also give important information about the number of escaped salmon, and especially about how many actually find their way up the rivers. This will, of course, be the most critical aspect in relation to escapes due to the consequences of interbreeding or damage to the spawning nests. A national programme to monitor escaped salmon in rivers will give important information for the future. In Norway such a programme is up and running and salmon rivers are monitored regularly for escaped salmon. After a known escape episode, monitoring of fish in local rivers close to the site of escape starts immediately in order to identify how many of the escapees actually migrate up the rivers.

8.8.2 Reasons for escape

There are a number of reasons for escape of fish from sea cage farms and several ways to categorize these reasons, for example: escape occurring during special working procedures; escape occurring during ordinary work; escape occurring due to bad weather conditions; or escape occurring due to special conditions. Included in the last group can be factors such as collisions with an object that destroys the cages (boats, icebergs, driftwood) and interaction with marine mammals that destroy the net bags.

A 5-year Norwegian study of reason for escape shows that escape under special working procedures was most significant [202]. In this category, the highest incidences of escape occurred in relation to transport of fish and delousing of fish. Bad management or routine failure seems to be an important reason for escape. Improvement of management and routines are therefore a major factor to reduce escapes. In a study in Chile the following reasons for escape were listed: severe weather conditions (29%), theft (21%), structural failure of sea cages and deficient handling of incidents (18%) [203]. In another Norwegian study in the period 2010–2016, the major reason for escape was holes in the net bags, both regarding number of fish and number of incidents, and 76% of the escapes were related to this [204]. Other significant reasons were the net bags dropping below the water surface (16%), leakage from tanks on land and escape in relation to internal movement of fish on the farm. The major reason for holes in the net bag was abrasion by the support system such as the bottom ring or weights, either during ordinary operation or especially in connection with operations including the handling of the support system. Problems with the cage collar during operation or breakdown of the cage collar were the major reasons for escape of fish due to lowering of the net below the water surface.

Even when there is a hole in the net bag, the salmon will not necessary swim out and escape such as is the case when farming cod. However, they will find their way out through the hole when they are crowded by hauling on the net bag.

When transporting salmon, fish need to be moved in or out of the cage and this is a situation where several escapes have been registered [204]. Handling of salmon and the net bags in relation to delousing is also a critical situation regarding escape, but this operation typically also includes moving fish in and out of the cages. Special care should therefore be taken in relation to delousing procedures and when performing handling procedures in bad weather.

However, individual incidents caused by extreme weather conditions and technical breakdown of sea cage farms can be important regarding escape of salmon, even if such incidents are occurring more seldomly, because when they occur the number of fish that escape can be very high. These incidents are responsible for the largest escape episodes, for example, a large escape in Chile in 2018 where approximately 800,000 salmon weighing over 3 kg each escaped [197]. There have also been a number of other episodes with escapes of more than 200,000 individuals due to bad weather conditions and farm breakdown. In bad weather, recapture operations will also be difficult to perform.

8.8.3 Escape prevention

There are a number of efforts that can be made to prevent escapes but the most important is to improve the management of the working procedures in relation to handling of fish and equipment.

Technically, improvement and strengthening of the collars, net bags and support system can of course be possible but this will also increase the investment and running costs of salmon production, and therefore the advantages must be weighed against these costs. Use of thicker threads in the net bags or the use of double nets are two possible solutions, but have the negative impacts of reduced water exchange inside the net bags and increased environmental loads on the farm, again requiring a stronger mooring system. Another possibility is to use stronger materials in the nets and mooring lines, such as Kevlar or in-woven metal threads. However, the strength of the net bags and mooring system should be adapted to the environmental conditions on the site and if this is not adequate, higher safety factors should be used both on environmental loads and on material strength and the regulations must be updated.

Sensors that can detect failure in components of the cage construction, such as mooring lines or net bags, could be one possibility for early warning of escapes. Electric wires weaved into the threads of the mesh in the net bag can be used for monitoring and can detect failures or holes in the bag based on electric signals. Adding sensors to monitor environmental forces acting on the cage structure, including net bags and mooring lines, can also be used to signal when high environmental forces are affecting the cages so special attention can be paid in such cases. In addition, this can be used in the long run to help to produce better designs and optimal selection of materials.

Use of closed floating sea cages is also proposed as a method to reduce escape. But, as for open sea cages, this is again dependent on the strength of the construction and most importantly on management of procedures requiring fish handling. The advantage of using closed sea cages, however, is that they can be located in less weather-exposed sites (though of course they require purification of wastewater) and therefore the risk of bad weather-related escapes could be reduced. Certain procedures such as sea lice treatment will not be required and so the number of handling operations where escape may occur will be reduced.

Land-based ongrowing is, of course, a way to reduce the risk of escape, but here efforts must still be taken to reduce risk. There have been registered escapes of smolt from land-based farms, either through effluent pipelines or due to construction breakdown. By using double gratings in the effluent pipeline and a concrete wall around the tank area this can be largely avoided. However, the major drawback with a land-based production facility for ongrowing salmon is the higher production cost, and also the use of energy and therefore the carbon footprint. Countries that have natural environmental advantages for sea cage farming may not have the same advantages if producing on land, where the geographical distance to the consumer market is an important advantage.

Recapture efforts should start immediately after episodes of escape to try to reduce the impacts of the breach. Protocols for this must be in place and the necessary equipment, such as fishing nets, must be ready and fit for use. Having an agreement with the local fishing community to perform recapture fishing will be an important precaution. Paying anglers for fish caught may also be a possibility to increase recapture.

8.8.3.1 Tracking of farmed fish

Easy identification of farmed fish that escape can be a tool to make it easier to identify from which sea cage farm the fish have escaped and possibly to recapture escaped fish. In relation to recapture fishing after escape episodes, it will be easier for local fishers to identify escaped salmon and avoid catching wild salmon. The same is the case for fishing in rivers. It may also be possible to build grading equipment in salmon ladders or to establish hurdles so only wild salmon are able to travel up rivers (Figure 8.28). Tagging the salmon so it is possible to identify which site the salmon have escaped from can elevate tracking escaped salmon to a higher level, and will make it easier to identify the reasons for escape and to give penalties and fines to the responsible farm.

Figure 8.28 Equipment installed in a salmon ladder for identifying escaped farmed salmon and preventing them continuing up the river.

There are several methods of fish identification that can be used (for instance, reviewed by [205], [206]): adipose fin clipping, freeze marking, external tags, internal tags, chemical marks and use of natural marks (gene, growth, scales and so on). There are, however, two major problems with large-scale marking of farmed salmon: the related cost, both including the marking and the reading of the marking, and most importantly how it affects animal welfare.

Requirements on how to reduce the possibility of escape of salmon are typically included in the national regulations. Such regulations will affect the management and working routines at sea cage farms. One such regulation (from Norway) requires that the farm shall have an emergency plan describing the efforts to be made when there is a major escape and how recapture is to be carried out. There must be regular monitoring by fishing with a net 20 m away from the farm in the period 1 October to 31 April. Any escape must be reported at once, and recapture of escaped salmon must start immediately (regulated by law). A number of other regulations have also been introduced to prevent the escape of fish, including the Nytek Regulations that also affect the management of the farm. The Nytek Regulations [57] set a minimum technical standard for floating fish farms and the main components that are included in such establishments. The site has to be classified by an independent bureau regarding wind, waves and current so that the correct equipment and mooring is used. There are also regulations regarding the equipment used, and farmers can only use new sea cages and main components that are certified according to NS 9415 [207]. Existing farms must have a capability certificate showing that the operational requirements of NS 9415 are followed. Certified companies must deliver the mooring systems and must have performed calculations to prove that the mooring systems tolerate the environmental forces on the actual site. A user manual must accompany all new main components, and the manual must also contain a maintenance section, including cage, moorings and net bag, that the farmer must follow.

8.8.3.2 Sterile or partly sterile fish

Another way to reduce the consequences of any escape is to use sterile or partly sterile fish, so they are not able to reproduce if they escape. Triploid fish or fish without functional germ cells are examples (as described in Chapter 5).

8.9 Number, weight and biomass control

As on smolt farms, it is of major importance to have control over the number of fish, the average weight and the standing biomass in each sea cage. Therefore, the fish placed into the cages and the dead fish removed from cages must be counted, and hopefully there will be no escapes or other reasons for missing fish. However, in practice this seems very difficult due to the high number of individuals, with above 100,000 per cage. It is important to keep track of numbers when grading/splitting or removing lice (when there might be increased mortality) and always count the fish when they are returned to the cage.

Several methods are available for counting fish. Manual counting is not an alternative because of the work involved with the high number of fish. Several automatic principles are available including concave v-shaped channels with light diodes, conductivity measurements by infrared light and camera technology. Typically, well boats are equipped with a fish counting device and so counting can be done in conjunction with procedures which involve well boats.

8.9.1 Weight sampling

Taking representative weight samples is challenging in large volume sea cages [208]. It can be done manually or by use of instruments. Weight sampling will also give estimations of the weight distribution which is important when planning any size grading and slaughter, in addition to finding the average weight of the fish in the cage.

Manual weight sampling is tricky in sea cages. When using a dip net, a typical problem is that smaller-sized fish may be at a different depth from larger fish, and therefore it is important to sample throughout the entire water column. It is also too easy to sample fish that are obviously not growing like the average fish, such as pin heads, mature fish and so on. It is important to use the same procedure each time. It will also be important to compare the weight samples with the slaughter data which gives the correct results, so enabling mistakes in weight sampling procedures to be detected or to assist in finding multiplying factors to add to the measured weight to obtain more correct weight data.

Below one procedure for weight sampling is described as an example (after [209]).

- Concentrate (crowd) the fish by:
  - drawing up the entire net bag, reducing the entire volume
  - drawing up a 'pocket' of the net bag
  - using a cast (seine) net.

- Use a dip net to take out the fish, by dragging the dip net from the bottom and upwards.
- Weigh 100 fish or at least 0.5% of the fish in the drawn-up pocket, release the fish and repeat the procedure once more.

8.9.2 Weight sampling intervals

How often the weight samples should be taken varies. Ideally monthly average samples are taken from each cage, but this can be too labour intensive and also disturbs the fish, resulting in stress. However, a number of weight samples ought to be taken throughout the year, and more often in the summer. At times of high growth rate weight samples should be taken at least every second month, and this may be combined with counting of sea lice. The sampling intervals may be decided in relation to the growth rate of the fish, therefore more often in periods with good growth and less often in other periods. This also means that sampling can be done less often when the fish are larger, but close to slaughter time it is important to know the weight and, if possible, gain an idea of the weight distribution. When the water temperature drops below 5°C it is easy to cause wounds on the salmon when taking weight samples and manual weight samples are not recommended.

Manual sampling can be done at an individual level or at a group or bulk level. Individual sampling is best, and gives most information, but requires much more work. In addition, it requires anaesthesia of the fish, and weighing of individual fish using an electronic weight. If the length is also measured, the condition factor can be calculated.

Individual weight sampling can be carried out in combination with sea lice counting, but it is most common to carry out bulk sampling. This can be done by counting the fish in or out of a large bucket filled with some water. Archimedes' law, which states that the increase in water volume is equal to the weight gain (1 l equalling approximately 1 kg), can then be used to find the total weight of all the fish. The method is not so accurate but much faster. If the difference in average fish weight between two throws from the same cage is above 10%, additional throws are recommended.

Weight sampling of the total fish weight by manual systems is not practical due to the quantity. Systems that allow automatic weight measuring of salmon in the cage are available [19,210,211]. In the future, this will be the method for monitoring the average weight and the fish biomass in the cage. Two different principles are used for measuring fish size. It can be carried out either in combination with pumping of fish or by a special device that is placed into the cage. If carried out in combination with pumping, a device that counts fish and measures approximate fish size is installed on the well boat.

The devices placed into the cage work by different principles. One uses a frame with light diodes (a linear camera) that the fish swim through [212]. Another uses a two-camera solution (a stereo camera) where the distance from each fixed camera is different to a given point on the fish, and by knowing the distance between the two cameras a virtual picture of the fish can be created and the size calculated. There is rapid development in this area, and new and improved camera technology is continually being developed so the accuracy of such methods will certainly be improved. Acoustic methods may also be used for biomass estimation, sometimes in combination with camera vision [213].

The total biomass in each cage can be calculated by multiplying the number of fish by the average fish size. It is therefore vital to count the number of dead fish in order to keep control of the total salmon biomass in each cage. The total biomass is an important value for calculating feed supply, controlling fish growth and calculating FCR.

8.10 Harvest/slaughter

8.10.1 Location of harvest/slaughter

The salmon are traditionally slaughtered at special slaughtering plants with customized infrastructure. Slaughter of aquaculture animals at the sea cage farm is generally not permitted. A special licence for establishment of a slaughtering plant is also necessary.

8.10.2 Delivery of fish for harvest/slaughter

Before delivery, it is important to have knowledge of biomass and size distribution to know what size and number of salmon to slaughter and sell to the consumer market. Some farms grade before slaughter. Slaughter can be carried out to order or the salmon can be sold on the market after slaughter. Information about the size and number to slaughter should be sent to the slaughterhouse a week or so in advance, so they can be prepared for the correct quantity of salmon. For control over slaughter, a standardized form should be filled out and sent to the authorities.

Before slaughter, the salmon must be starved for a time period. The recommended length of time is 7 days in summer and 12–14 days in the winter (this may also be regulated) and is dependent on the water temperature. The reason for starving is to empty the digestive system before slaughter to ensure hygiene (to satisfy regulations), but it will also improve fitness and slaughter yield [214–216] and will be beneficial for the water quality during transport in the well boat because metabolic rate and excretion will be reduced [217,218]. Gut evacuation time varies with temperature, which is the reason for an increased starvation time at lower water temperatures (see Chapter 10). The length of the starvation period may, however, be considered for benefits other than gut evacuation. Positive effects on stress tolerance have been demonstrated for longer starvation periods of up to 5 weeks [214,215]. However, this will also reduce fish weight which is why a longer starvation period is not used. Questions about welfare during starvation of salmon have been raised but up to 2 weeks' starvation seems to have no negative impacts [217,219]. It is also important not to stress the fish unnecessarily before slaughtering as stress will reduce the quality of the fish [215,220] (Figure 8.29). The well boat has equipment for loading fish from the sea cage and on to the boat, usually by vacuum. Other systems are possible, however, including bleeding of the salmon at the cage farm and transport of the bled fish on ice to the slaughterhouse. This can be done on specially equipped bleeding boats. Recently a full slaughtering boat has been introduced, which carries out the entire slaughtering process on board. This boat may then take the fish directly from the cages on the fish farm and deliver the fish to the consumer market as freshly slaughtered fish on ice.

Figure 8.29 Well boat collecting fish for slaughter at the sea cage farm to deliver to the slaughterhouse.

References

1. MOWI. *Salmon Farming Industry Handbook*. 2020.
2. Lekang, O.-I. and Fjæra, S.O. *Teknologi for Akvakultur*. 1997. Landbruksforl.
3. FAO. National Aquaculture Legislation Overview (NALO). http://www.fao.org/fishery/collection/nalo/en.
4. Dale, T., Cusack, C. and Ruiz, M. Aquaculture site selection report. 2017. https://ec.europa.eu/research/participants/documents/downloadPublic?documentIds=080166e5b4d2021d&appId=PPGMS.
5. Murray, A.G. and Gubbins, M. Aquaculture zoning, site selection and area management in Scottish marine finfish production, in *Aquaculture Zoning, Site Selection and Area Management under the Ecosystem Approach to Aquaculture: a Handbook,* J. Anguilar-Manjarrez, D. Soto, and R. Brummett (eds). 2017. FAO: Rome, pp. 358–367.
6. Stelzenmüller, V. et al. Aquaculture site-selection and marine spatial planning: the roles of GIS-based tools and models, in *Aquaculture Perspective of Multi-use Sites in the Open Ocean*. 2017. Springer, Dordrecht, pp. 131–148.
7. Longdill, P.C., Healy, T.R. and Black, K.P. An integrated GIS approach for sustainable aquaculture management area site selection. *Ocean and Coastal Management*. 2008. 51(8–9): 612–624.

8. Falconer, L. et al. Use of geographic information systems for aquaculture and recommendations for development of spatial tools. *Reviews in Aquaculture*. 2020. 12(2): 664–677.
9. Nath, S.S. et al. Applications of geographical information systems (GIS) for spatial decision support in aquaculture. *Aquacultural Engineering*. 2000. 23(1–3): 233–278.
10. Benetti, D.D. et al. Site selection criteria for open ocean aquaculture. *Marine Technology Society Journal*. 2010. 44(3): 22–35.
11. Aguilar-Manjarrez, J., Soto, D. and Brummett, R. *Aquaculture Zoning, Site Selection and Area Management under the Ecosystem Approach to Aquaculture. A Handbook*. 2017. FAO: Rome.
12. Halide, H. et al. Developing a decision support system for sustainable cage aquaculture. *Environmental Modelling and Software*. 2009. 24(6): 694–702.
13. Myroms.org. Regional ocean modeling systems (ROMS). http://www.myroms.org/.
14. Albretsen, J. et al. NorKyst-800 Rapport nr. 1: Brukermanual og tekniske beskrivelser. 2011.
15. Egbert, G.D. and Erofeeva, S.Y. Efficient inverse modeling of barotropic ocean tides. *Journal of Atmospheric and Oceanic Technology*. 2002. 19(2): 183–204.
16. Thefishsite. A new tool for site selection? 2019. https://thefishsite.com/articles/a-new-tool-for-site-selection.
17. Ervik, A. et al. Regulating the local environmental impact of intensive marine fish farming I. The concept of the MOM system (Modelling-Ongrowing fish farms-Monitoring). *Aquaculture*. 1997. 158(1–2): 85–94.
18. Australian Institute of Marine Science. A Cage Aquaculture Decision Support Tool (CADS_TOOL). 2009. https://apps.aims.gov.au/metadata/view/bd1a43f0-fafc-11dc-b047–00008a07204e.
19. Lekang, O.-I., *Aquaculture Engineering*. 2020. John Wiley and Sons: Chichester.
20. Lekang, O.-I., Aquaculture hatchery water supply and treatment systems, in *Advances in Aquaculture Hatchery Technology*. 2013. Elsevier: Amsterdam, pp. 3–22.
21. Hvas, M. et al. Assessing swimming capacity and schooling behaviour in farmed Atlantic salmon *Salmo salar* with experimental push-cages. *Aquaculture*. 2017. 473: 423–429.
22. Hvas, M. et al. The effect of thermal acclimation on aerobic scope and critical swimming speed in Atlantic salmon, *Salmo salar*. *Journal of Experimental Biology*. 2017. 220(15): 2757–2764.
23. Noble, C. et al. Welfare Indicators for farmed Atlantic salmon: tools for assessing fish welfare. 2018. http://hdl.handle.net/11250/2575780.
24. Kapetsky, J.M., Aguilar-Manjarrez, J. and Jenness, J. A global assessment of offshore mariculture potential from a spatial perspective. FAO fisheries and aquaculture technical paper, 2013(549): I.
25. Solstorm, F. et al. Not too slow, not too fast: water currents affect group structure, aggression and welfare in post-smolt Atlantic salmon *Salmo salar*. *Aquaculture Environment Interactions*. 2016. 8: 339–347.
26. Priede, M., Biology of salmon, in *Handbook of Salmon Farming*, S. Stead and L. Laird (eds). 2002. Springer-Praxis: Amsterdam.
27. Folkedal, O. et al. Food anticipatory behaviour as an indicator of stress response and recovery in Atlantic salmon post-smolt after exposure to acute temperature fluctuation. *Physiology and Behavior*. 2012. 105(2): 350–356.
28. Oppedal, F., Dempster, T. and Stien, L.H. Environmental drivers of Atlantic salmon behaviour in sea-cages: a review. *Aquaculture*. 2011. 311(1–4): 1–18.
29. Grefsrud, E.S. et al. Risikorapport norsk fiskeoppdrett 2018. 2018.
30. Svåsand, T. et al. Risikovurdering norsk fiskeoppdrett 2016. 2016.
31. ISO-1107. Fishing nets–Netting–Basic terms and definitions. 2017. International Organization for Standardization.

32. Sægrov, H. and Urdal, K. Rømt oppdrettslaks i sjø og elv; mengd og opphav. *Rådgivende Biologer Rapport*. 2006. 947: 21.
33. Hansen, T., Svar på bestilling–undersøkelse av sammenheng mellom maskevidde og smoltstørrelse. *Havforskningsinstituttet, arkivnr*. 2007. 352.
34. Harboe, T. and Skulstad, O.F. *Undersøkelse av Maskeåpning og Smoltstørrelse*. 2013, Havforskningsinstituttet.
35. Herrmann, B. et al. Prediction of selectivity from morphological conditions: methodology and a case study on cod (*Gadus morhua*). *Fisheries Research*. 2009. 97(1–2): 59–71.
36. Sistiaga, M. et al. Prediction of size-dependent risk of salmon smolt (*Salmo salar*) escape through fish farm nets. *Aquacultural Engineering*. 2020. 89: 102061.
37. Johansson, D. et al. The interaction between water currents and salmon swimming behaviour in sea cages. *PLOS One*. 2014. 9(5): e97635.
38. Solstorm, F. et al. Fast water currents reduce production performance of post-smolt Atlantic salmon *Salmo salar*. *Aquaculture Environment Interactions*. 2015. 7(2): 125–134.
39. Remen, M. et al. Critical swimming speed in groups of Atlantic salmon *Salmo salar*. *Aquaculture Environment Interactions*. 2016. 8: 659–664.
40. Marras, S. et al. Fish swimming in schools save energy regardless of their spatial position. *Behavioral Ecology and Sociobiology*. 2015. 69(2): 219–226.
41. Herskin, J. and Steffensen, J.F. Energy savings in sea bass swimming in a school: measurements of tail beat frequency and oxygen consumption at different swimming speeds. *Journal of Fish Biology*. 1998. 53(2): 366–376.
42. Nilsen, A. et al. The importance of exercise: Increased water velocity improves growth of Atlantic salmon in closed cages. *Aquaculture*. 2019. 501: 537–546.
43. Norwegian Regulations. Operation of aquaculture facilities. FOR-2018–04–19–673, l. 2018.
44. Skretting. Feed catalog (in Norwegian). 2012.
45. European Union. Regulation (EC) no. 1774/2002 of the European Parliament and of the Council of 3 October 2002 laying down health rules concerning animal by-products not intended for human consumption. *Off J Eur Communities* L. 2002. pp. 1–95.
46. Tatterson, I.N. and Windsor, M.L. Fish silage. *Journal of the Science of Food and Agriculture*. 1974. 25(4): 369–379.
47. Raa, J., Gildberg, A. and Olley, J.N. Fish silage: a review. *Critical Reviews in Food Science and Nutrition*. 1982. 16(4): 383–419.
48. Arruda, L.F.d., Borghesi, R. and Oetterer, M. Use of fish waste as silage: a review. *Brazilian archives of Biology and Technology*. 2007. 50(5): 879–886.
49. Remen, M. et al. The oxygen threshold for maximal feed intake of Atlantic salmon post-smolts is highly temperature-dependent. *Aquaculture*. 2016. 464: 582–592.
50. Stien, L.H. et al. Salmon Welfare Index Model (SWIM 1.0): a semantic model for overall welfare assessment of caged Atlantic salmon: review of the selected welfare indicators and model presentation. *Reviews in Aquaculture*. 2013. 5(1): 33–57.
51. Turnbull, J. et al. Stocking density and welfare of cage farmed Atlantic salmon: application of a multivariate analysis. *Aquaculture*. 2005. 243(1–4): 121–132.
52. Adams, C.E. et al. Multiple determinants of welfare in farmed fish: stocking density, disturbance, and aggression in Atlantic salmon (*Salmo salar*). *Canadian Journal of Fisheries and Aquatic Sciences*. 2007. 64(2): 336–344.
53. Johansson, D. et al. Effect of environmental factors on swimming depth preferences of Atlantic salmon (*Salmo salar* L.) and temporal and spatial variations in oxygen levels in sea cages at a fjord site. *Aquaculture*. 2006. 254(1–4): 594–605.

54. Akvaplan-Niva. Sleping av oppdrettsmerd. Faktaark 7.
55. Gansel, L.C. et al. Drag forces and deformation of aquaculture cages – full-scale towing tests in the field. *Aquacultural Engineering.* 2018. 81: 46–56.
56. ISO-1107. Fishing nets — Netting — Basic terms and definitions. 2017: International Organization for Standardization.
57. Norwegian Regulation. Regulations on claims for technical standard on floating fishfarming installations. FOR-2011–08–16–849. 2011.
58. Bannister, J. et al. Biofouling in marine aquaculture: a review of recent research and developments. *Biofouling.* 2019. 35(6): 631–648.
59. Fitridge, I. et al. The impact and control of biofouling in marine aquaculture: a review. *Biofouling.* 2012. 28(7): 649–669.
60. Hellio, C. and Yebra, D. *Advances in Marine Antifouling Coatings and Technologies.* 2009. Elsevier Science: Amsterdam.
61. Railkin, A.I.. Marine Biofouling: Colonization Processes and Defenses. 2003. CRC Press: Boca Raton, FL.
62. Flemming, H.C. et al. *Marine and Industrial Biofouling.* 2010. Springer: Berlin.
63. Madin, J., Chong, V.C. and Hartstein, N.D. Effects of water flow velocity and fish culture on net biofouling in fish cages. *Aquaculture Research.* 2010. 41(10): e602–e617.
64. Klebert, P. et al. Hydrodynamic interactions on net panel and aquaculture fish cages: A review. *Ocean Engineering.* 2013. 58: 260–274.
65. Bi, C.-W. et al. Drag on and flow through the hydroid-fouled nets in currents. *Ocean Engineering.* 2018. 161: 195–204.
66. Cornejo, P. et al. Hydrodynamic effect of biofouling in fish cage aquaculture netting. *Aquaculture.* 2020: 735367.
67. Bi, C.-W. et al. Numerical study on wave attenuation inside and around a square array of biofouled net cages. *Aquacultural Engineering.* 2017. 78: 180–189.
68. Gansel, L.C. et al. Drag of clean and fouled net panels–measurements and parameterization of fouling. *PLOS One.* 2015. 10(7): e0131051.
69. Milne, P.H. *Fish Farming: a Guide to the Design and Construction of Net Enclosures.* 1970. Dept. of Agriculture and Fisheries for Scotland, Edinburgh.
70. Beveridge, M.C.M. *Cage Aquaculture.* 2008. Wiley: Chichester.
71. Tan, C.K.F., Nowak, B.F. and Hodson, S.L. Biofouling as a reservoir of *Neoparamoeba pemaquidensis* (Page, 1970), the causative agent of amoebic gill disease in Atlantic salmon. *Aquaculture.* 2002. 210(1–4): 49–58.
72. Vladkova, T. *Surface Modification Approach to Control Biofouling, in Marine and Industrial Biofouling.* 2009. Springer: Dordrecht.
73. Crab Project, CRAB. Collective Research on Aquaculture Biofouling, Final Activity Report. 2006. European Union, Commission.
74. Bhoj, Y., Tharmavaram, M. and Rawtani, D. A comprehensive approach to antifouling strategies in desalination, marine environment, and wastewater treatment. *Chemical Physics Impact.* 2020: 100008.
75. Iversen, A. et al. Kostnadsutvikling og forståelse av drivkrefter i norsk lakseoppdrett. Faglig sluttrapport, in *Nofima rapportserie.* 2019. Nofima AS.
76. Nilsen, A. et al. Effective protection against sea lice during the production of Atlantic salmon in floating enclosures. *Aquaculture.* 2017. 466: 41–50.
77. Bricknell, I.R. et al. Effect of environmental salinity on sea lice *Lepeophtheirus salmonis* settlement success. *Diseases of Aquatic Organisms.* 2006. 71(3): 201–212.

78. Ljungfeldt, L.E.R. et al. A pedigree-based experiment reveals variation in salinity and thermal tolerance in the salmon louse, *Lepeophtheirus salmonis. Evolutionary Applications.* 2017. 10(10): 1007–1019.
79. Overton, K. et al. Thermal delousing with cold water: effects on salmon lice removal and salmon welfare. *Aquaculture.* 2019. 505: 41–46.
80. Samsing, F. et al. Salmon lice (*Lepeophtheirus salmonis*) development times, body size, and reproductive outputs follow universal models of temperature dependence. *Canadian Journal of Fisheries and Aquatic Sciences.* 2016. 73(12): 1841–1851.
81. Norwegian Regualtion. Norwegian Regulation, treatment of sea lice in aquaculture facilities (in Norwegian). FOR-2012–12–05–1140, 2012.
82. Barrett, L.T. et al. Prevention not cure: a review of methods to avoid sea lice infestations in salmon aquaculture. *Reviews in Aquaculture,* 2020. 12(4): 2527–2543.
83. Costello, M. Review of methods to control sea lice (Caligidae: Crustacea) infestations on salmon (*Salmo salar*) farms. *Pathogens of Wild and Farmed Fish: Sea Lice.* 1993. 219: 252.
84. Overton, K. et al. Salmon lice treatments and salmon mortality in Norwegian aquaculture: a review. *Reviews in Aquaculture.* 2019. 11(4): 1398–1417.
85. Holan, A.B. et al. Beste praksis for medikamentfrie metoder for lakseluskontroll (MEDFRI)-Faglig sluttrapport. Nofima rapportserie, 2017.
86. Aaen, S.M. et al. Drug resistance in sea lice: a threat to salmonid aquaculture. *Trends in Parasitology.* 2015. 31(2): 72–81.
87. Denholm, I. et al. Analysis and management of resistance to chemotherapeutants in salmon lice, *Lepeophtheirus salmonis* (Copepoda: Caligidae). *Pest Management Science: formerly Pesticide Science.* 2002. 58(6): 528–536.
88. Bui, S. et al. Assessment of long-term implementation of sea lice prevention technologies: efficiency in reducing infestations and impact on fish welfare. Rapport fra havforskningen, 2018.
89. Stien, L.H. et al. Skirts on salmon production cages reduced salmon lice infestations without affecting fish welfare. *Aquaculture.* 2018. 490: 281–287.
90. Grøntvedt, R.N., Kristoffersen, A.B. and Jansen, P.A. Reduced exposure of farmed salmon to salmon louse (*Lepeophtheirus salmonis* L.) infestation by use of plankton nets: estimating the shielding effect. *Aquaculture.* 2018. 495: 865–872.
91. Stien, L.H. et al. Skirt around a salmon sea cage to reduce infestation of salmon lice resulted in low oxygen levels. *Aquacultural Engineering.* 2012. 51: 21–25.
92. Oppedal, F. et al. Sea lice infestation levels decrease with deeper 'snorkel' barriers in Atlantic salmon sea-cages. *Pest Management Science.* 2017. 73(9): 1935–1943.
93. Stien, L.H. et al. 'Snorkel' sea lice barrier technology reduces sea lice loads on harvest-sized Atlantic salmon with minimal welfare impacts. *Aquaculture.* 2016. 458: 29–37.
94. Bui, S. et al. Salmon lice survive the straight shooter: a commercial scale sea cage trial of laser delousing. *Preventive Veterinary Medicine.* 2020: 105063.
95. Howell, M. Is thermal delousing welfare-friendly? The Fish Site. 2019. https://thefishsite.com/articles/is-thermal-delousing-welfare-friendly.
96. Gismervik, K. et al. Thermal injuries in Atlantic salmon in a pilot laboratory trial. *Veterinary and Animal Science.* 2019. 8: 100081.
97. Nilsson, J. et al. Sudden exposure to warm water causes instant behavioural responses indicative of nociception or pain in Atlantic salmon. *Veterinary and Animal Science.* 2019. 8: 100076.
98. Wright, D.W., Oppedal, F. and Dempster, T. Early-stage sea lice recruits on Atlantic salmon are freshwater sensitive. *Journal of Fish Diseases.* 2016. 39(10): 1179–1186.

99. Powell, A. et al. Use of lumpfish for sea-lice control in salmon farming: challenges and opportunities. *Reviews in Aquaculture*. 2018. 10(3): 683–702.
100. Overton, K. et al. Sea lice removal by cleaner fish in salmon aquaculture: a review of the evidence base. *Aquaculture Environment Interactions*. 2020. 12: 31–44.
101. Imsland, A.K. et al. The use of lumpfish (*Cyclopterus lumpus* L.) to control sea lice (*Lepeophtheirus salmonis Krøyer*) infestations in intensively farmed Atlantic salmon (*Salmo salar* L.). *Aquaculture*. 2014. 424: 18–23.
102. Brooker, A.J. et al. Sustainable production and use of cleaner fish for the biological control of sea lice: recent advances and current challenges. *Veterinary Record*. 2018. 183(12): 383–383.
103. Lam, C.T. et al. Sea lice exposure to non-lethal levels of emamectin benzoate after treatments: a potential risk factor for drug resistance. *Scientific Reports*. 2020. 10(1): 1–8.
104. Espedal, P.G. et al. Emamectin benzoate resistance and fitness in laboratory reared salmon lice (*Lepeophtheirus salmonis*). *Aquaculture*. 2013. 416: 111–118.
105. Leclercq, E. et al. Dietary supplementation with a specific mannan-rich yeast parietal fraction enhances the gut and skin mucosal barriers of Atlantic salmon (*Salmo salar*) and reduces its susceptibility to sea lice (*Lepeophtheirus salmonis*). *Aquaculture*. 2020. 529: 735701.
106. Gautam, R. et al. Evaluating bath treatment effectiveness in the control of sea lice burdens on Atlantic salmon in New Brunswick, Canada. *Journal of Fish Diseases*. 2017. 40(7): 895–905.
107. Haya, K. et al. A review and assessment of environmental risk of chemicals used for the treatment of sea lice infestations of cultured salmon, in *Environmental Effects of Marine Finfish Aquaculture*. 2005. Springer: Dordrecht, pp. 305–340.
108. Jones, M.W., Sommerville, C. and Wootten, R. Reduced sensitivity of the salmon louse, *Lepeophtheirus salmonis*, to the organophosphate dichlorvos. *Journal of Fish Diseases*. 1992. 15(2): 197–202.
109. Aaen, S.M. *Chemotherapeutants Against Salmon Lice Lepeophtheirus salmonis: Screening of Efficacy*. 2016. Norwegian University of Life Sciences.
110. Bakke, M.J. et al. Deltamethrin resistance in the salmon louse, *Lepeophtheirus salmonis* (Krøyer): Maternal inheritance and reduced apoptosis. *Scientific Reports*. 2018. 8(1): 1–14.
111. Kaur, K. et al. Mechanism behind resistance against the organophosphate azamethiphos in salmon lice (*Lepeophtheirus salmonis*). *PLOS One*. 2015. 10(4): e0124220.
112. Thomassen, J.M. and Lekang, O.I. Hydrogenperoksyd mot lakselus (hydrogenperoxide against sealice) (in Norwegian). ITF report, N.u.o.l. Sciences. 1990. Realtek: NMBU.
113. Thomassen, J.M. and Lekang, O.I. Method for controlling aquatic parasites. 1994. Google Patents.
114. Bravo, S. et al. Effectiveness of hydrogen peroxide in the control of *Caligus rogercresseyi* in Chile and implications for sea louse management. *Aquaculture*. 2010. 303(1–4): 22–27.
115. Overton, K. et al. The use and effects of hydrogen peroxide on salmon lice and post-smolt Atlantic salmon. *Aquaculture*. 2018. 486: 246–252.
116. Gebauer, P. et al. Lethal and sub-lethal effects of commonly used anti-sea lice formulations on non-target crab *Metacarcinus edwardsii* larvae. *Chemosphere*. 2017. 185: 1019–1029.
117. Urbina, M.A. et al. Effects of pharmaceuticals used to treat salmon lice on non-target species: Evidence from a systematic review. *Science of the Total Environment*. 2019. 649: 1124–1136.
118. Fang, J. et al. Acute toxic effects of hydrogen peroxide, used for salmon lice treatment, on the survival of polychaetes *Capitella* sp. and *Ophryotrocha* spp. *Aquaculture Environment Interactions*. 2018. 10: 363–368.
119. Frantzen, M. et al. Effects of the sea lice bath treatment pharmaceuticals hydrogen peroxide, azamethiphos and deltamethrin on egg-carrying shrimp (*Pandalus borealis*). *Marine Environmental Research*. 2020. 159: 105007.

120. Raynard, R.S. et al. Development of vaccines against sea lice. *Pest Management Science: formerly Pesticide Science*. 2002. 58(6): 569–575.
121. Carpio, Y. et al. Novel gene isolated from *Caligus rogercresseyi*: a promising target for vaccine development against sea lice. *Vaccine*. 2011. 29(15): 2810–2820.
122. Jensen, P.M. Vaccine cuts lice loads by 97% in lab tests. Fishfarming Expert. 2018. https://www.fishfarmingexpert.com/article/vaccine-cuts-lice-loads-by-97-in-lab-tests/.
123. Contreras, M. et al. Vaccination with ectoparasite proteins involved in midgut function and blood digestion reduces salmon louse infestations. *Vaccines*. 2020. 8(1): 32.
124. Norwegian Veterinary Institute. Louse Off 2 – Cost effective sea lice vaccine; development and efficacy evaluation. 2020. https://wwweng.vetinst.no/research-and-innovation/ongoing-research-projects/louse-off-2-cost-effective-sea-lice-vaccine-development-and-efficacy-evaluation.
125. Thefishsite. Searching for an in-feed sea lice vaccine. 2019. https://thefishsite.com/articles/searching-for-an-in-feed-sea-lice-vaccine.
126. Gharbi, K. et al. The control of sea lice in Atlantic salmon by selective breeding. *Journal of the Royal Society Interface*. 2015. 12(110): 20150574.
127. Tsai, H.-Y. et al. Genomic prediction of host resistance to sea lice in farmed Atlantic salmon populations. *Genetics Selection Evolution*. 2016. 48(1): 1–11.
128. Lhorente, J.P. et al. Quantitative genetic basis for resistance to *Caligus rogercresseyi* sea lice in a breeding population of Atlantic salmon (*Salmo salar*). *Aquaculture*. 2012. 324: 55–59.
129. Glover, K.A. et al. Variation of Atlantic salmon families (*Salmo salar* L.) in susceptibility to the sea lice *Lepeophtheirus salmonis* and *Caligus elongatus*. *Aquaculture*. 2005. 245(1): 19–30.
130. Kolstad, K. et al. Genetic variation in resistance of Atlantic salmon (*Salmo salar*) to the salmon louse *Lepeophtheirus salmonis*. *Aquaculture*. 2005. 247(1–4): 145–151.
131. S.S.P.O. Limited. Seawater lochs, in *Code of Good Practice, Scottish Finfish Aquaculture*. 2015.
132. Jevne, L.S. and Reitan, K.I. How are the salmon lice (*Lepeophtheirus salmonis* Krøyer, 1837) in Atlantic salmon farming affected by different control efforts: a case study of an intensive production area with coordinated production cycles and changing delousing practices in 2013–2018. *Journal of Fish Diseases*. 2019. 42(11): 1573–1586.
133. Guarracino, M., Qviller, L. and Lillehaug, A. Evaluation of aquaculture management zones as a control measure for salmon lice in Norway. *Diseases of Aquatic Organisms*. 2018. 130(1): 1–9.
134. Díaz, P.A. et al. Impacts of harmful algal blooms on the aquaculture industry: Chile as a case study. *Perspectives in Phycology*. 2019. 6(1–2): 39–50
135. Montes, R.M. et al. Quantifying harmful algal bloom thresholds for farmed salmon in southern Chile. *Harmful Algae*. 2018. 77: 55–65.
136. Haigh, N. and Esenkulova, S. Economic losses to the British Columbia salmon aquaculture industry due to harmful algal blooms, 2009–2012. PICES Scientific Report, 2014(47): 2.
137. Purcell, J.E. and Angel, D.L. *Jellyfish Blooms: New Problems and Solutions*. 2015. Springer: Dordrecht.
138. Burridge, L.E. et al. Lethality of microalgae to farmed Atlantic salmon (*Salmo salar*). *Aquaculture*. 2010. 308(3–4): 101–105.
139. Småge, S.B. et al. Concurrent jellyfish blooms and tenacibaculosis outbreaks in Northern Norwegian Atlantic salmon (*Salmo salar*) farms. *PLOS One*. 2017. 12(11): e0187476.
140. Ferguson, H.W. et al. Jellyfish as vectors of bacterial disease for farmed salmon (*Salmo salar*). *Journal of Veterinary Diagnostic Investigation*. 2010. 22(3): 376–382.
141. Baxter, E.J. et al. Gill damage to Atlantic salmon (*Salmo salar*) caused by the common jellyfish (*Aurelia aurita*) under experimental challenge. *PLOS One*. 2011. 6(4): e18529.
142. Rapid Tables. Online calculators and tools. https://www.rapidtables.com/calc/light/how-watt-to-lumen.html.

143. Thimijan, R.W. and Heins, R.D. Photometric, radiometric, and quantum light units of measure: a review of procedures for interconversion. *HortScience*. 1983. 18(6): 818–822.
144. Apogee Instruments. Conversion – PPFD to Lux. https://www.apogeeinstruments.com/conversion-ppfd-to-lux/.
145. Chambers, E.G. Lighting radiation conversion. http://www.egc.com/useful_info_lighting.php.
146. FSC. Light in ponds and lakes. https://www.lifeinfreshwater.org.uk/Web%20pages/ponds/Abiota%20Ponds/Light.htm.
147. Bernal, C.E. Light transmission in the ocean. Waterencyclopedia. http://www.waterencyclopedia.com/.
148. Moser, P.M. Spectral transmission of light through seawater. 1992, Pacific-Sierra Research Corporation Warminster United States.
149. Gupta, S.K. Spectral transmission studies of ocean water under different sea conditions. *Defence Science Journal*. 1984. 34(1): 19–28.
150. Wozniak, B. and Dera, J. *Light Absorption in Sea Water*. 2007. Springer: New York.
151. Shilling, C. *The Underwater Handbook: A Guide to Physiology and Performance for the Engineer*. 2013. Springer: New York.
152. Gordon, H.R. Can the Lambert-Beer law be applied to the diffuse attenuation coefficient of ocean water? *Limnology and Oceanography*. 1989. 34(8): 1389–1409.
153. Bowers, D.G. and Thomas, D. *Introducing Oceanography*. 2012. Dunedin Academic Press.
154. Holmes, J.J. Ocean electromagnetics, in *Springer Handbook of Ocean Engineering*. 2016. Springer: Dordrecht, pp. 177–196.
155. Morel, A. et al. Optical properties of the "clearest" natural waters. *Limnology and Oceanography*. 2007. 52(1): 217–229.
156. Arnone, R.A. and Parsons, A.R. Real-time use of ocean color remote sensing for coastal monitoring, in *Remote Sensing of Coastal Aquatic Environments*. 2007. Springer: Dordrecht, pp. 317–337.
157. Smith, R.C. and Baker, K.S. Optical properties of the clearest natural waters (200–800 nm). *Applied Optics*. 1981. 20(2): 177–184.
158. Hansen, T.J. et al. Effects of light source and intensity on sexual maturation, growth and swimming behaviour of Atlantic salmon in sea cages. *Aquaculture Environment Interactions*. 2017. 9: 193–204.
159. Orrego, R. Effect of LED lighting on growth and development of Atlantic salmon. Fishfarmingexpert. 2015. www.fishfarmingexpert.com/.
160. Boeuf, G. and Le Bail, P.-Y. Does light have an influence on fish growth? *Aquaculture*. 1999. 177(1–4): 129–152.
161. Qiu, D. et al. Effects of spectral composition, photoperiod and light intensity on the gonadal development of Atlantic salmon *Salmo salar* in recirculating aquaculture systems (RAS). *Chinese Journal of Oceanology and Limnology*. 2015. 33(1): 45–56.
162. Taranger, G.L. et al. Control of puberty in farmed fish. *General and Comparative Endocrinology*. 2010. 165(3): 483–515.
163. Endal, H.P. et al. Effects of continuous additional light on growth and sexual maturity in Atlantic salmon, *Salmo salar*, reared in sea cages. *Aquaculture*. 2000. 191(4): 337–349.
164. Porter, M.J.R. et al. The use of cage lighting to reduce plasma melatonin in Atlantic salmon (*Salmo salar*) and its effects on the inhibition of grilsing. *Aquaculture*. 1999. 176(3–4): 237–244.
165. Stien, L.H. et al. Low intensity light of different colours modifies Atlantic salmon depth use. *Aquacultural Engineering*. 2014. 62: 42–48.
166. Saunders, R.L. and Harmon, P.R. Extended daylength increases postsmolt growth of Atlantic salmon. *World Aquaculture*. 1988. 19(4): 72–73.

167. Kråkenes, R. et al. Continuous light increases growth rate of Atlantic salmon (*Salmo salar* L.) postsmolts in sea cages. *Aquaculture*. 1991. 95(3–4): 281–287.
168. Juell, J.-E. and Fosseidengen, J.E. Use of artificial light to control swimming depth and fish density of Atlantic salmon (*Salmo salar*) in production cages. *Aquaculture*. 2004. 233(1–4): 269–282.
169. Juell, J.E. et al. Submerged light increases swimming depth and reduces fish density of Atlantic salmon *Salmo salar* L. in production cages. *Aquaculture Research*. 2003. 34(6): 469–478.
170. Frenzl, B. et al. Manipulation of farmed Atlantic salmon swimming behaviour through the adjustment of lighting and feeding regimes as a tool for salmon lice control. *Aquaculture*. 2014. 424: 183–188.
171. Bui, S. et al. Modifying Atlantic salmon behaviour with light or feed stimuli may improve parasite control techniques. *Aquaculture Environment Interactions*. 2013. 3(2): 125–133.
172. Wright, D.W. et al. Atlantic salmon *Salmo salar* instantaneously follow vertical light movements in sea cages. *Aquaculture Environment Interactions*. 2015. 7(1): 61–65.
173. Malpaux, B. et al. Biology of mammalian photoperiodism and the critical role of the pineal gland and melatonin. *Journal of Biological Rhythms*. 2001. 16(4): 336–347.
174. Migaud, H. et al. A comparative ex vivo and in vivo study of day and night perception in teleosts species using the melatonin rhythm. *Journal of Pineal Research*. 2006. 41(1): 42–52.
175. Berthelsen, K. Hunting the optimal light for aquaculture (in Norwegian). 2017. www.kyst.no.
176. Amano, M. et al. Roles of melatonin in gonadal maturation of underyearling precocious male masu salmon. *General and Comparative Endocrinology*. 2000. 120(2): 190–197.
177. Hansen, T., Stefansson, S. and Taranger, G.L. Growth and sexual maturation in Atlantic salmon, Salmon, salar L., reared in sea cages at two different light regimes. *Aquaculture Research*. 1992. 23(3): 275–280.
178. Leclercq, E. et al. The use of continuous light to suppress pre-harvest sexual maturation in sea-reared Atlantic salmon (*Salmo salar* L.) can be reduced to a four month window. *Aquaculture Research*. 2010. 41(10): e709–e714.
179. Leclercq, E. et al. The potential of alternative lighting-systems to suppress pre-harvest sexual maturation of 1+ Atlantic salmon (*Salmo salar*) post-smolts reared in commercial sea-cages. *Aquacultural Engineering*. 2011. 44(2): 35–47.
180. Oppedal, F. et al. Photoperiod in seawater influence seasonal growth and chemical composition in autumn sea-transferred Atlantic salmon (*Salmo salar* L.) given two vaccines. *Aquaculture*. 2006. 254(1–4): 396–410.
181. Vera, L.M. et al. Differential light intensity and spectral sensitivities of Atlantic salmon, European sea bass and Atlantic cod pineal glands ex vivo. *General and Comparative Endocrinology*. 2010. 165(1): 25–33.
182. Falcon, J. et al. Current knowledge on the melatonin system in teleost fish. *General and Comparative Endocrinology*. 2010. 165(3): 469–482.
183. Migaud, H. et al. The effect of spectral composition and light intensity on melatonin, stress and retinal damage in post-smolt Atlantic salmon, *Salmo salar*. *Aquaculture*. 2007. 270(1–4): 390–404.
184. Vera, L.M. and Migaud, H. Continuous high light intensity can induce retinal degeneration in Atlantic salmon, Atlantic cod and European sea bass. *Aquaculture*. 2009. 296(1–2): 150–158.
185. Bowmaker, J.K. *Retinal Structure of Fishes. The Visual System of Fish*. 1990. Chapman and Hall: London.
186. Nakano, N. et al. Color vision, spectral sensitivity, accommodation, and visual acuity in juvenile masu salmon *Oncorhynchus masou masou*. *Fisheries Science*. 2006. 72(2): 239–249.
187. Cheng, C.L. et al. Photoreceptor layer of salmonid fishes: transformation and loss of single cones in juvenile fish. *Journal of Comparative Neurology*. 2006. 495(2): 213–235.

188. Flamarique, I.N. The ontogeny of ultraviolet sensitivity, cone disappearance and regeneration in the sockeye salmon *Oncorhynchus nerka*. *Journal of Experimental Biology*. 2000. 203(7): 1161–1172.
189. Thorstad, E.B. et al. Incidence and impacts of escaped farmed Atlantic salmon *Salmo salar* in nature. *NINA Special Report*, 2008. 36(6).
190. Keyser, F. et al. Predicting the impacts of escaped farmed Atlantic salmon on wild salmon populations. *Canadian Journal of Fisheries and Aquatic Sciences*. 2018. 75(4): 506–512.
191. Glover, K.A. et al. Half a century of genetic interaction between farmed and wild Atlantic salmon: status of knowledge and unanswered questions. *Fish and Fisheries*. 2017. 18(5): 890–927.
192. Jonsson, B. and Jonsson, N. Cultured Atlantic salmon in nature: a review of their ecology and interaction with wild fish. *ICES Journal of Marine Science*. 2006. 63(7): 1162–1181.
193. Karlsson, S. et al. Widespread genetic introgression of escaped farmed Atlantic salmon in wild salmon populations. *ICES Journal of Marine Science*. 2016. 73(10): 2488–2498.
194. Bolstad, G.H. et al. Gene flow from domesticated escapes alters the life history of wild Atlantic salmon. *Nature Ecology and Evolution*. 2017. 1(5): 1–5.
195. NASCO. *State of the North Atlantic Salmon*. 2019. North Atlantic Salmon Conservation Organization (NASCO): Edinburgh.
196. Dizerud, O.H. et al. Genetisk påvirkning av rømt oppdrettslaks på ville laksebestander–oppdatert status 2019. 2019. Norsk Institutt for Naturforskning (NINA).
197. Quinones, R.A. et al. Environmental issues in Chilean salmon farming: a review. *Reviews in Aquaculture*. 2019. 11(2): 375–402.
198. Arismendi, I. et al. Differential invasion success of salmonids in southern Chile: patterns and hypotheses. *Reviews in Fish Biology and Fisheries*. 2014. 24(3): 919–941.
199. Fiskeridirektoratet. Rømmingsstatistikk – antall og art (in Norwegian). 2020. https://www.fiskeridir.no/Akvakultur/Tall-og-analyse/Roemmingsstatistikk/Roemmingsstatistikk-antall-og-art.
200. Scotland`s Aquaculture. Fish escapes. 2020. http://aquaculture.scotland.gov.uk/data/fish_escapes.aspx.
201. Myrhol, K.F. Flest fisk rømmer på grunn av hull i not (In Norwegian). 2019. https://www.fhf.no/nyheter/nyhetsarkiv/flest-fisk-roemmer-paa-grunn-av-hull-i-not/.
202. Fiskeridirektoratet. Omstendigheter ved rømming 2014–2019 (In Norwegian). 2019. https://www.fiskeridir.no/Akvakultur/Tall-og-analyse/Roemmingsstatistikk/Omstendigheter-ved-roemming-2014–2019.
203. Sepúlveda, M. et al. Escaped farmed salmon and trout in Chile: incidence, impacts, and the need for an ecosystem view. *Aquaculture Environment Interactions*. 2013. 4(3): 273–283.
204. Føre, H.M. and Thorvaldsen, T. Årsaker til rømming av oppdrettslaks og ørret i perioden 2010–2016. 2017. SINTEF Ocean.
205. Uglem, I. et al. Evaluation of large-scale marking methods in farmed salmonids for tracing purposes: impact on fish welfare. *Reviews in Aquaculture*. 2020. 12(2): 600–625.
206. Mortensen, A., Puvanendran, V. and Hansen, Ø.J. Fysisk merking av oppdrettslaks. Test og evaluering av forskjellige merkemetoder. 2013. Nofima.
207. Standards Norway. Marine fish farms-Requirements for site survey, risk analyses, design, dimensioning, production, installation and operation. 2009.
208. Nilsson, J. and Folkedal, O. Sampling of Atlantic salmon *Salmo salar* from tanks and sea cages is size-biased. *Aquaculture*. 2019. 502: 272–279.
209. Kittelsen A. et al. Drift og røkt, akvakultur. 2002. Gan ForlagL Oslo.
210. Li, D., Hao, Y. and Duan, Y. Nonintrusive methods for biomass estimation in aquaculture with emphasis on fish: a review. *Reviews in Aquaculture*. 2019. 12: 1390–1411.

211. Antonucci, F. and Costa, C. Precision aquaculture: a short review on engineering innovations. *Aquaculture International*. 2020. 28(1): 41–57.
212. Difford, G.F. et al. Validating non-invasive growth measurements on individual Atlantic salmon in sea cages using diode frames. *Computers and Electronics in Agriculture*. 2020. 173: 105411.
213. Puig-Pons, V. et al. Automatic bluefin tuna (*Thunnus Thynnus*) biomass estimation during transfers using acoustic and computer vision techniques. *Aquacultural Engineering*. 2019. 85: 22–31.
214. Einen, O., Waagan, B. and Thomassen, M.S. Starvation prior to slaughter in Atlantic salmon (*Salmo salar*): I. Effects on weight loss, body shape, slaughter-and fillet-yield, proximate and fatty acid composition. *Aquaculture*. 1998. 166(1–2): 85–104.
215. Mørkøre, T., Tahirovic, V. and Einen, O. Impact of starvation and handling stress on rigor development and quality of Atlantic salmon (*Salmon salar* L). *Aquaculture*. 2008. 277(3–4): 231–238.
216. Einen, O. and Thomassen, M.S. Starvation prior to slaughter in Atlantic salmon (*Salmo salar*): II. White muscle composition and evaluation of freshness, texture and colour characteristics in raw and cooked fillets. *Aquaculture*. 1998. 169(1–2): 37–53.
217. Robb, D. Welfare of fish at harvest. *J Fish Welfare*. 2008: 217–242.
218. Hjeltnes, B. and Waagbø, R. Transportation of fish within a closed system. Opinion of the Panel on Animal Health and Welfare. Norwegian Scientific Committee for Food Safety. Norwegian, 2008. 3.
219. Waagbø, R. et al. Short-term starvation at low temperature prior to harvest does not impact the health and acute stress response of adult Atlantic salmon. *PeerJ*. 2017. 5: e3273.
220. Skjervold, P.O., Fjæra, S.O. and Østby, P.B. Rigor in Atlantic salmon as affected by crowding stress prior to chilling before slaughter. *Aquaculture*. 1999. 175(1–2): 93–101.

CHAPTER 9

Production planning and management

9.1 Introduction

The first step involved in initiating and projecting plans for a fish farm is to have an adequate production plan. With this information the required infrastructure, rearing units, water treatment and heating/cooling requirements can be estimated according to the type of farm. When the farm is up and running the production plan is of major importance for the economics of the farm. A production plan is not a static document; it must respond to changes, for example, if larger-sized smolt are delivered from the smolt farm, or if the fish grow much faster than expected, the plan will need to react.

Setting a production plan is like setting a budget; in place of financial variables a production plan includes expectations based on biological factors such as mortality and growth. These may vary considerably, which makes estimation of future developments more difficult, especially if the background information is limited. Care must also be taken to set environmental factors at a level that ensures attainment of the requirements for optimal performance of the individual fish, including optimal growth and welfare. Environmental factors that could affect mortality and likelihood of disease outbreaks must be within manageable limits. This includes factors such as water quality, necessary space/volume (kilogram of fish per cubic metre), water current and light. After ensuring the environmental factors, other factors such as optimizing the use of volume, hydraulic self-cleaning of tanks and grading must be taken into consideration. The values in the production plan, when turned into economic values, predict whether or not the planned production will produce a profit.

The plan may be set up manually (Table 9.1) or using a computer program. The use of computer programs is usually the best alternative. This permits the simulation of several alternatives and allows the selection of the best scenario depending on the production characteristics of the individual farm. It is, however, important to understand how a plan is set up, and how the computer program works.

It must also be emphasized that it is a plan, an estimation. Just as for an economic budget, it depends on the assumptions on which it is based. If there are uncertainties in the assumptions, the same will be the case for the budget, and the same uncertainties will then also be transferred to the production plan. It is not like an accounting process where the attained values are

Table 9.1 Example production plan set-up for a smolt farm.

Date	Jan	Feb	Mar	Apr	May	Jun	Jul	Aug	Sep	Oct	Nov	Dec
Raw water temperature (°C)												
Egg water temperature (°C)												
0^+ temperature (°C)												
1^+ temperature (°C)												
Weight 0^+ (g)												
Weight 1^+ (g)												
Number of eggs												
Number of 0^+												
Number of 1^+												
Total number												
Biomass eggs												
Biomass 0^+ (kg)												
Biomass 1^+ (kg)												
Total biomass (kg)												
Water requirement eggs												
Water requirement 0^+ (l)												
Water requirement 1^+ (l)												
Total water required (l)												
Hatch trays												
Tank volume 0^+												
Tank volume 1^+												
Total tank volume (l)												

documented. There have been a number of cases where biological assumptions that were too positive have been used when making the production plans for the establishment of new farms, with poor farm economics and bankruptcy as the result. This is due to the great variations in the biological assumptions that can be used. It is therefore possible to choose positive inputs, for example, assumption of expected fish growth or mortality, which result in a positive prediction, but if more moderate assumptions had been used a negative prediction would have been the result. Sensitive analyses for major biological factors should therefore be used when setting up production plans, especially when using the production plan as a basis for economic planning and future budgets.

Owing to the usual structure of salmon farms, it is typical to have two individual production plans, one for the smolt production farm and one for the ongrowing farm, which is normally a sea cage farm.

To have some idea of what is needed for salmonid production, some very rough estimated values are listed below, but these may vary a lot depending on water temperature and fish size:

- water requirement per kilogram of fish produced: 70 m^3
- oxygen requirements per kilogram of fish produced: 0.3 kg
- oxygen requirement per kilogram of feed supplied: 0.25–0.5 kg
- water requirement per kilogram of fish in the farm: 0.8–1 l/min.

- yearly amount of slaughtered fish (up to double the amount of fish in the farm when the biomass is maximal)
- dry feed requirements per kilogram of fish in the farm: 0.8–1.2 kg
- production of a salmon smolt of 75 grams: 3000–4000 D°C

9.2 Biological expectations

9.2.1 When salmon reach the different life stages

The estimated time involved in the growth of salmonids mainly depends on water temperature (Table 9.2). This is also the reason for using the term day degrees, which is the sum of daily temperatures over all days in the period. Other factors, such as water quality and fish density, may also be significant, but to a lesser degree as long as they are within acceptable levels. Access to enough feed is, of course, a requirement.

The day degrees (D°C), also called accumulated temperature units (ATU), can be calculated by summing the water temperature over the days in the period using the following equation where n represents the number of days and T the water temperature.

$$D°C = \sum_{1}^{n} T$$

A more simplified estimation of day degrees may be calculated by multiplying the average water temperature over a period by the number of days that temperature has occurred. For example, 300 D°C may be 30 days at 10°C, 100 days at 3°C or any other multiple that results in 300. In practice the water temperature must be within the acceptable range for salmon to use this method, and in addition the salmon have distinct life stages. By setting the water temperature, estimates for when the salmon are going to reach the different life stages based on day degrees can be produced (Table 9.2).

However, it is not just an increase in the water temperature that reduces the time the salmon take to move from one life stage to the next. There are both optimal temperatures and lethal temperatures at the different life stages. For incubation of eggs, water temperatures below 8°C should be used to avoid skeletal deformations, and temperatures for start feeding are typically between 12°C and 14°C. For large fish, the range of temperature tolerance is quite large, where the lower

Table 9.2 Necessary day degrees for the salmon to reach the different life stages (approx.), and recommended water temperatures within each life stage.

Life stage	Day degree	Normal temperatures (°C)
Eyed egg	220–230	5–8
Hatching	450–520	8–12
Start feeding	570–620	7–14
End of start feeding	1200–1400	4–14

Note: Start feeding when 80–90% of yolk sac is consumed.

end is 0–7°C and the upper end is 22–28°C (review in [1]). The thermal optimum for growth is not documented but it is thought to be in the area of 16–20°C. With water temperatures above 16°C it becomes difficult to maintain satisfactory water quality, including supply of an optimal amount of oxygen to the individual fish in the production unit.

Box 9.1 provides an example calculation using day degrees.

Box 9.1

A salmon is stripped on 1 November, and the water temperature of the eggs is 8°C. *When will* hatching take place?

The life stage is estimated to be reached after 450 D°C (Table 9.2)

Number of days: 450/8 = 56

The time for hatching is 1 November plus 56 days – which is 25 December

9.2.2 Mortality

Owing to the large number of individuals there will always be some natural mortality. The level of mortality is dependent on life stage. The highest mortality occurs in the early life stages, when the weakest individuals will die. It is important to consider that there can be a large variation in mortality and that this is dependent on various factors, such as feed quality, water quality and breeding stock. When making a production plan the mortality numbers have to be estimated by making a qualified guess. Some estimates are given in Table 9.3. For management, it is more advantageous to have an overestimation of the mortality rather than an underestimation to ensure that there will be enough surviving fish. This is at least correct at the start of the growing period, as numbers are easier to handle at this stage. If there is a higher level of survival than estimated, some of the smaller fish may be removed at size grading to optimize the production. When transferring the smolt to the sea, due to the change in environment the mortality will typically be higher just after transfer than later in the ongrowing period.

Incidences of disease or parasites will increase the mortality rate. Today, for example, the sea lice problem in sea cage farms is causing much higher mortality than what would be expected

Table 9.3 Rough estimation of proposed mortality at different life stages.

Life stage	Mortality (%) Atlantic salmon
Eyed eggs	5–15
Hatching	5
Start feeding	5–15
Ongrowing including seawater stage (% per month)	0.1–1
Total survival	50–80

naturally. This will always create a problem regarding the estimate to include in the production plan, and again it must be emphasized that it is a prognosis.

It is important to be able to calculate the number of fish that are left after a given period with an expected mortality rate. This could, for instance, be used if the smolt farm has a contract to deliver of a given amount of smolt. The following equation can be used for this purpose when estimating the mortality number per month:

$$N_1 = N_0 (1 - m/100)^t$$

where N_1 = the number of fish after a given period; N_0 = the initial number of fish; m = mortality per month; t = number of months.

This estimation is also necessary in order to calculate the number of fish at the different life stages in the production plan (Box 9.2).

Box 9.2

An order of 500,000 smolt is to be satisfied. How many fry must be left after start feeding, given that the period needed to produce the smolt is 12 months and the mortality rate is set to be 0.15% of the stock per month after start feeding?

500,000 = x × (1 – 0.15/100)12
500,000 = x × 0.98
x = N_0 = 509,088

9.2.3 Salmon growth

Growth can be calculated according to fish weight, the length, or both. The first is most common. However, knowing the condition factor for the actual length of the fish allows the weight to be found based on fish length (see Chapter 4). To be able to calculate the expected growth of the fish it is necessary to know the initial weight of the fish before commencing feeding. A newly hatched salmon fry will have a weight of 0.18–0.2 g, while the weight when ready for start feeding will be 0.15–0.18 g.

The growth rate of the salmon is based on heritage (the genetic origin of the fish) and environment. Breeding has been a major factor in the success of the salmon farming industry and the growth rate has been one of the major breeding goals and as a result the growth rate is improving from generation to generation. It is therefore important to use the latest generation of salmon to get the best genetic material possible and thereby a fish with the best growth characteristics possible. The other factor of major importance for growth is the environmental conditions, and here water temperature is the single most important parameter, assuming the water quality is optimal. Important water quality parameters that can inhibit growth under practical farming conditions are too low a dissolved oxygen concentration in general, and too high a carbon dioxide concentration with high fish densities and with the use of RAS. It is, of course, a prerequisite that feed composition is optimal in order to maximize growth, but today this is the case with salmon feed bought from the large feed producers.

Several models are available for estimation of salmon growth [2], including specific growth rate (SGR), growth factor (GF), growth index (GI) and average daily growth rate (ADG) (see also Chapter 10). It is again important to be aware that these are all models that give estimates, not exact values. The basis for the development of the growth models has been estimation of how much feed to deliver to a fish group the following day, and this value can be found by expecting an FCR of 1.

The SGR model estimates growth rate based on water temperature and the size or life stage of the fish; small fish have a higher specific growth rate per kilogram of fish than larger fish (see also Chapter 10). For future production planning it is necessary to estimate the growth of the fish in relation to the expected temperature. For this purpose, growth tables can be used (Table 9.4, Table 9.5) [3]. The tables give estimates of daily specific growth rates for different fish sizes in relation to the water temperature in freshwater or seawater, and show the percentage weight gain that fish of a certain size will have during a given time interval at a particular water temperature. A model based on the tables has also been proposed [4] (limited to seawater and a size range from 50 g to 3000 g).

$$\mathrm{SGR} = 0.9\ \mathrm{T}^{-0.97} \times \mathrm{w}^{-0.34}$$

where T = temperature; w = weight.

The thermal growth coefficient (TGC), normally simply called the growth factor (GF), may also be used for estimation of future growth (Table 9.6), and is the estimated growth according

Table 9.4 Achieved specific growth rate (SGR) for salmon in freshwater (based on [3, 6–8]).

Weight (g)	Temperature (°C)						
	4	6	8	10	12	14	16
>1	2	2.5	3.2	4	4.5	5	5.5
1–5	1.5	2.3	2.7	3.5	4	4.5	4.7
5–10	1.4	2	2.3	2.9	3.5	4	4.3
10–25	1.2	1.8	2	2.6	2.8	3.6	3.9
25–40	1	1.3	1.6	1.9	2.2	2.7	3.2
40–60	0.9	1.1	1.3	1.8	2.1	2.4	2.8
60–100	0.6	0.9	1.2	1.6	1.9	2.2	2.7

Table 9.5 Achieved specific growth rate (SGR) for salmon in seawater (based on [3, 6–8]).

Weight (g)	Temperature (°C)							
	2	4	6	8	10	12	14	16
30–150	0.35	0.6	0.7	1	1.3	1.7	2	2.1
150–600	0.3	0.5	0.7	0.9	1.2	1.4	1.6	1.7
600–2000	0.2	0.4	0.5	0.6	0.8	1	1	1
>2000 g	0.15	0.3	0.35	0.5	0.5	0.6	0.6	0.6

Table 9.6 Examples of achieved thermal growth coefficients (TGC) or growth factors (GF) for salmonids of different sizes (based on [9, 10]).

Size	TGC/GF
Start feeding (0.15–1 g)	1.0
Fry (1–10 g)	1.5
Juvenile smolt (>10 g)	2.0
Sea water – ongrowing	2.7–3.5

to water temperature, time and fish size. By calculating the GF it is possible to see how much of growth potential is achieved. It also gives an opportunity to make a partial comparison of growth at different sites, unlike the SGR that only permits an overall view of the production at one specific site with the actual temperature regime. For example, one site may have a good SGR due to a high water temperature while another site may have a lower water temperature and for this reason a lower SGR. Using TGC will compensate for this and therefore may be better suited for comparisons. However, comparing sites only using TGC may have a limited value because factors other than fish size and water temperature will affect growth rate. TGC may also vary with the season, being higher in spring, but it can be used as an estimate for future growth [5].

TGC can be calculated over a period based on initial weight (W_0), the end weight (W_n), and the number of day degrees in the period (DGR):

$$TGC = 1000\,((W_n^{1/3} - W_0^{1/3}) / DGR)$$

where W_0 = start weight (gram); W_n = end weight after n days; DGR = the number of day degrees in the period = $T_{avg} \times D$.

Reversing the equation can estimate the end weight and thereby is used as a tool for predicting future growth:

$$((TGC \times DGR) / 1000) + W_0^{1/3} = W_n^{1/3}$$

$$W_n^{1/3} = W_0^{1/3} + (TGC \times DGR) / 1000))$$

$$W_n = (W_0^{1/3} + (TGC \times DGR) / 1000)))^3$$

There are, as mentioned, a number of other growth indexes available that can be used for estimation of future growth, even though they require more input data and therefore more assumptions when predicting the future. One is the Ewos growth index (EGI), which is an index using real growth values for farms as a basis [2]. It has shown quite good correlation to real growth for the ongrowing phase [2]. The model uses day length in addition to fish size and water temperature, and other factors may also be included since it is based on a regression model. The data as is the basis for this model shows that the most important factor for growth is fish size, followed by day length and then water temperature. The index value is 100 if the growth is as expected for the site with the given conditions. An EGI below 100 indicates that the full growth potential has not been achieved. Another similar index is the relative growth index from Skretting

(RGI). Estimation of future growth may also be based on the energy requirement for growth (see Chapter 10) but this is not normally used when setting up production plans.

9.2.3.1 Specific growth rate calculation examples

For future planning it is usual to use SGR. It can be used to predict weight after a specified period with the following equation.

$$V_1 = V_0 \times (1 + d/100)^t$$

where V_1 = final weight at the end of the period; V_0 = initial weight at the start of the period; d = daily specific growth rate (SGR) from Table 9.4; t = number of days.

Box 9.3

In a farm the weight of a juvenile salmon is 10 g. The water temperature is 10°C. What is the weight of the salmon after 1 month of growth?

The daily growth rate is set to 2.1 or slightly below optimal growth rate at the given temperature (2.6–2.9 from Table 9.4). This means that it will have a weight gain of 2.1% each day if it gets enough feed and that the environment is satisfactory in general.

$$\text{New weight} = 10 \times (1 + 2.1 / 100)^{30}$$
$$\text{New weight} = 18.65 \text{ g}$$

A reversed calculation can also be used to find the SGR for a given period, but this requires the initial and final weight of the salmon to be known. This is a method suited for future control of growth. The following equation may be used

$$SGR = 100 \times \frac{(\ln V_1 - \ln V_0)}{t}$$

where V_1 = final weight at the end of the period; V_0 = initial weight at the start of the period; t = number of days.

The actual equation is as follows, but it is simplified above:

$$SGR = 100 \times (e^{\frac{\ln v_1 - \ln v_0}{t}} - 1)^t$$

Box 9.4

Find the SGR when two weight samples have been taken within a 30 day interval. The initial weight is 10 g and the final weight is 18.65 g.

$$SGR\ (\%) = 100 \times (\ln (18.65) - \ln (10)) / 30$$
$$SGR\ (\%) = 2.1$$

Box 9.5

Calculate the increase in biomass for a group of fish during one month for a certain fish size.

Number of fish	10,000
Average size	10 g
Initial biomass	100 kg
Mortality per month	1%
Water temperature	10°C
SGR is set to	2.1%

Number of fish at the end of the period = Initial number – Mortality

$$= 10{,}000 - \frac{10{,}000 \times 1}{100}$$

$$= 9900$$

New fish size: $10 \times (1 + 2.1 / 100)^{30} = 18.65$

New total biomass: $9900 \times 18.65 = 184.6$ kg

For one month the total weight (biomass) for the fish group will increase to 184.6 kg.

9.2.3.2 Growth factor calculation examples

Growth factor may, as discussed, also be used but it requires more information to be estimated. The following equation can be used for calculation of the growth factor:

$$GF = 1000\ ((W_n^{1/3} - W_0^{1/3})\ /\ D°C)$$

Expected end weight (W_n) after n days can be calculated as:

$$W_n = (W_0^{1/3} + (GF / 1000) \times D°C)^3$$

where W_0 = initial weight (gram); W_n = end weight after n days; D°C = the number of day degrees in the period.

Day degrees can be calculated by using the expected daily water temperatures over the whole period or by multiplying the expected average temperature over the period with the number of days in the period.

Box 9.6

There is an intake of smolt of 120 g on 1 June and the growth up to 1 January needs to be estimated. GF is set to 3.5 and expected temperatures are given in Table 9.7.

The fish have grown up to 2032 kg by 1 January.

Table 9.7 Expected water temperatures.

Month	Average water temperature (°C)	Day degrees (D°C)	Weight (g, 1st of each month)
June	13	390	120
July	15	465	240.6
August	15	465	462.5
September	13	390	780
October	11	330	1114.5
November	10	310	1436.5
December	8	248	1770
January			2032

2.4 Feed requirements

Once growth is estimated the volume of fee required can be estimated based on expected FCR. Generally, values here are within the range 1–1.3, where the highest values are for the ongrowing stage. Even though it is possible to go below 1 on smolt farms and down to 1 on sea cage farms, such values are not recommended when planning and budgeting. These values can also be used for economic budgeting and for estimation of required space for feed storage.

Box 9.7

The fish have grown from 1 kg to 4 kg; in one case 3 kg feed has been used and in another case 4.5 kg feed has been used to achieve this growth. Calculate the feed conversion rate in the two cases. In both cases the weight increase has been 4 kg–1 kg = 3 kg.

FCR first case

$$\text{FCR} = \frac{\text{Used amount of feed}}{\text{Weight gain}} = \frac{3\text{ kg}}{3\text{ kg}} = 1$$

FCR second case

$$\text{FCR} = \frac{\text{Used amount of feed}}{\text{Weight gain}} = \frac{4.5\text{ kg}}{3\text{ kg}} = 1.5$$

9.3 Environmental requirements for different production systems

9.3.1 Oxygen requirements

Salmon require oxygen to digest food to produce energy for normal body functions (aerobic metabolism). The oxygen requirement depends on the metabolic rate. The metabolic rate depends on a number of factors, of which the important ones are water temperature and fish size. A higher temperature increases the metabolic rate and therefore the oxygen requirement, and an increased fish size reduces the oxygen requirement per kilogram weight of fish.

The metabolic rate also depends on the level of activity, whether or not feed is consumed and the growth rate if feed is supplied. It may be divided into standard metabolism and maximum metabolism [11]. Owing to the different factors affecting metabolism, giving exact values for estimation of oxygen consumption is difficult.

Standard metabolism may also be named maintenance metabolism; it is the metabolism of fasting fish. Oxygen consumption is then low and only sufficient to ensure normal body function, with no growth. The maximum metabolic rate is the maximum metabolism when the fish is active, for example, high speed swimming, but this is not a normal long-term behaviour so if this value is used a general overestimation of the oxygen requirement is produced. When fish are stressed oxygen consumption increases a lot compared to standard metabolism. The growth metabolism or feeding metabolism includes oxygen consumption for digestion and growth and lies in between the standard metabolism and the maximum metabolism (see also Chapter 10). The level will, of course, depend on the growth rate and how much of the growth potential is achieved. Furthermore, feed composition will play a role due to the differing oxygen requirements for digestion of each particular feed. The regularity of feeding is also important, as less frequent feeding results in increased oxygen consumption due to increased activity.

It is of major importance to know the oxygen requirements of the fish, because if these are not met, production will be affected negatively. Low levels of dissolved oxygen in the water will cause reduced growth, and if levels are too low, stress and possibly mortality will be the result. Stress due to low oxygen content increases the risk for disease outbreak and will also present a fish welfare issue.

Fish oxygen requirements may be met either by the available oxygen in the natural water source, preferably aerated water, or by supersaturation of the water with pure oxygen gas, so manipulating the gas concentration in the water. However the oxygen is provided, the oxygen requirements of the salmon must be known. Different models for estimating the oxygen requirements of salmonids have been developed. These are either calculated by equations or directly transferred into graphs or tables and usually show the oxygen requirements for different fish sizes and water temperatures, but other factors, such as swimming velocity or water velocity, may also be included. The oxygen consumption can also be calculated based on the metabolic rate of the fish or estimated by the feed consumption of the fish, because a known amount of oxygen is required to digest the feed, even though this also varies based on the feed composition.

A number of experiments have been carried out regarding oxygen consumption in salmonids. Liao [12] was one of the first to perform a detailed study and developed an equation to estimate oxygen requirements for salmonids based on fish size and water temperatures. Values attained by those equations seem too low today, mainly because fish now grow much faster. Practical experience has shown that in general, Liao's oxygen requirements can be used if multiplied by 1.5 for small fish (0.2–100 g) and 2 for large fish (>100 g) [13] (Table 9.8).

Liao's original equation to calculate oxygen requirement is [12]:

$$O_2 = \frac{K}{0.144}(32 + 1.8T)^n \times (0.454W)^m$$

where O_2 = oxygen requirement (mg O_2/kg fish/minute); T = temperature (°C); W = weight (kg); K, m, n = constants necessary for calculating the oxygen requirement (Table 9.9).

Another equation developed to calculate oxygen requirement is [14]:

$$O_2 = 5.5 \times W^{-0.2} \times e^{0.07T}$$

where O_2 = oxygen requirement (mg O_2/kg fish/minute); T = temperature (°C); W = weight (g). This equation shows an approximate doubling of the values calculated by Liao's original equation. Other studies concur [15].

For land-based post-smolt farms a model has been developed that also uses water current in the equation as below [16]:

$$O_2 = 1.92 \times W^{-0.27} \times T \times 10^{0.01C}$$

Table 9.8 Oxygen requirements in milligram per kilogram live weight per minute with varying water temperature and fish weight (based on [12, 13]).

Fish weight (g)	Water temperature (°C)								
	2	4	6	8	10	12	14	16	18
0,2	3.09	4.22	5.58	7.23	9.18	10.59	12.15	12.15	
1	2.27	3.08	4.08	5.28	6.72	7.74	8.90	8.90	11.43
5	1.66	2.26	2.99	3.87	4.91	5.67	6.51	6.51	8.36
10	1.44	1.97	2.61	3.38	4.29	4.95	5.69	5.69	7.31
50	1.06	1.44	1.91	2.48	3.14	3.63	4.16	4.16	5.35
100	0.93	1.26	1.67	2.16	2.75	3.17	3.65	3.65	4.68
500	0.90	1.23	1.63	2.11	2.68	3.09	3.55	3.55	4.56
1000	0.80	1.08	1.42	1.84	2.34	2.70	3.10	3.10	4.00
2500	0.66	0.90	1.19	1.55	1.96	2.26	2.60	2.60	3.34
5000	0.58	0.78	1.04	1.36	1.72	1.98	2.28	2.28	2.92

Table 9.9 Constants in Liao's equation [12].

Temperature	K	m	n
<10°C	7.2×10^{-7}	−0.194	3.2
>10°C	4.9×10^{-5}	−0.194	2.12

where O_2 = oxygen requirement (mg O_2/kg fish/minute); T = temperature (°C); W = weight (g); C = water current (cm/second).

A number of other equations have been developed, some newer, which estimate a slightly higher oxygen consumption (for instance, reviewed for post-smolt [10]). Therefore, when performing calculations, it is important to be aware that they only give an estimate of the oxygen consumption because there are so many influencing factors. Consumption will also depend on the production system used, and further factors that have an effect are the water velocity on smolt farms and the currents on the sea cage farming site.

The oxygen requirement may, as mentioned, also be estimated based on the feed consumed because this shows the metabolic rate. Values between 0.25–0.5 kg oxygen per kilogram feed digested have been reported [10,16,17].

Selective breeding is a factor that results in improved growth rate, and the faster growth achieved will require more oxygen for feed digestion for each kilogram of fish.

As has been discussed, the purpose of the values in this chapter are for use in further production planning; therefore, estimates can be used. In order to cover all eventualities, values in the highest range can be used or safety factors can be included. Fluctuations in oxygen requirements, for instance, during food consumption or resulting from stress, may indicate that safety factors should be included. When using values for the necessary supply of oxygen it is also important to understand that this represents the requirements of the individual fish, and so optimal distribution of the oxygen in the production volume to ensure that each individual fish can obtain enough oxygen will be important.

9.3.1.1 Minimum and maximum oxygen concentration levels in the water

Oxygen is supplied to the fish by the water. If the concentration becomes too low, hypoxia, it will be harder for the fish to obtain sufficient oxygen, and the result will be reduced growth. The required outlet saturation values, both in closed units and in open sea cages, are a matter of discussion. In the past, 5 mg/l was proposed, but it was seen that improved growth was achieved when levels were raised to 7 mg/l. Replacing milligram per litre with oxygen gas pressure in the water (% oxygen saturation) as a measure has also been proposed because oxygen is 'pushed' into the fish due to the partial pressure gradient (see Table 9.10). The measurements of milligram per litre and percentage oxygen saturation are, of course, related to a degree. If the saturation pressure drops, the salmon take up less oxygen. By increasing the oxygen pressure gradient over the gills, the oxygen transfer will increase. Theoretically an oxygen saturation pressure below 100% will result in growth depression. However, above 80% oxygen saturation the effect will be limited and above 90% it is difficult to observe any effect on growth.

Even though a very limited growth increase is found above 90% saturation in the outlet water, such values will still present problems. There will be very little oxygen available if using naturally aerated water and in all cases there will be a requirement to add pure oxygen gas to the water. Before choosing to go above 70–80% saturation in outlet water, a cost–benefit analysis for the specific production system used should be performed. Recommendations for high oxygen concentrations in the outlet of the production unit will also increase the challenges associated with good, even distribution of water and oxygen in the unit. Using 75–80% saturation is quite common today when calculating oxygen supplies and requirements to add pure oxygen gas to the water. The water temperature also has an effect; at lower temperatures a lower saturation may be used but

Table 9.10 Dissolved oxygen content at equilibrium with different water salinity and temperature (mg/l).

T° / Sal ‰	0	5	10	15	20	25	30	35	40
0	14.6	14.1	13.6	13.1	12.7	12.3	11.9	11.5	11.1
1	14.2	13.7	13.3	12.8	12.4	12.0	11.6	11.2	10.8
2	13.8	13.4	12.9	12.5	12.1	11.7	11.3	10.9	10.5
3	13.4	13.0	12.6	12.2	11.8	11.4	11.0	10.6	10.6
4	13.1	12.7	12.3	11.9	11.5	11.1	10.7	10.4	10.0
5	12.8	12.3	11.9	11.6	11.2	10.8	10.5	10.1	9.8
6	12.4	12.0	11.6	11.3	10.9	10.6	10.2	9.9	9.6
7	12.1	11.7	11.4	11.0	10.7	10.3	10.0	9.7	9.4
8	11.8	11.5	11.1	10.7	10.4	10.1	9.8	9.4	9.1
9	11.5	11.2	10.8	10.5	10.2	9.8	9.5	9.2	8.9
10	11.3	10.9	10.6	10.3	9.9	9.6	9.3	9.0	8.7
11	11.0	10.7	10.3	10.0	9.7	9.4	9.1	8.8	8.6
12	10.8	10.4	10.1	9.6	9.5	9.2	8.9	8.6	8.4
13	10.5	10.2	9.9	9.6	9.2	9.0	8.7	8.4	8.2
14	10.3	10.0	9.7	9.4	9.1	8.8	8.6	8.3	8.0
15	10.1	9.8	9.5	9.1	8.9	8.5	8.4	8.1	7.8
16	9.9	9.6	9.3	9.0	8.7	8.5	8.2	8.0	7.7
17	9.7	9.4	9.1	8.8	8.6	8.3	8.1	7.8	7.6
18	9.6	9.2	8.9	8.6	8.4	8.1	7.9	7.7	7.4
19	9.3	9.0	8.7	8.5	8.2	8.0	7.7	7.5	7.3
20	9.1	8.8	8.6	8.3	8.1	7.8	7.6	7.4	7.2
21	8.9	8.6	8.4	8.1	7.9	7.7	7.5	7.2	7.0
22	8.7	8.5	8.2	8.0	7.8	7.5	7.3	7.1	6.9
23	8.6	8.3	8.1	7.8	7.6	7.4	7.2	7.0	6.8
24	8.4	8.2	7.9	7.7	7.5	7.3	7.1	6.9	6.7
25	8.2	8.0	7.8	7.6	7.4	7.1	7.0	6.8	6.6

can still achieve a maximal level of feed intake (DO_{maxFI}) [18]. Too low a dissolved oxygen concentration in the water will affect the salmon negatively in several ways, including reduced growth, poor feed conversion rates, behavioural changes and excessive respiratory ventilation.

Too high a concentration of oxygen in the water, hyperoxia, is not recommended either. Natural water contains a certain amount of oxygen and manipulation of the oxygen gas saturation in the water can result in levels that are too high, which may become toxic for the salmon. The hyperoxia that occurs when salmon are exposed to higher oxygen concentrations than the normal partial pressure (supersaturation) may cause severe gill damage due to oxidation of the gill tissue by the pure oxygen gas bubbles in the water. Supersaturation of oxygen gas above 140–150% saturation may result in reduced growth and increased mortality [19,20]. With this amount of oxygen, a reduction of nitrogen gas below saturation is needed to avoid problems with overall gas supersaturation (TGP) [21]. From a welfare perspective, any supersaturation of oxygen in the water is not recommended within the production unit. However, in practice it may not be a problem to have above 100% oxygen saturation in the inlet water by adding pure oxygen gas to the water. The kinetics of dissolving oxygen in supersaturated water is important here. If adding super-oxygenated water to a volume of water with a saturation below 100%, it

Box 9.8

How much oxygen is available to fish in water with 95% oxygen saturation when the water temperature is 10°C and the required outlet concentration is set to 7 mg/l? (See also Table 9.10).

Fully saturated water of 10°C contains 11.3 mg oxygen/litre (Table 9.10).
Oxygen in 95% saturated water: $0.95 \times 11.3 = 10.74$ mg/l
Available oxygen = Inlet concentration – Outlet concentration
10.74 mg/l – 7 mg/l = 3.74 mg/l

Box 9.9

The temperature in two water bodies is 5°C and 15°C, and the dissolved oxygen saturation in the water is 75% of the value at equilibrium (Table 9.10). Calculate the concentration of dissolved oxygen in milligrams per litre for the two situations.

Using Table 9.10, the equilibrium concentration (100% saturation) is found to be 12.8 mg/l at 5°C and 10.1 mg/l at 15°C. Calculating 75% saturation gives 9.6 mg/l at 5°C and 7.6 mg/l at 15°C.

This also shows that there are differences for the two values regarding milligrams per litre and also that saturation in percentage is more important than the actual value in milligrams per litre.

depends how long the superoxygenation in the water remains before it becomes a problem. Most probably the oxygen will be dissolved very fast, presenting no problem for the fish. In general, a supersaturation of the incoming water with oxygen gas above 120–130% is not recommended.

9.3.2 Water requirements

Water supply is an essential requirement when growing salmon but how it best supplied and in which form depends on the life stage of the salmon and on the production systems used. The water supply is needed to supply oxygen to the fish, as it is the oxygen carrier. It is also necessary to supply water to dilute and remove the metabolic waste products released from the fish. There are differences depending on whether eggs, fry, juvenile fish or ongrowing fish are produced, and also depending on whether closed systems, such as tanks, or open systems, such as sea cages, are used. With closed systems it will be possible to alter the content of the incoming water more easily, for instance by adding pure oxygen gas. It is also beneficial to consider the differences between water use in flow through systems (FT) and recycling aquaculture systems (RAS) for production in closed production units. In a RAS the new water requirement depends on the ratio of the new incoming water flow compared to total water used in the production unit and the efficiency of the water treatment units placed in the reuse cycle.

9.3.2.1 Water requirements for egg incubation

A hatchery's water requirement will be highly dependent on the type of closed system that is used for storing the eggs. Here the systems can be separated into those in which the eggs can hatch or those in which the eggs must be removed before hatching, usually at eyed egg stage. The second system is common on broodstock farms where the eggs are transferred to smolt farms at eyed egg stage (see section 6.2.1). The supply of new raw water will also be dependent on whether the water is used once or multiple times, that is, recirculated or not. Today, systems where the new water is reused in the hatchery are used. The advantage compared to RAS in fish farms is that fewer metabolic products are released from the eggs compared to a growing fish, and therefore a simpler RAS may be used.

In a hatchery, therefore, the amount of new water is not normally calculated according to any tabled values, but by the operational requirements of the hatching system used. However, rough values are in the area of 0.3–1 l water supplied per minute per litre of eggs. Regardless of the system used, it is important that the supply of water is accurate and that the eggs are not lifted by the incoming water flow. The velocity of the water passing the eggs is important and a velocity of 100 cm/h has been proposed for salmon in the wild [22]. This ensures adequate oxygen reaches the eggs and that metabolic waste products released by the eggs are removed.

9.3.2.2 Water requirement for closed production units

Closed production systems can be used from fry to ongrowing salmon and typically consist of tanks or basins on land, but may also include ponds and closed sea cages (closed contaminant systems). The water requirement here will vary to some degree with the production system, but mainly depends on the fish density and growth rate.

Tanks are most commonly used in intensive salmon farming and the required amount of water to supply to the tank will depend on how the farm and tank are operated. The water supplied can be used to provide the entire oxygen consumption of the fish within the tank without any supersaturation of the inlet water with oxygen gas. However, the inlet water may also be supersaturated with oxygen gas and the amount of water supplied should then only be enough to ensure adequate dilution of metabolic waste products for optimal growth of the fish in the production unit. The amount of water to add to the tank may also be used to keep the tank free of waste products and to some degree biofouling (hydraulic self-cleaning), but then the amount of new incoming water has to be above a minimum amount and the tank, including inlet and outlet, must be designed correctly.

Systems that partly or almost completely reuse the water are available, but then water purification equipment must be included in the reuse circuit to avoid problems with accumulation of metabolic waste products (RAS system).

9.3.2.2.1 Water requirements in flow through systems without oxygenation

In what is called a flow through (FT) system, the water is sent through the production unit once and not several times as in a RAS. All the oxygen that the salmon in the closed production unit need can be supplied with the incoming water (the raw water) or alternatively, additional oxygen gas may be directly added, either to the incoming water or directly with diffusors within the production unit. The amount of oxygen gas that can be dissolved in water under natural conditions depends on water temperature, water pressure and water salinity (for more info [21]). Under natural conditions, the concentration of dissolved oxygen will be in equilibrium with the oxygen concentration in the air above the water surface, but there can also be a reduced saturation in the water, meaning that the oxygen level is lower than the equilibrium level (shown in Table 9.10). Therefore, aerators should always be used to raise the oxygen saturation to 100% of the level possible at equilibrium. This also reduces any supersaturation of nitrogen gas to 100%, as fish do not tolerate values above 100% nitrogen saturation. As previously discussed, the oxygen concentration can be raised above 100% saturation (supersaturation) by the addition of pure oxygen gas to the water. The salinity of the water will also affect the dissolved gas content, with higher salinity decreasing the gas content. For example, the available dissolved oxygen content will be reduced when the salinity increases (Table 9.10).

When the oxygen requirement of the fish in the production unit, the concentration of oxygen in the inlet water and the required concentration in the outlet water to ensure optimal growth are all known, the water supply requirement and water exchange in the production unit can be calculated. If the entire oxygen requirement is to be supplied with the incoming water without any supersaturation with additional oxygen gas, the oxygen saturation in fully aerated water is typically set to 95% of the equilibrium concentration, since it can be difficult to achieve 100% saturation. The following equation may then be used:

$$Q = \frac{O_{2-req} \times m}{C_{in} - C_{out}}$$

where Q = water flow requirement – l/minute; $O_{2\text{-}req}$ = specific oxygen requirement of the actual fish size – mg/minute/kg fish; m = total mass of fish – kg; C_{in} = oxygen concentration in the inlet water – mg/l; C_{out} = oxygen concentration in the outlet water – mg/l.

Based on the oxygen requirement of the salmon and the available oxygen in the water, tables showing the necessary supply of water can be produced. These tables will then give the water requirements in litres per minutes per kilogram of fish, and the values will depend on the levels set for oxygen saturation in the inlet water and the required saturation in the outlet water (Table 9.11, Table 9.12).

9.3.2.2.2 Water requirements in flow through systems with superoxygenation

The oxygen content in the inlet water may, as discussed, be raised above the natural equilibrium concentration with air by adding oxygen gas directly to the water. The closed production unit can still be operated as a flow through system with supersaturation of oxygen gas in the inlet water. This will reduce the required supply of new water to the closed production unit because each volume unit of inflowing water contains more oxygen that can be utilized by the fish. It may also be possible to add oxygen gas with diffusors directly within the closed production unit, but this may disturb the flow pattern in the tank resulting in suboptimal water distribution, though this depends on the type of closed production unit used. Diffusers installed at the bottom of the tank are typically used as emergency systems in case of incidents such as water stoppages.

Box 9.10

The amount of water required for oxygen fulfilment in a production tank must be found. There is 1000 kg of fish with an average weight 100 g in the tank, the water temperature is 15°C and normal growth is expected. The oxygen concentration is set to 100% in the inlet water and the requirement is set to 80% in the outlet water to avoid reduction in growth. How much water must be supplied?

Calculate first concentration in milligram per litre from 80% saturation using Table 9.10. 100% saturation is 10.1 mg/l, 80% saturation is 8.08 mg/l.

The specific oxygen requirement for the fish using Table 9.8 is found to be 3.65 mg/kg/minute.

Using the equation:

$$Q = \frac{O_{2-req} \times m}{C_{in} - C_{out}}$$

$$Q = \frac{3.65 \times 1000}{10.1 - 8.08}$$

$$Q = 1816 \text{ l of water}$$

Table 9.11 Freshwater requirements of fish over a range of fish weights and water temperatures, given in litres per minutes per kilogram of fish (for 95% saturation of intake water and 7 mg/l in outlet water) (based on tables 9.8 and 9.9 [12,13]).

Fish weight (g)	Water temperature (°C)								
	2	4	6	8	10	12	14	16	18
0.2	0.51	0.78	1.14	1.68	2.48	3.21	4.19	5.53	
1	0.37	0.57	0.83	1.23	1.82	2.35	3.07	4.05	5.72
5	0.27	0.42	0.61	0.90	1.33	1.72	2.24	2.96	4.18
10	0.4	0.36	0.53	0.79	1.16	1.50	1.96	2.59	3.66
50	0.17	0.27	0.39	0.58	0.85	1.10	1.43	1.90	2.68
100	0.15	0.23	0.34	0.50	0.74	0.96	1.26	1.66	2.34
500	0.15	0.23	0.33	0.49	0.72	0.94	1.22	1.62	2.28
1000	0.13	0.20	0.29	0.43	0.63	0.82	1.07	1.42	2.00
2500	0.11	0.17	0.24	0.36	0.53	0.68	0.09	1.18	1.67
5000	0.10	0.14	0.21	0.32	0.43	0.60	0.79	1.03	1.46

In closed production systems when growing salmon with normal fish densities and water temperatures it is common today to add pure oxygen gas. In smolt farms this is typically due to a lack of sufficient freshwater on site. If it is necessary to pump the water to the farms, such as for land-based farms for ongrowing, it is usually more economically beneficial to perform superoxygenation rather than to pump enough water to supply the entire oxygen requirement.

Table 9.12 Seawater requirements of fish over a range of fish weights and water temperatures given in litres per minutes per kilogram of fish (for 95% saturation of intake water and 7 mg/l in outlet water) (based on Table 9.8 and 9.9 [12, 13]).

Fish weight (g)	Water temperature (°C)							
	2	4	6	8	10	12	14	16
100	0.28	0.45	0.73	1.14	1.72	2.64	4.06	6.90
500	0.27	0.44	0.71	1.11	1.68	2.25	3.94	6.73
1000	0.24	0.39	0.62	0.97	1.46	2.25	3.44	5.90
2500	0.20	0.32	0.52	0.82	1.23	1.88	3.89	4.93
5000	0.18	0.28	0.45	0.72	1.08	1.65	2.53	4.30

Box 9.11

How much freshwater is required for a fish farm under the following conditions?

Water temperature:	15°C
Fish size:	100 g
Total biomass:	1 tonne
Oxygen in outlet water:	7 mg/l

Use Table 9.11 to find the specific water requirements: 1.46 l/minute/kg fish

Total water required: 1.46 l/minute/kg fish × 1000 kg = 1460 l/minute.

Owing to the large variation in requirements, it is normal to add a safety factor in production planning, for instance 30%.

The water requirement is then 1460 l/minute × 30% = 1898 l/minute.

If all oxygen is to be supplied with the inlet water without supersaturation of oxygen and both fish densities and water temperature are high, the amount of water needed will be very high and special precautions must be taken to avoid the water velocity in the tank becoming too high for the fish. A flow so strong that the fish will be pushed backwards against the drain may be created. However, consideration must also be taken when adding pure oxygen gas to the water to avoid the oxygen concentration in the incoming water reaching a concentration or a saturation degree that reduces welfare and may, in the worst case, be directly toxic to the fish. However, some supersaturation can be used without any problems. The same equation as that used for calculating water requirement without supersaturation can be used, but with the difference that the inlet concentration will be higher.

$$Q = \frac{O_{2-req} \times m}{C_{in} - C_{out}}$$

where Q = water flow requirement – l/minute; $O_{2\text{-req}}$ = specific oxygen requirement of the actual fish size – mg/minute/kg fish; m = total mass of fish – kg; C_{in} = oxygen concentration in the inlet water – mg/l – this will now be above 100% saturation; C_{out} = oxygen concentration in the outlet water – mg/l.

There are several ways to add oxygen gas to the inlet water, and these are typically divided into high pressure and low pressure oxygenation (review in [21]). One single inlet supplying highly oxygenated water may also be used [21]. However, by adding pure oxygen gas to the new raw water supply, the volume of new water into the tank will be reduced, which in turn reduces the water exchange rate thereby reducing the dilution caused by the new incoming water; this can result in the concentration of metabolic waste products released from the fish rising above levels required for good welfare and optimal growth. A reduced water exchange rate will also increase the hydraulic retention time within the production unit. In tanks this may be reduced so much that it results in settling of particles inside the tank, leading to a requirement for manual cleaning as the velocity and force of the inlet water will not be high enough to ensure hydraulic self-cleaning of the tank (described more in section 9.3.2.2.4).

9.3.2.2.3 Water requirement for dilution of metabolic waste products

When the supply of water into the production unit is low, the volume of new water may be too small to dilute the metabolic waste products produced by the salmon to acceptable limits within

Box 9.12

The same values as Box 9.11 are used: water temperature 15°C, fish weight 100 g, total biomass 1000 kg and oxygen requirement in outlet water 7 mg/l. The oxygen concentration in the inlet water is now raised to 120% saturation (supersaturation with oxygen). How much water must now be supplied to the production tank to fulfil the oxygen requirement?

With 120% supersaturation the oxygen concentration in the inlet water is 12.12 mg/l based on Table 9.10. The oxygen requirement for the actual fish size and with the given water temperature is taken from Table 9.8 to be 3.65 l/kg fish/minute.

$$Q = \frac{O_{2-req} \times m}{C_{in} - C_{out}}$$

$$Q = \frac{3.65 \times 1000}{12.12 - 7}$$

$$Q = 712 \text{ l/minute}$$

Compared to the prior example the water requirement can therefore be halved by adding pure oxygen gas, but this requires that the metabolic waste products are still diluted to satisfactory levels and that it is within the interval for hydraulic self-cleaning if this is required (described more in section 9.3.2.2.4).

the tank. This must be considered when superoxygenating the inlet water and having high fish densities in the production unit at the same time, especially if the temperature and growth rate is high.

The metabolic waste released by the fish are the product of digesting food. The amount and composition of the metabolic waste products are based on the feed composition, particularly the protein, fat and carbohydrate ratios (further described in Chapter 10). In addition, any feed loss and non-digested feed contributes to the volume of waste.

The amount of water required to dilute the metabolic waste to the required concentrations can be calculated using the following equation:

$$Q = \frac{MT_{released} \times m + MT_{inlet}}{(MT_{outlet\ max} - MT_{inlet})}$$

where $MT_{released}$ = metabolic waste products released (mg per kg fish per minute); m = mass of fish (kg); MT_{inlet} = concentration of metabolic waste products in inlet water ($\frac{mg}{l}$); $MT_{outlet\ max}$ = max concentration in outlet (mg/litre).

The first problem that typically occurs from an accumulation of metabolic waste is too high a concentration of dissolved carbon dioxide gas in the production unit. Carbon dioxide gas is one of the end products of fish metabolism. A simplified calculation ratio that can be used is that the amount of carbon dioxide produced is equal to the amount of oxygen consumed by the fish in mol. More accurate is a ratio of 0.9 mol CO_2 released per mol oxygen used. Converting to milligrams, a ratio of 1.1 mg CO_2 is released per 1 mg O_2 consumed. A linear regression model has been developed to calculate the carbon dioxide production where d represents the oxygen used in milligrams per litre [23, 24]:

$$CO_2\ (mg/l) = d \times 1.02 - 0.18$$

However, this ratio depends on the feed composition (based on energy metabolism and Krebs cycle, see Chapter 10) and growth rate, and therefore is only used to find approximate values, but does assist in calculating the concentration of carbon dioxide in the outlet water based on the oxygen consumption. Calculation of carbon dioxide release can also be carried out based on feed consumption instead of using oxygen consumption, but of course those values are correlated. The concentration in the outlet water ought to be below 10–15 mg/l to achieve optimal growth and to ensure good fish welfare. Current recommendations are 10 mg of carbon dioxide per litre of water, but this can be hard to fulfil. If values in the outlet water rise above this value (see Box 9.13), action must be taken. Too high a concentration of carbon dioxide in the water may be removed by degassing the water, but this cannot be done inside the closed unit because a large quantity of air has to bubble through the tank, creating a high gas to liquid ratio, which will affect the fish negatively. Degassing of carbon dioxide must therefore be done outside the tank in an external water circuit. This results in the water cycle being partly closed and the system created may be defined as a simple RAS (low degree of reuse, described later). Water now enters the production unit either as raw water or as water that has been returned to the production unit after degassing. The latter will be reused water containing all metabolic waste products apart from carbon dioxide. Degassing of carbon dioxide is a process that requires a supply of energy and a high amount of air to be supplied (review in [21]).

When the fish release carbon dioxide into the water a chemical reaction will occur, and the pH will drop (carbonate system, explained, for instance, in [21]). How fast and how much the pH drops will depend on the alkalinity of the water. In highly alkaline water the pH drop will be limited while the opposite will occur in less alkaline water. Seawater is an example of water with high alkalinity and here the drop in pH due to release of carbon dioxide will be lower. For the quality of water generally used, the drop in pH will not be so high that it results in problems with too low a pH for the fish. The drop in pH can be calculated based on carbon dioxide production and the alkalinity of the water [21]. When removing the produced carbon dioxide from the water by degassing, the pH of the water will rise as H^+ is removed.

However, carbon dioxide is not the only metabolic waste product released by fish that accumulates and presents a problem: there is also ammonium (NH_4^+). In general, when adding NH_4^+ (ion) to water, a chemical reaction will occur and some will be converted into ammonia (NH_3) (dissolved gas in water); there will be an equilibrium between ammonium and ammonia. Normally it is presented as total ammonia nitrogen (TAN) which represents the sum of

Box 9.13

How much water is required for the fish tank under the same conditions as the prior examples (water temperature 15°C, fish weight 100 g, total biomass 1000 kg) to dilute the water so the carbon dioxide levels can be kept below 10 mg/l?

First use Table 9.8 to find the specific oxygen consumption: 3.65 mg/kg live weight/minute.

Total oxygen requirement: 3.65 mg/kg live weight/minute × 1000 kg = 3650 mg/minute.

Set carbon dioxide to oxygen ratio to 1:1.1 meaning the carbon dioxide production is 3650 mg/minute × 1.1 = 4015 mg/minute.

Using the following equation and setting the inlet concentration of carbon dioxide to 0 mg/l (not exactly correct as there will always be some present due to carbon dioxide in the atmosphere) and the maximum amount in the outlet to be <10 mg/l, will give:

$$Q = \frac{CO_{2-rel} \times m + CO_{2-inlet}}{CO_2 - \text{outlet max}}$$

$$Q = \frac{4015 + 0}{10}$$

$$Q = 401.5 \text{ l/minute}$$

ammonium and ammonia, but only the nitrogen component. The toxicity is caused by ammonia as this damages the gills at high concentrations. Values for NH_3-N below 0.012 to 0.03 mg/l have been reported safe for salmonids [25–27], but this may be affected by fish size and interaction with other water quality parameters. What value this represents in the TAN is, of course, dependent on water temperature and pH, but once the pH is below 7 almost all TAN will be the ionic form.

When starting to remove gaseous carbon dioxide from the water, the pH will increase and therefore more of the TAN will be transferred to the more toxic ammonia form. It is important to remember this when installing aerators for removal of gaseous carbon dioxide from the water to avoid creating concentrations that are too high. Measurement of TAN and pH will then be important to ensure that a new problem is not created.

It is possible to roughly calculate the concentration of TAN and ammonia based on how much oxygen the fish consume. An estimate says that for each milligram of oxygen consumed, 0.04 to 0.14 mg of TAN is produced [27,28]. By using this, the concentration in the outlet water can be found based on oxygen consumption. This also applies when using carbon dioxide aerators to remove carbon dioxide gas. Which form the TAN is in can be calculated based on the pH of the water.

Instead of using the oxygen consumption to find the TAN production it may also be calculated from the feeding rate. As mentioned before, the oxygen consumption depends on the growth rate. One equation that gives the TAN production based on feeding rate is as follows [17]:

$$P_{TAN} = F \times PC \times 0.092$$

where F = feeding rate (kg/day); PC = percentage protein in the feed.

Box 9.14

How much water is required for the fish tank under the same conditions as the prior examples (water temperature 15°C, fish weight 100 g, total biomass 1000 kg) to dilute the water so the TAN levels can be kept below 1 mg/l? The pH is set to 6.8, so almost all TAN is in ammonium form.

First use Table 9.8 to find the specific oxygen consumption: 3.65 mg/kg live weight/minute.

Total oxygen requirement: 3.65 mg/kg live weight/minute × 1000 kg = 3650 mg/minute.

Set TAN to oxygen ratio to 0.1:1 meaning that the TAN production is 3650 mg/min × 0.1 = 365 mg/minute.

Using the equation and setting inlet concentration of TAN to 0 mg/l and the maximum amount in the outlet to be <1 mg/l will give:

$$Q = \frac{MT_{released} \times m}{(MT_{outlet\ max} - MT_{inlet})}$$

$$Q = \frac{TAN_{-rel} \times m}{(TAN_{outlet\ max} - TAN_{inlet})}$$

$$Q = \frac{365}{1}$$

$$Q = 365 \text{ l/minute}$$

When starting to remove carbon dioxide, problems with suspended solids may also present a problem, because there may be too low an amount of raw water to dilute them to safe concentrations. Problems with suspended solids (TSS) start to occur if the concentration rises above 25 mg/l. The same calculation as above can be used here to find the necessary requirements to dilute them to safe outlet concentrations. A dual drain in the tank may, however, be used to avoid this. When using a dual drain the particles are taken out in a separate outlet that is directed out of the system together with 1–20% of the outlet water flow.

Based on the above equation, tables can be produced to show how much raw water or dilution water must be supplied to ensure safe levels of carbon dioxide, TAN and TSS in relation to water temperature. Raw water requirements in litres per minute per kilogram of fish in the production unit for different temperatures can then be presented. However, because carbon dioxide is part of the carbonate system the alkalinity of the water will also have an influence and has to be taken into consideration; this equation is not as simple and requires more chemical knowledge (see, for instance, [21]). As a basis and for normal alkalinity the above equation will at least give a recommendation.

9.3.2.2.4 Water requirement for the hydraulic self-cleaning of tanks

The addition of pure oxygen gas means that the supply of water to the same amount of fish can be reduced. This means that the flow of water into tanks or other closed units can be reduced; however, the inlet water can also be used to keep the tank free of particles and biofouling (hydraulic self-cleaning) (reviewed in [21]). The water flow in the tank will wash the tank walls and bottom and carry the effluent to the outlet of the tank. Hydraulic self-cleaning can be achieved if the tank shape, inlet and outlet are correctly designed and if the flow into the tank is within acceptable limits. If the retention time in the tank becomes too low, below 10 l/minute/m^3 tank volume, it will be difficult to attain hydraulic self-cleaning with a normal tank inlet and outlet design. This value corresponds with water retention time in the tank of over 100 minute, which means that with the chosen volume flow rate it will take over 100 minutes to fill one tank volume. If a theoretical retention time of 100 minutes is exceeded, either a specially designed inlet and outlet must be used, or mechanical cleaning of the tank must be carried out in addition to the now-limited hydraulic self-cleaning.

Too high a water flow into tanks may, however, also create problems due to too short a retention time in tanks, creating turbulence in the tank, and this therefore has to be controlled. It is not recommended to have an inlet water flow of over 30 l/minute/m^3 tank volume unless using a specially designed tank inlet and outlet. This corresponds to not having a theoretical water retention time in the tank of less than 30 minute, or not taking less than 30 minutes to fill the tank volume; this water flow is fairly typical, however, when fish density and growth rates are high in order to avoid carbon dioxide problems, so therefore a specially design inlet and outlet must be used.

Box 9.15

A fish tank contains 10 m^3 of water, and the water supply is 300 l/minute. Calculate the theoretical water retention time in the tank.

$$\text{Water retention time} = \frac{\text{Tank volume}}{\text{Water flow rate}}$$

$$\text{Water retention time} = \frac{10{,}000 \text{ l}}{300 \text{ l/minute}} = 33.3 \text{ minute}$$

Box 9.16

A tank with a volume of 50 m^3 contains 1000 kg salmon with an average size of 100 g. The water temperature is 10°C. Find the minimum inlet flow of water to ensure hydraulic self-cleaning, and the amount of extra oxygen that has to be added when the minimum inlet water flow is supplied.

Necessary water inlet: 50 m^3 × 10 l/minute/m^3 = 500 l/minute

Specific oxygen requirement from Table 9.8: 2.75 mg/minute/kg fish

Total oxygen requirement: 2.75 mg/minute/kg fish × 1000 kg fish = 2 750 mg/minute

Oxygen content in the water at 10°C and 100% saturation (Table 9.10): 11.3 mg/l

Oxygen supplied with the inlet water: 11.3 mg/l – 7 mg/l = 4.3 mg/l

Total supplied: 4.3 mg/l × 500 × l/minute = 2150 mg/minute

Additional pure oxygen requirement: 2750 mg/minute – 2150 mg/minute = 600 mg/minute

Required supersaturation of water in the inlet water:

$$\text{Additional oxygen per l} = \frac{\text{Additional oxygen requirement}}{\text{Incoming water flow}}$$

$$= \frac{600 \text{ mg/minute}}{500 \text{ l/minute}} = 1.2\text{mg/l}$$

$$\text{Oxygen supersaturation} = \frac{11.3 \text{ mg/l} + 1.2 \text{ mg/l}}{11.3 \text{ mg/l}} \times 100 = 110.6\%$$

9.3.2.2.5 Fish density and growth rate

As seen above, the water requirement of closed production units can be calculated in different ways, but it is important to be aware that there can be situations where the maximum density that can be kept in a closed production unit is limited by other variables than water requirement. This occurs particularly with high growth rates. A water velocity that is too high can be created in the production units if supersaturated water is not used, resulting in turbulence which pushes the fish backwards. In such cases the only solution is to use units with a specially designed inlet and outlet. If only supplying enough water to dilute metabolic waste products to safe levels, the supersaturation of oxygen in the inlet water to fulfil the oxygen requirement of the fish will be too high. Toxic levels of oxygen gas may then be reached.

Box 9.17

Use the same values as in the prior examples (water temperature 15°C, fish weight 100 g, total biomass 1000 kg) but a fish density of 50 kg/m^3, meaning the tank volume is 1000 kg/50 kg/m^3 = 20 m^3. Calculate the water retention time for the following three cases: (1) all oxygen supplied with incoming water; (2) inlet water supersaturated with 120% oxygen; (3) only enough water to dilute carbon dioxide concentration to levels <10 mg/l.

In the first case all oxygen is supplied with the inlet water, so the amount of water to supply will then be 1460 l/minute:

$$\text{Water retention time} = \frac{\text{Tank volume}}{\text{Water flow rate}}$$

$$\text{Water retention time} = \frac{20{,}000}{1460}$$

$$\text{Water retention time} = 13.7 \text{ minute}$$

This is a retention time far below recommended values and a special design must be used to avoid turbulence in the tank.

In the second case inlet water is supersaturated with oxygen, water supply is 712 l/minute:

$$\text{Water retention time} = \frac{20{,}000}{712}$$

$$\text{Water retention time} = 28.1 \text{ minutes}$$

Recommended retention time is almost within recommended values.

The third case uses only enough water to dilute carbon dioxide levels to <10 mg/l, the water supply is 401 l/minute:

$$\text{Water retention time} = \frac{20{,}000}{401}$$

$$\text{Water retention time} = 49.8 \text{ minutes}$$

Even though the water retention time is within the recommended values, this will not be the case for the oxygen supersaturation in the inlet water, which will be far above recommended values:

Total oxygen requirement: 3.65 mg/minute/kg fish × 1000 kg fish = 3650 mg/minute

Oxygen content in the water at 15°C and 100% saturation (Table 9.10): 10.1 mg/l

Oxygen supplied with the inlet water: 10.1 mg/l – 7 mg/l = 3.1 mg/l

Total supplied: 3.1 mg/l × 401 × l/minute = 1243 mg/minute

Additional pure oxygen requirement: 3650 mg/minute – 1243 mg/minute = 2407 mg/minute

Required supersaturation of water in the inlet water:

$$\text{Additional oxygen per litre} = \frac{\text{Additional oxygen requirement}}{\text{Incoming water flow}}$$

$$= \frac{2407 \text{ mg/minute}}{401 \text{ l/minute}} = 6.0 \text{ mg/l}$$

$$\text{Oxygen supersaturation} = \frac{10.1 \text{ mg/l} + 6.0 \text{ mg/l}}{10.1 \text{ mg/l}} \times 100$$

$$= 159.4\%$$

9.3.2.2.6. Water requirement with reuse of water – RAS

One of the major factors that influences the development of salmon farming is the access to water, both regarding quality and quantity. Any system that can include reuse and conservation of water is therefore very useful, especially for freshwater since it is a limited resource, but also when seawater must be pumped or if a large amount of costly water treatment is necessary on the raw water to achieve the required water quality for salmon. Optimal salmon growth is dependent on water temperature and by keeping the temperature stable at an optimum point, increased growth can be achieved. This will, however, typically involve the use of energy to heat the water. Therefore, systems which conserve energy are also of interest, which will be the case if the water is reused.

Water reuse or water recirculation systems, RAS, involve reuse of varying degrees of waste water/outlet water from closed rearing systems, such as tanks, closed sea cages, raceways or ponds. Greater water reuse requires less new water (raw water or dilution water). The disadvantage of a lower volume of new water compared to the volume of reused water is that more water treatment of the reused waste water is required in the cycle.

When reusing water, the diluting effect of the inlet water will be reduced and accumulation of metabolic waste will occur. The retention time for all substances in the water in the system will increase, and therefore the fish will be exposed to all substances not removed from the RAS for a longer period of time. This is the same situation as when the amount of inlet water to a given production unit is reduced, resulting in problems with too high a concentration of carbon dioxide and TAN; exactly the same will happen in a RAS system. Compared to a traditional flow through farm, the retention time from when the new water enters the production unit until it leaves is much higher, on another scale, and problems with first carbon dioxide and then TAN occur. However, today there is a great interest in RAS systems in salmon farming, including smolt farms and land-based ongrowing farms, due to lack of both freshwater and good sites for sea cage farming. By introducing RAS technology for ongrowing, areas other than those naturally suited in both the northern and southern hemispheres can be used for ongrowing salmon production. Disease problems in the sea, particularly sea lice, are also an important factor influencing the interest in utilization of land-based facilities. RAS will also give possibilities for a higher degree of environmental control, allowing year-round growth at optimum temperatures with collection of waste products. The drawback is the higher investment and running costs.

Finding the requirement for new water or dilution water for a RAS system depends on a number of factors, including how the RAS system is designed, the fish density, the growth rate, the temperature and so on. In addition, several terms are used when describing RAS systems and their performance which also makes this area slightly confusing. In general, the new/dilution water plus the water that has gone through the RAS circuit with the purification units ensures that the overall water quality in the production unit is satisfactory.

When discussing RAS, as mentioned a number of factors are used for calculations. One is the degree of recirculation, which means how large a proportion of the water is reused in relation to the total water consumption. This can be given as time independent or time dependent. Most commonly, time independent is used. This allows use of the equation below:

$$\text{Degree of recirculation (\%)} = \frac{(1 - \text{Amount of reused water})}{\text{Total water consumption}} \times 100$$

Box 9.18

The total water consumption is 750 l/minute. Of this amount, 37 l/minute is new. Calculate the degree of recirculation:

$$(1 - \frac{37}{750}) \times 100 = 95\%$$

Box 9.19

In a tank with 1.5 tonnes of fish with a weight of 100 g. The temperature is 10°C. Calculate the need for addition of new water with a degree of 90% recirculation.

From (Table 9.11) the water requirement of these fish at the given temperature is found to be 0.5 l/minute/kg fish.

Total need for water: 1500 kg × 0.5 l/minute/kg fish = 750 l/minute

Need for new water: 750 – (0.90 × 750) = 75 l/minute

New water requirements can, however, also be calculated as time-dependent water exchange rate over a given period, for instance, how much of the water in the overall water volume, including fish tanks, piping and the purification system, is replaced during an interval of time, or how much water is supplied to the system compared to the internal flow in the system over the same time, for instance, a day. Another way to define this is the time needed to fill up the entire system with only the dilution water, or the theoretical retention time for the system.

Water exchange in RAS is typically on a continuous basis, meaning some new water is continuously supplied with the same amount of water flowing out of the system. However, a batch exchange may also be used where a batch of water in the system is exchanged, for example once a day, similar to an aquarium. However, the latter results in degeneration of water quality until the next water exchange, so the water quality will be variable.

The values described above are of limited interest alone because they do not factor in the fish density, fish species and growth rate. If these values are given, much more can be learned about the system and its possible performance. Another way to quantify the performance of a RAS system, therefore, is to give the new water supply per kilogram of fish or perhaps how much feed could be supplied per litre of new water added.

With regard to the production plan, the use of RAS will result in reduced raw water requirements. How to estimate these requirements will depend on the RAS degree, the components in the RAS and their achieved purification degree, and the amount of released metabolic products from the fish (for further description see [21]). The following equation can be used to estimate the additional concentration (C) due to operating the farm as a RAS rather than a flow through.

$$C = \frac{1}{(1 - R + (R \cdot RE))}$$

where R = degree of reuse; RE = removal efficiency of component in RAS circuit.

Box 9.20

The recycling degree is 97% and the removal efficiency of carbon dioxide in the carbon dioxide aerator is set to 80%. If running the system as a flow through the carbon dioxide concentration would be 10 mg/l. What will the outlet concentration be when using the RAS?

$$C = \frac{1}{(1 - R + (R \times RE))}$$

$$C = \frac{1}{(1 - 0.97 + (0.97 \times 0.8))} = 1.24$$

The concentration of carbon dioxide in the tank outlet will then be 1.24 × 10 mg/l = 12.4, but at the same time the necessary amount of new water/raw water will only be 3% of the requirement of the flow through to dilute to acceptable values.

When using RAS the water coming into the production unit will be a mix of new raw water and water coming from the recycling unit. What differs from a flow through system, as stated above, is that the water may contain products that are not removed fully in the purification steps in the recycling unit. These can be metabolic waste, such as carbon dioxide, suspended solids, or ammonia. When calculating necessary production units this has to be taken into account. More water will need to be supplied to the production units to dilute these substances to satisfactory levels.

Moreover, new substances may be added in the RAS that also have to be diluted below toxic levels by raw water. A biological filter is a normal component in the RAS and it might be that the degree of reuse only requires a nitrification process. This is a process where bacteria 'convert' ammonia to nitrate, which is much less toxic for the fish. However, this substance may also be toxic if the concentration becomes too high.

9.3.2.3 Water requirement for sea cages

The water requirement of a production unit is, as described, dependent on the unit and how it is operated. For open units, such as sea cages, the requirement can be defined based on the number of fish, fish weight, water temperature and whether pure oxygen is added or not. However, it is actually of little interest to define a water quantity requirement for open systems because it only depends on the natural water currents of the site. Adding additional water currents on the site,

Box 9.21

How much water is required for a fish tank under the following conditions: water temperature 15°C, fish weight 100 g, total biomass 1000 kg, to dilute the water so the carbon dioxide levels can be kept <10 mg/l (same as in earlier examples). The fish tank is a part of a RAS system where carbon dioxide is removed but only with a 50% removal rate, meaning that the inlet concentration from the RAS is 5 mg/l. The RAS degree is set to 99%.

Almost all water comes from the RAS, with only 1% raw water, so in the later calculations 5 mg/l CO_2 is approximated in the inlet water to the fish tank.

First use Table 9.8 to find the specific oxygen consumption: 3.65 mg/kg live weight/minute.

Total oxygen requirement: 3.65 mg/kg live weight/minute × 1000 kg = 3650 mg/minute

Set carbon dioxide to oxygen ratio to 1:1.1 giving carbon dioxide production of 3650 mg/min × 1.1 = 4015 mg/minute

Using the equation and setting inlet concentration of carbon dioxide to 5 mg/l and the maximum amount in the outlet to be below 10 mg/l will give:

$$Q = \frac{CO_{2-rel} \times m + CO_{2-inlet}}{CO_2 - \text{outlet max}}$$

$$Q = \frac{4015 + 0}{10 - 5}$$

$$Q = 803 \text{ l/minute}$$

Compared to using water without carbon dioxide gas (as in the earlier example) the amount of new water to supply will be increased.

for instance, by the use of large propellers, is not of interest due to cost. Therefore, sites with no or very low water current are not suited for salmon production in sea cages. Although there will always be some current in the sea created by wind, tide, oceanic currents and other locally created currents, the degree will depend on each actual site. Currents that are too high will be a challenge both for the technical equipment and for the salmon. Currents above 1 m/second result in high environmental loads on the sea cage construction and on the mooring system, and in this case special considerations must be made. Specially designed offshore cages including a heavier mooring system can be used in such cases, but the salmon will also have problems if the water currents become too high. The current tolerance or swimming speed of the salmon is often given by fish length/body length per second, and this decreases with fish size. It is not the maximum swimming speed that is of interest, but the velocity that the salmon can tolerate over longer periods without negatively affecting fish welfare. Maximum swimming speed or critical swimming speed of post-smolt salmon is 2–4 body lengths per second, while welfare may be negatively affected by velocities over 1.5 body lengths per second [29]. An increase in water velocity on the site can be visually observed by the behaviour of the salmon, as the fish change behaviour from swimming in the traditional circular pattern to swimming against the water current. However, it is important to be aware that the current velocity outside the cages is higher than inside the cages because the net panel in the net bag reduces the velocity, and the more cages and net panels that the current has to pass, the more the current will reduce. The recommendation for current velocity inside the cage is over 0.2 m/second. Smaller fish will tolerate less current than a larger salmon, and therefore large salmon may be kept on more current-exposed sites than newly transferred smolt. Sites may therefore be divided into sites suitable for smolt and sites adapted for larger-sized salmon.

In open cages supply of pure oxygen gas will also be a possibility, utilizing special equipment for dissolving oxygen inside the cages. This is still not common, but interest is increasing especially due to increasing seawater temperatures caused by climate change. With high seawater temperatures, the available dissolved oxygen content will be reduced and at the same time the oxygen requirements of the salmon will increase. In Chile, oxygenation of sea cages has been introduced in commercial ongrowing farms.

9.3.2.4 Requirements for heating and cooling of water

Heating and cooling of the incoming water is possible in closed production units. It is then possible to have higher or lower water temperatures in the production units than the temperature of the raw water. This is an important tool in production planning and for optimizing production, but requires the farm to have the means to heat or chill the raw water. Equipment for temperature regulation also has a given capacity, meaning it cannot provide temperature regulation above its capacity. For smolt farming, heating of water is usually a necessity to speed up the growth in wintertime. This is also the case for RAS farms but here the heating requirement is reduced because it is mainly just the inlet water/dilution water that needs to be heated. However, on RAS farms there may be the need to chill the water in summertime depending on location and summer temperatures.

However, it is costly to both heat and chill water, and this has to be taken into consideration when making production plans that include requirement of temperature regulation. For each kilogram of water that is heated or chilled, 4.18 kJ (1.16 Wh) must be added or removed per degree Celcius of change.

Direct heating of the raw water can be performed by electric or gasoline boilers, but due to the large amount of heat that is needed heat pumps will normally be used. Quite high performance coefficients (COP), above 5, can be achieved because the temperature change will be quite low, but still it will be costly. Therefore, when heating water, reusing heat either through heat exchangers or by use of water reuse systems (RAS) will be a requirement. If using heat exchangers on the outlet water, improved overall COP values can be achieved. If possible, use of other temperate water sources will be an advantage, for instance, seawater can be used for heating freshwater on smolt farms in wintertime by using heat exchangers because the water temperature is higher in the sea in winter. Overall COP for the entire heating system, including heat pumps and several heat exchangers, can reach over 30, meaning that for each 1 kW of electric energy supplied, over 30 kW will be added to the water.

Box 9.22

A water flow of 100 kg/second with freshwater must be heated from 2°C to 12°C. Calculate the energy requirement:

$$E = \dot{m}C_p\,(T_{out} - T_{in})$$

$$E = 1000\,\frac{kg}{minute} \times 4.18\,\frac{kJ}{kg°C} \times (12°C - 2°C)$$

$$E = 4180\ kJ/second = 4180\ kW$$

Box 9.23

Instead of direct heating, a heat pump is installed to the example in Box 9.22. The heat pump has a COP of 5. The energy supplied with direct heating is 4180 kW; how much direct energy must be supplied when using the heat pump?

COP = Delivered energy / Supplied direct energy
Supplied direct energy = Delivered energy / COP
Supplied direct energy = 4180 kW / 5
Supplied direct energy = 836 kW

Chilling of water is even more energy consuming than heating and the only practical way to do this is by utilization of refrigeration plants. A refrigeration plant will have a lower COP than a heat pump for the same temperature difference. Refrigeration plants and heat pumps utilize the same technology but based on how they work, lower COPs will be achieved by refrigeration plants. The possibilities for using heat exchangers and chilling water with available cold water sources are therefore of great interest where chilling of water is a requirement in summer. A good COP for aquaculture refrigeration plants is around 4.

9.4 Requirements for production volume

To farm eggs and fish a production volume is required. This will be dependent on the production systems used, and the volume requirement will differ between eggs and fish. For fish it depends on the fish density that can be kept in the production systems while still securing optimal growth and fish welfare.

9.4.1 Egg incubation

As discussed in Chapter 5 on broodstock farms the size of the individual egg varies, but typically the volume requirement is in the area of 4000 to 6500 eggs per litre. The production volume requirement for incubation of eggs in the hatchery will, however, depend on whether the eggs are to hatch in the system or not. If it is only designed to produce eyed eggs, such as on broodstock farms that send their eyed eggs to smolt farms, this sets one requirement. But if the eggs are to hatch, the unit used must be suitable for this.

For storage of eggs from fertilization until eyed egg stage, it is typical to use hatching cylinders where the eggs are stacked in thick layers. These units can only be used for storing eggs, not hatching. The size of cylinder varies; one type can store 30 l of eggs (diameter 0.5 m and height 0.8 m) while others have a storage capacity of up to 120 l/m^2. But although the sizes vary, in simple terms it is only a large bucket, and larger ones can be used if the production volume is high.

A popular system for storing and hatching eggs is the California system that includes troughs with trays. The size and number of trays varies but a much-used type has a tray dimension of 40 cm × 40 cm and four to seven trays are put in each trough. The number of eggs in each tray is 1.5–2 l depending on egg size, but there are typically two layers of eggs in each tray. For Atlantic salmon, therefore, 6000 to 13,000 eggs are placed in each tray, depending on egg size. Two or three troughs may be positioned above each other in a room.

Box 9.24

Ffty thousand salmon eggs are to be placed into the hatchery, and a California system is used. How many troughs and trays must be used? How much water is needed?

> 5000 eggs per litre gives 10 l of eggs to be placed in the system. With a capacity of 1.5 l per tray the requirement will be 6.67 trays = 7 trays. Choosing a trough with seven trays requires a water supply of 12 l per minute.

A more area-intensive system is also used in the form of cabinets. There are a few different systems and designs available. In one system 4 l of eggs are stored per drawer with a water supply of 3–7 l, and eight double drawers with a size of 0.8 m × 1.6 m are placed on top of each other, with a total capacity of 360,000 eggs. This allows for over 280,000 eggs per square metre of floor area.

9.4.2 Tank requirement

On smolt farms and land-based ongrowing farms, tanks or basins are the typical closed units for holding salmon although other closed units can also be used, such as closed sea cages.

When the total biomass of the production is known, the necessary tank requirement can be calculated. When setting up tanks, larger tanks are cheaper per cubic metre of farming volume and therefore represent a lower initial investment. However, better expertise is required to operate large tanks and it is more difficult to monitor the individual fish. A number of tanks will be required in order to have the ability to separate salmon of different weights and sizes in different tanks when size grading for moving the fish and for vaccination. It will also be necessary to have some free tanks when grading the fish. The operational cost will be lower with a few large tanks compared to a larger number of smaller volume tanks, but the increased biomass in each tank will increase the consequences of any failure in a tank. For the start-feeding stage, some smaller tanks are recommended. In the past, a 1–2 m tank diameter was used and the tanks were shallow, under 1 m, but today with increased knowledge of Atlantic salmon farming and improved start feed, tanks greater than 5 m in diameter with a depth of over 2 m are used. Normal tank size for ongrowing is larger and typically ranges 5–25 m in diameter and 2–5 m in depth. Recommended height to diameter ratio of the tank is in the area 0.25:0.5.

The maximum fish density depends on the life stage, as discussed in earlier chapters (Table 9.13). However, there is some discussion about these values with recent findings and an increased focus on animal welfare (reviewed in [29,30]). For larger salmon, good growth has been achieved with salmon densities up to 100 kg/m^3 [31,32]. Effects on skin properties show no problem with a density of 100 kg/m^3, while over 100 kg/m^3 show indications of cellular stress [33]. Experiments on post-smolt with emphasis on production performance and welfare show negative effects with stocking densities at 100 kg/m^3 or above [34]. The problems that occur when operating with high densities is adequate supply of water and removal of metabolic waste so that optimal metabolism of the individual fish can be achieved. Access to feed for all individuals may also be a challenge, and efforts must be taken to ensure good feed distribution. For fry and parr, experiments have shown reduced performance at densities above 146 kg/m^3 [35] while densities from 21 to 86 kg/m^3 showed no negative growth effects [36]. For parr grown in RAS, densities below 100 kg/m^3 are suggested, at least if the temperature is above 13°C [37]. For post-smolt even lower densities may be suggested, but this is due to the problem of keeping carbon dioxide concentrations below recommended values.

The discussion about these values will probably continue in the future due to the high importance of fish density in relation to the profitability of the farm. The high initial investment required in the production units, such as for land-based ongrowing salmon farms, mean a constantly high biomass per volume unit is required to make them economically profitable. However, high densities create challenges, as more security is required and the response time to correct failure will be reduced, representing both additional cost and increased risk.

Table 9.13 Recommended maximum fish densities.

	Salmon density (kg/m3)
Start feeding number per m^3	8000–10 000
Early ongrowing <20 g	5–20
Ongrowing until smolt	15–40
Ongrowing sea cages	5–25
Ongrowing land based	40–100

9.4.3 Sea cage requirements

In sea cages densities up to 25 kg/m^3 are recommended, which is also the legislative limit in some countries [38]. There may also be limitations on the number of individuals in one cage unit independent of stocking density, like in Norway which has a limit of 200,000 individuals, due to the increased consequences of any disease.

Box 9.25

One hundred thousand fry are to be start fed. Calculate the need for production tanks.

$$100{,}000/10{,}000 = 10\ m^3$$

Use of a circular tank with a diameter of 2.8 m and a height of 1.6 m will fulfil this requirement.

Box 9.26

A farm is to deliver 1 million smolt of an average size of 125 g. Calculate the need for tanks when setting a maximum density of 35 kg/m^3.

Total biomass: 1,000,000 × 0.125 kg = 125,000 kg
Specific density: 35 kg/m^3

$$\text{Total tank volume: } \frac{25{,}500}{35} = 3571\ m^3$$

Use circular tanks 15 m in diameter and 5 m deep. Each tank has a volume of 883 m^3. Therefore:

$$\frac{3571}{883} = 4.044$$

In practice four tanks of 15 m diameter will be needed, even though the density will be slightly above 35 kg/m^3.

9.5 Internal transport and size grading

When making a production plan, ideally requirements for internal fish transport and size grading should be included. However, to include this in practice will be an additional challenge, because growth and mortality are estimated and it will be difficult to obtain exactly the same values. If including internal transport and size grading in the plan, the plan will be much more detailed and will then include setting up the individual production units and following the growth and mortality in each individual unit. When size grading, the fish size groups are divided into individual units. A timeline for each production unit including which fish group is in each unit for a given time must then be set up. Making a plan with this level of detail is very labour intensive and in addition will have a higher degree of uncertainty. Therefore, this is not normally done, but it is of major importance that it is well thought out, and is focused on below.

9.5.1 Reasons for moving fish

The number of fish that can be produced on fish farms depends on the fish density; higher fish density will result in higher production as long as it does not negatively affect growth, mortality or welfare. The more expensive the production volume, the greater the production per volume unit that is required (kilogram of fish per kilogram of production volume). As an example, a land-based fish farm for intensive farming with a high investment cost per cubic metre of farming volume requires a continuously high fish density. This requires frequent intake and output of fish, in addition to frequent movement or reallocation of fish inside the farm. The frequency of movement is mainly decided by the growth rate, with a higher growth rate requiring more frequent movement. A high average fish density which requires regular intake and output of fish is often seen in RAS smolt farms. It is a necessity to have both an eyed egg supply and to transfer fish to the sea all year round. For land-based farming of market size salmon, the same will be the case. This can be visualized with the following example where the maximum density for large salmon is 80–100 kg/m^3 and all environmental factors are optimal. The economic requirement for the farm is to keep an average density in the production volume per year of 30–45 kg/m^3. What will then happen if the fish are not going to be moved during the entire production cycle? The input density cannot rise above 1.5–2 kg/m^3. In Table 9.14 the number of months between each movement of fish with different SGR and different densities are presented in order to keep different average densities.

When fish are moved from the original production unit and handled, they become stressed, which of course is suboptimal. Nevertheless, many normal work routines on the farm, such as weight sampling, size grading and vaccinating, involve handling. All handling procedures and equipment should therefore be operated by qualified personnel and designed to minimize the stress response in the fish and the possibility of injury to the fish. Animal welfare recommendations must be followed, and it is important to have suitable conditions for the fish to recover as fast as possible after handling.

Table 9.14 Required moving frequency in months with different growth rates (SGR) to keep 60 kg/m^3 and 70 kg/m^3 average fish density in the production unit.

Density (min., avg, max.)	Specific growth rate					
	0.3	0.5	0.7	0.9	1.1	1.3
40, 70, 100	10–11	ca. 6	4–5	3–4	ca. 3	2–3
40, 60, 80	ca. 8	ca. 5	3–4	2–3	ca. 2	1–2

9.5.2 Reasons for size grading

In a large fish group, there will be a variation in the individual growth based on heritage which will result in a weight distribution. This can be approximated to be a normal distribution (Table 9.15). Size grading is the process of sorting the fish group into different weight or size classes. Fish need to be graded to avoid formation of a hierarchy within the group and to fulfil the growth potential of the entire group. This also enables smaller, less dominant individuals to get access to food and grow. It also prevents cannibalism, even though this does not present a problem in salmon farming as long as the fish have access to enough feed.

In large fish groups normal distribution can be used to estimate the weight distribution of individual fish. By knowing the average fish weight (x mean = $\bar{x}$) of the individuals in the group, the individual weight variation within the group, measured as standard deviation (Std = δ). The

Table 9.15 Normal distribution table.

X	0.000	0.01	0.02	0.03	0.04	0.05	0.06	0.07	0.08	0.09
0	0.500	0.504	0.508	0.512	0.516	0.520	0.524	0.528	0.532	0.536
0.1	0.540	0.544	0.548	0.552	0.556	0.560	0.564	0.567	0.571	0.575
0.2	0.579	0.583	0.587	0.591	0.595	0.599	0.603	0.606	0.610	0.614
0.3	0.618	0.622	0.626	0.629	0.633	0.637	0.641	0.644	0.648	0.652
0.4	0.655	0.659	0.663	0.666	0.670	0.674	0.677	0.681	0.684	0.688
0.5	0.691	0.695	0.698	0.702	0.705	0.709	0.712	0.716	0.719	0.722
0.6	0.726	0.729	0.732	0.736	0.739	0.742	0.745	0.749	0.752	0.755
0.7	0.758	0.761	0.764	0.767	0.770	0.773	0.776	0.779	0.782	0.785
0.8	0.788	0.791	0.794	0.797	0.800	0.802	0.805	0.808	0.811	0.813
0.9	0.816	0.819	0.821	0.824	0.826	0.829	0.831	0.834	0.836	0.839
1.0	0.841	0.844	0.846	0.849	0.851	0.853	0.855	0.858	0.860	0.862
1.1	0.864	0.867	0.869	0.871	0.873	0.875	0.877	0.879	0.881	0.883
1.2	0.885	0.887	0.889	0.891	0.893	0.894	0.896	0.898	0.900	0.901
1.3	0.903	0.905	0.907	0.908	0.910	0.911	0.913	0.915	0.916	0.918
1.4	0.919	0.921	0.922	0.924	0.925	0.926	0.928	0.929	0.931	0.932
1.5	0.933	0.934	0.936	0.937	0.938	0.939	0.941	0.942	0.943	0.944
1.6	0.945	0.946	0.947	0.948	0.949	0.951	0.952	0.953	0.954	0.954
1.7	0.955	0.956	0.957	0.958	0.959	0.960	0.961	0.962	0.962	0.963
1.8	0.964	0.965	0.966	0.966	0.967	0.968	0.969	0.969	0.970	0.971
1.9	0.971	0.972	0.973	0.973	0.974	0.974	0.975	0.976	0.976	0.977
2.0	0.977	0.978	0.978	0.979	0.979	0.980	0.980	0.981	0.981	0.982
2.1	0.982	0.983	0.983	0.983	0.984	0.984	0.985	0.985	0.985	0.986
2.2	0.986	0.986	0.987	0.987	0.987	0.988	0.988	0.988	0.989	0.989
2.3	0.989	0.990	0.990	0.99	0.990	0.991	0.991	0.991	0.991	0.992
2.4	0.992	0.992	0.992	0.992	0.993	0.993	0.993	0.993	0.993	0.994
2.5	0.9940	0.9940	0.9942	0.9943	0.9945	0.9946	0.9949	0.9949	0.9951	0.9952
2.6	0.9953	0.9955	0.9956	0.9957	0.9959	0.9960	0.9962	0.9962	0.9963	0.9964
2.7	0.9965	0.9966	0.9967	0.9968	0.9969	0.9970	0.9972	0.9972	0.9973	0.9974
2.8	0.9974	0.9975	0.9976	0.9977	0.9977	0.9979	0.9980	0.9980	0.9980	0.9981
2.9	0.9981	0.9982	0.9982	0.9983	0.9984	0.9984	0.9986	0.9985	0.9986	0.9986
3.0	0.9987	0.9987	0.9988	0.9988	0.9988	0.9989	0.9990	0.9989	0.9990	0.9990

Box 9.27

In a sea cage with 10,000 individual fish the average fish weight is measured as 4 kg and the standard deviation is measured to be 1 kg. Find the number of fish in the weight class of 1 kg using the normal distribution table (Table 9.15).

First the number of fish <5 kg is x + δ.

From the normal distribution table, use the value of δ = 1 which gives us 0.841.

This means that 84.1% of the fish have a weight <5 kg and therefore 15.9% must be larger than this weight. So, of the 10,000 fish, 1590 are >5 kg.

Similarly, the number of fish >3 kg can be found as x – δ.

This means that 1590 fish are also <3 kg, since the value of δ = 1.

In the area between 3 kg and 5 kg there will therefore be 10,000 – 1590 – 1590 = 6820 fish.

With half in each weight class, there are 3410 fish in weight class 3–4 and class 4–5 respectively.

Next, the number of fish >6 kg and <2 kg is x + 2δ.

2δ = 2, which in the normal distribution table gives 0.977.

97.7% of the fish are >6 kg and <2 kg, so 2.3% of the population will be >6 kg and 2.3% of the population will be <2 kg, which is equal to 230 fish.

distribution of fish in different weight classes can then be estimated based on the normal distribution table.

9.5.2.1 Knowledge about size distribution

The coefficient of variation (CV) shows the standard deviation (δ) in relation to the average value (x- mean) and is a way to illustrate the relative size variation in a fish group. In smolt production a coefficient of variation (CV) of up to 100% is measured in non-size-graded fish groups. For this reason, frequent grading ought to be performed. This is typically done at least three times during the production cycle on a smolt farm. If size grading is neglected during the freshwater phase, the size distribution will change from a normal distribution to a kind of bimodal distribution, with a group of 'winners' and a group of 'losers'. On ongrowing farms less grading is necessary, because larger fish need more time to grow and therefore more time is needed for the CV of the fish group to increase. A CV above 30% is seldom the case in ongrowing farms.

9.5.2.2 Advantages and disadvantages with size grading

There are several good reasons for size grading. As stated above, a bimodal size distribution with winners and losers may develop. But it is also a great advantage for regular production control to have a small size variation (low CV) within the fish group. Good production control requires representative weight sampling in fish groups to monitor the growth and feed conversion rate. However, in a fish group with a large CV it will be very difficult to obtain representative weight samples, and a large number of fish are needed to get a reliable result; it is therefore important to maintain homogenous-sized fish groups.

Low variance in the size of fish in a group allows for a higher yield of supplied feed due to more optimal feed particle size. At slaughter, it is also an advantage to have a limited size variation in order to more accurately estimate fish size prior to slaughter; this is an important parameter to report to the slaughterhouse. It is also an advantage to have a homogeneous size distribution at slaughter. For example, if the average individual fish weight of the group is 4 kg and the CV is 25% there will theoretically be 6 different weight classes. This situation makes it very difficult to manage and then grading before slaughtering will usually be required.

Box 9.28

A fish group contains 10,000 individuals; assuming that representative samples are withdrawn, how many single fish are needed in a sample if the average weight with 5% security is to be found (supposing normal distribution) with different CV in the fish group?

CV%	20	30	50	100
Number of Fish	64	144	400	1600

9.6 Example of smolt production plan

Box 9.29 presents a full production plan displaying the necessary calculations.

Box 9.29

A production plan for a smolt farm must be made. The farm is supplied with eyed eggs.

Conditions:

The farm must deliver 500,000 smolt
The smolt are to be 1 year old when delivered
The eyed eggs are delivered on 12 November

Mortality until hatching is set to 5%, from hatching to start feeding 10% and from start feeding to delivered smolt 10%
All oxygen must be supplied by the inlet water
The specific growth rate is given by table (here an older feed table is used [3])
Inlet water is heated to 8°C from eyed egg stage until beginning of start feeding, afterwards heated to 12°C until the raw water equals this temperature
The raw water has the following temperatures (°C): Jan 1, Feb 1, Mar 1, Apr 2, May 8, Jun 10, Jul 12, Aug 12, Sept 8, Oct 6, Nov 4, Dec 2

Solution

Hatching time

Eggs arrive 12 November. (This should normally be earlier in the year)
Water temperature: 8°C
Time from input to hatching is set to 500D°C.
Number of days from inlet to hatching: 500/8 = 62 days
Point of time for hatching: 12 November + 62 days = 15 January

Point of time for start feeding

From hatching to start feeding 210D°C
210/8 = 26 days
Start feeding will begin 15 January + 26 = 10 February

Point of time to finish start feeding

Start-feeding endurance: 500D°C
Water temperature: 12°C
500/12 = 42 days
10 February + 42 = 24 March

Calculation of growth

We expect that the fish weighing 0.18 g to start to grow from 15 February
SGR table value is 3.1 %. The weight of the fish on 1 March, after 13 days:

$$0.18 \times (1 + 3.1 / 100)^{13} = 0.27 \text{ g}$$

These same procedures must be followed further in the period.

Counting fish

The next step is to calculate the amount of eggs that must be taken in. Here we must do a retroactive calculation, and for this it is important to know the basis of the percentage calculation.

The number of fish required is 500,000. The mortality from the end of start feeding to smolt is 10%. The amount of fish at the end of start feeding (x_1) is thereby:

$$x_1 \times (100 - \frac{10}{100}) = 500{,}000$$

$$x_1 = 555{,}556$$

This mortality occurs over 14 months, which corresponds to 0.75% a month.

Then we calculate the number of fish ready for start feeding (x_2):

$$x_2 \times (100 - \frac{10}{100}) = 555{,}556$$

$$x_2 = 617{,}284$$

This mortality occurs over 42 days, which corresponds to 0.25% per day.

The number of eggs ready for hatching (x_3) is:

$$x_3 \times (100 - \frac{5}{100}) = 617{,}284$$

$$x_3 = 649{,}773$$

The period is approximately 26 days which is 0.2 % mortality per day.

The amount of eggs that have to be put in (x_4) is:

$$x_4 \times (100 - \frac{5}{100}) = 649{,}773$$

$$x_4 = 683{,}971$$

The mortality per day is calculated to 0.08%.

Equipment for eggs

Necessary volume of eggs 683,971 / 5000 eggs per litre = 137 l
Number of hatching trays with 2 l in each tray is 137 / 2 = 69
If we choose troughs with seven trays and we need ten troughs
Water requirement 12 l/minute × 10 = 120 l/minute

Water requirement for the fish

0^+ in June as an example
Size = 4.31 g and temperature of 12°C
Water requirement 1.8 l/minute/kg fish

Table 9.16 Complete production plan: temperature (°C), weight (grams), number (units), biomass (tonnes), water requirement (m3/min), tank volume (m3).

Date	Jan	Feb	Mar	Apr	May	Jun	Jul	Aug	Sep	Oct	Nov	Dec
Raw water temperature (°C)	1	1	1	2	8	10	12	12	8	6	4	2
Egg water temperature (°C)											8	8
0^+ temperature (°C)	8	8.0/12.0	12	12	12	12	12	12	8	6	4	2
1^+ temperature (°C)	1	1	1	2	8	10						
Weight 0^+ (g)	0.2	0.18	0.27	0.69	1.72	4.31	9.87	22.5	48.5	69.3	88.7	97.3
Weight 1^+ (g)	103	107	110	114	117	158						
Number of eggs											683,971	674,419
Number of 0^+	657,343	629,583	590,031	554,732	550,631	546,561	542,521	538,510	534,529	530,578	526,656	522,762
Number of 1^+	520,281	518,898	511,254	507,475	503,725	500,000						
Total number	1,177,624	1,143,790	1,101,285	1,062,189	1,050,285	1,046,561	542,521	538,510	534,529	530,578	526,656	522,762
Biomass eggs											1371	1371
Biomass 0^+ (kg)	131	113	158	383	949	2355	5353	12,135	25,897	36,766	46,730	50,887
Biomass 1^+ (kg)	53,739	55,020	56,332	57,675	58,991	78,924						
Total biomass (kg)	53,870	55,133	56,490	58,050	59,940	81,279	5353	12,135	25,897	36,766	46,730	50,887
Water requirement eggs											120	120
Water requirement 0^+ (l)	221	364	507	1228	2230	4239	8029	16989	15020	13604	11213	7633
Water requirement 1^+ (l)	6986	7153	7323	8651	29,496	58,404						
Total water required (l)	7207	7517	7830	9879	31,726	62,643	8029	16,989	15,020	13,604	11,213	7633
Hatch trays											69 trays	69 trays
Tank volume 0^+	69 trays	63 m^2	59 m^2	38	95	236	535	607	1295	1838	1557	1696
Tank volume 1^+	1791	1834	1878	1923	1966	2631						
Total tank volume (l)	1791	1834	1878	1961	2061	2867	535	607	1295	1838	1557	1696

Total water requirement
= Number of fish × Fish size × Specific amount of water
= 546,561 × 4.31 × 1.8 = 4239 l/minute
We calculate the corresponding values for each month and add for eggs, 0^+ and 1^+.

Tank volume
April for 1^+ as an example
Total weight of the fish is 56,675. Maximum density 30 kg/ m^3.

Necessary volume: $\frac{57,675}{30} = 1923\ m^3$

The total production plan is shown in Table 9.16.

References

1. Elliott, J. and Elliott, J.A. Temperature requirements of Atlantic salmon *Salmo salar*, brown trout *Salmo trutta* and Arctic charr *Salvelinus alpinus*: predicting the effects of climate change. *Journal of Fish Biology*. 2010. 77(8): 1793–1817.
2. Aunsmo, A. et al. Field validation of growth models used in Atlantic salmon farming. *Aquaculture*. 2014. 428: 249–257.
3. Austreng, E., Storebakken, T. and Åsgård, T. Growth rate estimates for cultured Atlantic salmon and rainbow trout. *Aquaculture*. 1987. 60(2): 157–160.
4. Forsberg, O.I. Empirical investigations on growth of post-smolt Atlantic salmon (*Salmo salar* L.) in land-based farms. Evidence of a photoperiodic influence. *Aquaculture*. 1995. 133(3–4): 235–248.
5. Jobling, M. The thermal growth coefficient (TGC) model of fish growth: a cautionary note. *Aquaculture Research*. 2003. 34(7): 581–584.
6. Lall,S.Atlanticsalmon,*SalmoSalar*.http://www.fao.org/fishery/affris/species-profiles/atlantic-salmon/atlantic-salmon-home/en/.
7. Fôr Skretting – fôrkatalog. 2012.
8. Ewos Innovation, Ewos growth index. 2014. Cargill.
9. Einen, O. and Mørkøre, T. Fôringslære for akvakultur. 1997. Landbruksforl.
10. Thorarensen, H. and Farrell, A.P. The biological requirements for post-smolt Atlantic salmon in closed-containment systems. *Aquaculture*. 2011. 312(1–4): 1–14.
11. Remen, M. The oxygen requirement of Atlantic salmon (*Salmo salar* L.) in the ongrowing phase in sea cages. 2012. The University of Bergen.
12. Liao, P.B. Water requirements of salmonids. *The Progressive Fish-Culturist*. 1971. 33(4): 210–215.
13. Kittelsen, A. and Fjæra, S.O. Oxygen and water requirement, in *Teknologi for akvakultur*, O.-I. Lekang and S.O. Fjæra (eds). 1997. Landbruksforl.
14. Christiansen, J.S., Jobling, M. and Jørgensen, E. Oksygen og vannbehov. Nye reviderte tabeller (In Norwegian), in *Norsk Fiskeoppdrett*. 1990, pp. 28–29.
15. Grisdale-Helland, B. et al. Influence of high contents of dietary soybean oil on growth, feed utilization, tissue fatty acid composition, heart histology and standard oxygen consumption of Atlantic salmon (*Salmo salar*) raised at two temperatures. *Aquaculture*. 2002. 207(3–4): 311–329.

16. Forsberg, O.I. The impact of varying feeding regimes on oxygen consumption and excretion of carbon dioxide and nitrogen in post–smolt Atlantic salmon *Salmo salar* L. *Aquaculture Research*. 1997. 28(1): 29–41.
17. Timmons, M.B., Ebeling, J.M., Wheaton, F.W., Summerfelt, S.T. and Vinci, B.J. *Recirculating Aquaculture Systems*, 2nd Edition. 2007. Cayuga Aqua Ventures.
18. Remen, M. et al. The oxygen threshold for maximal feed intake of Atlantic salmon post-smolts is highly temperature-dependent. *Aquaculture*. 2016. 464: 582–592.
19. Lygren, B., Hamre, K. and Waagbø, R. Effect of induced hyperoxia on the antioxidant status of Atlantic salmon *Salmo salar* L. fed three different levels of dietary vitamin E. *Aquaculture Research*. 2000. 31(4): 401–407.
20. Fridell, F. et al. Effect of hyperoxygenation and low water flow on the primary stress response and susceptibility of Atlantic salmon *Salmo salar* L. to experimental challenge with IPN virus. *Aquaculture*. 2007. 270(1–4): 23–35.
21. Lekang, O.-I., *Aquaculture Engineering*. 2020. John Wiley and Sons: Chichester.
22. Crisp, D.T. Environmental requirements of common riverine European salmonid fish species in freshwater with particular reference to physical and chemical aspects. *Hydrobiologia*. 1996. 323(3): 201–221.
23. Bergheim, A. and Fivelstad, S. Atlantic salmon (*Salmo salar* L.) in aquaculture: metabolic rate and water flow requirements, in *Salmon: Biology, Ecological Impacts and Economical Importance*, P.T.K. Woo and D.J. Noakes (eds). 2014. Nova Science Publishers: Hauppauge, NY, pp. 155–173.
24. Fivelstad, S. and Binde, M. Effects of reduced waterflow (increased loading) in soft water on Atlantic salmon smolts (*Salmo salar* L.) while maintaining oxygen at constant level by oxygenation of the inlet water. *Aquacultural Engineering*. 1994. 13(3): 211–238.
25. Summerfelt, S.T. et al. A partial-reuse system for coldwater aquaculture. *Aquacultural Engineering*. 2004. 31(3–4): 157–181.
26. Working, E.I.F.A.C. Water quality criteria for European freshwater fish: report on ammonia and inland fisheries. *Water Research*. 1973. 7(7): 1011–1022.
27. Ebeling, J.M. and Timmons, M.B. *Recirculating Aquaculture*. 2010. Cayuga Aqua Ventures.
28. Molleda, M.I., Thorarensen, H. and Johannsson, R. Water quality in recirculating aquaculture systems (RAS) for arctic charr (*Salvelinus alpinus* L.) culture. The United Nation University, Fisheries Training Programme, 2008.
29. Noble, C. et al. Welfare Indicators for farmed Atlantic salmon: tools for assessing fish welfare. 2018. http://hdl.handle.net/11250/2575780.
30. Thorarensen, H. and Farrell, A.P. The biological requirements for post-smolt Atlantic salmon in closed-containment systems. *Aquaculture*. 2011. 312(1): 1–14.
31. Berg, A.J. et al. Effect of stocking density, oxygen level, light regime and swimming velocity on the incidence of sexual maturation in adult Atlantic salmon (*Salmo salar*). *Aquaculture*. 1996. 143(1): 43–59.
32. Kjartansson, H. et al. Effects of different stocking densities on physiological parameters and growth of adult Atlantic salmon (*Salmo salar* L.) reared in circular tanks. *Aquaculture*. 1988. 73(1): 261–274.
33. Sveen, L.R. et al. Impact of fish density and specific water flow on skin properties in Atlantic salmon (*Salmo salar* L.) post-smolts. *Aquaculture*. 2016. 464: 629–637.
34. Calabrese, S. et al. Stocking density limits for post-smolt Atlantic salmon (*Salmo salar* L.) with emphasis on production performance and welfare. *Aquaculture*. 2017. 468: 363–370.
35. Soderberg, R.W., Meade, J.W. and Redell, L.A. Growth, survival, and food conversion of Atlantic salmon reared at four different densities with common water quality. *The Progressive Fish-Culturist*. 1993. 55(1): 29–31.

36. Hosfeld, C.D. et al. Effects of fish density on growth and smoltification in intensive production of Atlantic salmon (*Salmo salar* L.). *Aquaculture*. 2009. 294(3–4): 236–241.
37. Bæverfjord, G. et al. Fiskevelferd i RAS: Effekt av vannhastighet, tetthet og temperatur. 2012.
38. Oppedal, F., Dempster, T. and Stien, L.H. Environmental drivers of Atlantic salmon behaviour in sea-cages: a review. *Aquaculture*. 2011. 311(1): 1–18.

CHAPTER 10

Feeding

10.1 Introduction

Feed costs represent over 50% of the production cost for salmon, so feed and feeding is of major importance. The feed and its composition must be optimal; the nutrients in the feed must be transformed into muscle and the feed has to be adapted to the needs of the salmon to ensure growth, reproduction and health. Different life stages will require a different feed composition. As an example, the feed must contain ingredients (typically astaxanthin) that ensure the salmon meat is red. Salmon cannot digest much carbohydrate as they are a carnivorous species, adapted to eat animals. The size and development stage of the salmon fry when they have consumed the yolk sac means they do not require live feed for start feeding, unlike several fresh- and seawater species. Formulated dry feed is a good start feed for salmon.

Only complete diets are used in salmon farming today, meaning a single feed supplies all the needs of the salmon and there is no need for supplemental feeding.

The feed is not only important for optimal growth: if suboptimal feed is used, it can result in health problems. The waste production and waste composition also depends on the feed, with low feed digestibility increasing waste production. A rough rule of thumb for the salmon feed currently available is that around 20% of the weight of the feed ends up as waste on a dry matter basis.

When a feed particle is supplied to a salmon production unit a sequence of processes occur (partly based on [1]):

1. ingestion – the salmon takes up the feed particle
2. digestion – the particle is broken down into smaller molecules in the gut
3. absorption – the solute of the broken-down particles is transferred through the gut to the blood for further transport within the body of the fish
4. assimilation – the absorbed nutrients are utilized
5. egestion – the undigested parts and non-absorbed feed (non-metabolic waste) are discharged from the fish's body
6. the end products of metabolism (metabolic waste) are released in the form of gas, liquid and heat.

This chapter gives an overview of salmon nutrition. For more in-depth knowledge, a number of textbooks both on general fish nutrition and specially related to salmonids are available (as example [2–7]). Review articles/reports are also a good source of further knowledge (as example [1,8–10].

10.2 Fish metabolism

A basic understanding of fish metabolism is required in order to understand how fish grow, how feed is composed and how feed is taken up. In addition to water, the two major elements required for metabolism are food (nutrients, including both organic and inorganic elements) and oxygen. If the fish's body is compared with a factory these represent the three major inputs in the production process. Metabolism can be defined as a biochemical process that occurs in the salmon to maintain its growth and life. During metabolism, nutrients are broken down and energy is produced. A simplified explanation is that large molecules are divided into smaller ones and during this process energy is released. All main nutrients (protein, lipids and carbohydrates) contain a 'skeleton' including carbon bonds, where energy is stored. To break these down, oxygen is required; it is therefore an oxidative process, like a combustion process. Chemical energy is released; used to fuel processes inside the body of the salmon for growth (building up new substances based on these broken-down substances), activity and reproduction. Metabolism can also be defined as catabolism (the breakdown of substances) and anabolism (the build-up of substances) (Figure 10.1), while growth and reproduction are defined as separate processes, but here the transition is fluid. If too low an amount or poorly balanced nutrients are supplied, energy production, growth and fish health will all be affected. If no nutrients are supplied the processes will continue, but now the already established substances are used as fuel, meaning the salmon will break down its own muscle tissue and use it for production of energy to run its essential body functions. If no oxygen is supplied, the processes will change from aerobic to anaerobic and the result will be production of lactate. In high concentrations lactate is toxic and therefore this process can only occur for short time periods.

In addition to the nutrients, oxygen and water, the working conditions for metabolism must be within acceptable parameters: simply explained, there must be good working conditions in the factory. For instance, if a salmon fry is put into seawater, the conditions for production are not met (a fatal outcome). The same will be the case if the water temperature is too high. Suboptimal water quality or disease outbreak will have the same effect.

For metabolism to function, it is also necessary for salmon to be able to get rid of the produced waste products, as these are all toxic in high concentrations. An example here is the level

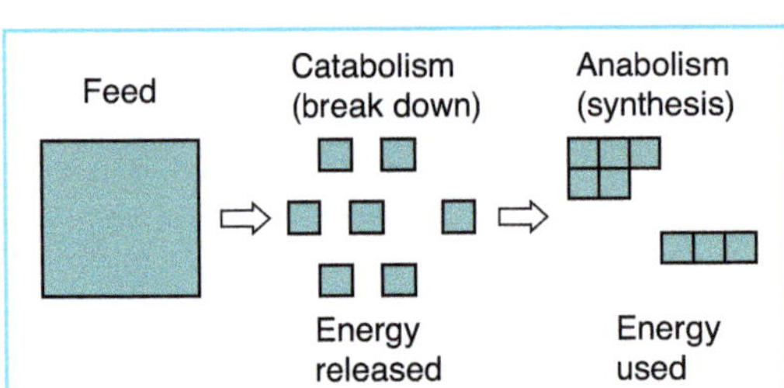

Figure 10.1 The feed is first decomposed into smaller molecules (catabolism), releasing energy, then some molecules are added together to create new substances (anabolism), a process that requires energy.

of carbon dioxide, a waste product from oxidation. If the concentration in the water around the gills is too high the salmon cannot get rid of the carbon dioxide produced through their metabolism and this will reach toxic levels for the fish.

When the necessary basic conditions for metabolism are met, the metabolic rate will depend on several factors. These also set the nutrient and oxygen requirements for the salmon under different conditions. Below are listed important factors affecting the metabolic rate; these are described more later (partly based on [11]):

- size of salmon: metabolic rate drops with increased size
- age of salmon: metabolic rate drops with increased age
- activity of salmon: metabolic rate increases with increased activity
- condition of the salmon: a fish in poor condition (low condition factor, thin fish) will have a higher metabolic rate
- surrounding environment of the salmon: higher temperatures and higher oxygen concentrations increase the metabolic rate within limitation
- stress level: a stressed fish will have a higher metabolic rate, for instance due to suboptimal water conditions, disease problems or handling.

10.3 Feed type and feed production

Feed can be divided into three types or categories: wet feed, moist feed and dry feed. The first two are made and stored on the farm while the last is made in large feed factories supplying the entire industry. Dry feed has a low water content (DM above 90%) and is stable in storage (with limitations described later), while the other two types contain much water, and must either be stored frozen or made regularly on the farm. In the past, wet and moist feed were partly used, especially in the early days of the salmon industry. Dry feed became commercially available early in the development of salmon farming and has been of major importance for the development and success of the industry. When wet feed was used it was mostly whole fish (trash fish), stored frozen, while moist feed contained milled fish mixed with binders and some additional nutrients, vitamins and minerals. Trials were performed using moist feed containing acid-preserved fish silage in the 1980s but this was not further developed [12]. More recently, there has also been some interest in using moist feed due to the possibility of a lower price and the utilization of local low-cost by-products from the traditional white fish industry. One such product includes alkaline fish silage (not from salmon) mixed with binders, vitamins and minerals [13], from which a soft pellet is made (Figure 10.2).

However, dry formulated feed totally dominates the salmon farming industry today. Dry feed has a DM content above 90% (92–93%) and is quite stable in storage, but cannot be kept too long partly due to possible oxidation of the lipids and breakdown of vitamins. The dry feed available today is a complete formulated diet containing all that the salmon need for optimal growth and welfare, and is their sole diet. It contains protein, fat, carbohydrates, vitamins and minerals, in addition to binders that hold the feed particles together to ensure technical quality.

Commercial dry feed for salmon used to be a pelleted feed made by mixing dry and wet raw materials into a dough, which was then placed into a pellet press, where it was pressed through

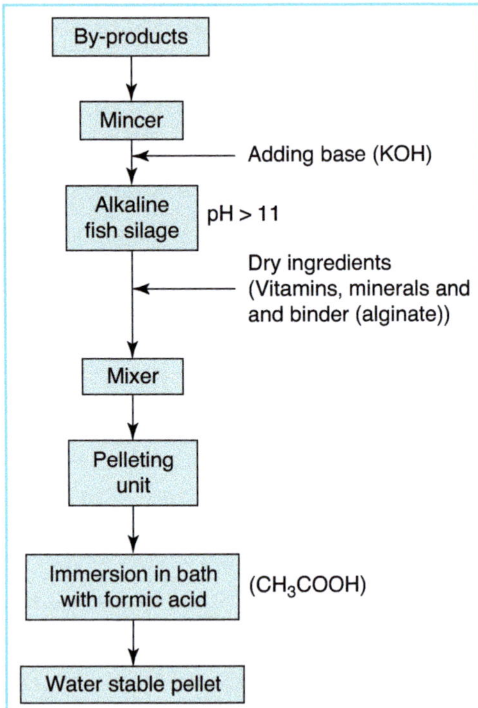

Figure 10.2 The production process for production of moist pellets based on by-products (partly based on[13]).

ring dies to produce the pellets. The pellets were then dried. Today, however, the pellet press has been exchanged for an extruder (Figure 10.4). One reason for this is the ability to obtain a higher fat content, which is required in today's energy-rich salmon feed. With a pellet press, a lipid content above 15% is not possible, even when utilizing preconditioners [12]. Another advantage of the extruder is that more pre-processing of the ingredients is possible. This can increase the digestibility of the raw materials/ingredients included in the feed, especially starch digestibility. Owing to the high temperatures in the extruder, a decrease in the amount of antinutrients and bacteria has been documented. The drawback is that heat-sensitive ingredients can be destroyed, but this is solved by coating the pellets with these after the extruder process.

In the extruder the ingredients are mixed and both heat (steam) and pressure are added. The raw material is melted and pressed through the die in the extruder that restructures it in a way that enhances the digestibility of the ingredients. The high pressure and temperature increases the pasteurization degree of the ingredients, improves starch gelatinization and may improve inactivation of antinutrients [9]. The high temperature combined with pressure can, however, have a negative effect on temperature sensitive ingredients, destroying them, so care should be taken. The density of the feed particle produced may be controlled by varying the porosity of the feed particle, determining whether it floats or sinks; typically, the latter is used for salmon. After leaving the extruder, the feed can be sent into a vacuum coater where additional fat can be added and sucked into the feed particle.

A pellet produced in an extruder will have a higher technical quality, and is less likely to break into finer particles and dust when exposed to forces compared to a pellet produced in a pellet press [12]. The water stability is also increased.

Figure 10.3 A production factory specializing in production of dry formulated salmon feed.

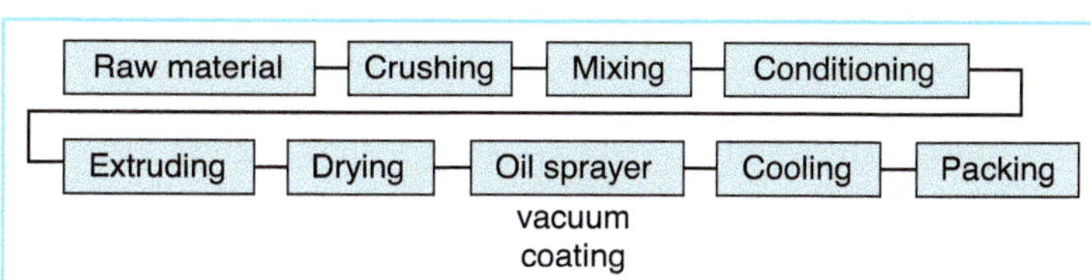

Figure 10.4 Production process for extruded feed.

A variety of feed particle sizes are used, depending on the size of the salmon. The size of the pellet is decided by the die in the extruder (this can be changed), while the length of the pellet can also be adapted. The pellet may also be crushed to produce crumbs of a size that is suitable for salmon fry. Micropellets can, however, also be used for fry and are a better solution. In general, particles with as large a size as possible should be used when feeding salmon. As a guide, the particle size of the produced pellet can be 20–30% of the fish mouth gape [14].

10.4 Composition of the complete feed

The composition of the feed is based on the requirements of the fish. Since it is a complete feed it must contain all essential ingredients needed for optimal growth. The nutrients are used to build muscle and body tissue (meat) and bones as the fish grows and as dead cells are replaced. The feed is also used as energy for metabolism: running the system or the 'factory'. If we consider the fish's body as an advanced processing factory it need inputs to run and to produce outputs. Some special nutrients (micronutrients, which are minerals and vitamins) are also necessary for certain processes within the fish, for example the building of cell walls. This can be visualized as oil to lubricate the machines in the factory. In addition, binders are added. They do not all have nutritional value, their purpose being to ensure that the feed particles are of good quality by binding the ingredients together to create a particle size that the salmon will eat effectively.

When the salmon eat the feed particles, the larger molecules have to be broken down throughout the digestive system so they can be absorbed and used by the fish. The materials are hydrolysed in the gut (the gastrointestinal [GI] tract), protein is broken down to amino acids and simple peptides, fat is broken down to free fatty acids and glycerols, while carbohydrates are broken down to monosaccharides, such as glucose.

10.5 Building new muscle

Most of the salmon's body is muscle, which is mainly protein, so the fish need protein or amino acids (the building blocks of protein) in the feed. The salmon's body also contains some fat, or lipids, and this must also be supplied with the feed. However, it is not essential to add carbohydrates as there are almost none of these in the salmon's body (although small amounts of lipids may be built from carbohydrates). The salmon cannot synthesize all necessary essential substances (muscle, organs and other necessary body parts) from the separate constituent elements (nitrogen, carbon, hydrogen, and oxygen). In addition, the salmon will require special additives (micronutrients) for specific purposes, for example phosphorus and calcium for bone formation.

In addition to supplying material for the construction of the body, energy for metabolism (or the running of the factory) must be supplied. Nutrients for metabolism are supplied by adding energy-rich nutrients to the feed. The study of how energy flows within a biological system is named bioenergetics, and is described more later [6,15,16]. What is known from the basic laws of energy is that there has to be an energy balance within a system. The fish does not make any energy itself, therefore all energy has to be supplied in the food and transformed to useable energy within the fish's body (catabolism). When the fish eat, the given amount of energy in the food is transformed to another form. For instance, when fat is broken down through metabolism, energy is released and this released energy can then be used for other purposes within the body where energy is required. All transformation of energy will result in production of heat to some degree.

The main ingredients included in the salmon feed are protein, fat and carbohydrates, which are classified as macronutrients based on the amount that the salmon need. In addition, the salmon need micronutrients (minerals and vitamins) but the required amounts are much lower. The main purpose of protein, as mentioned, is building muscle; however, if too large a fraction of protein is added it will be used for energy production (see section 10.11). The main purpose of fat is energy production but some will also be deposited, while the purpose of carbohydrates is purely energy production. The gut microbiota (population of microbes living in the gut) of salmon is not suited for digestion of carbohydrates [17]. It is mainly the non-starch part of

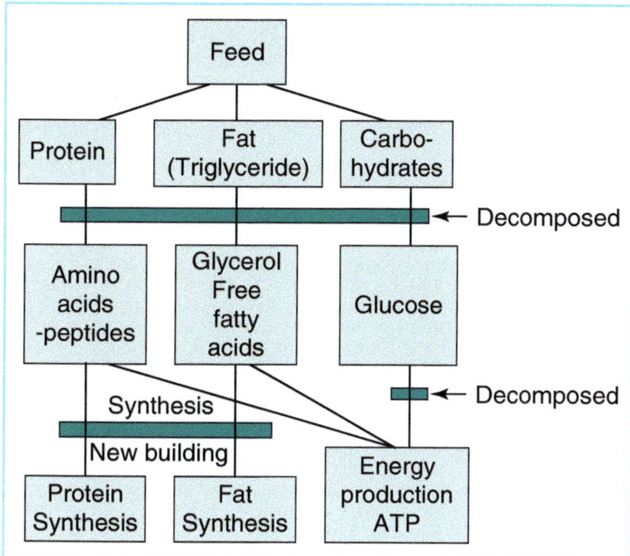

Figure 10.5 Flow chart of how the feed is utilized with the main nutrients: protein, fat and carbohydrates.

carbohydrates that salmon can digest (described later). It is necessary to add trace amounts of some vitamins and nutrients in the feed to ensure correct functioning of the body and normal growth. This is also important for reproduction.

In summary, the feed will be broken down to simpler components (catabolism); these are either used as building blocks in new components or for energy production (ATP, described later) (Figure 10.5). In addition, there will be waste products from the catabolism, which include substances not needed for building blocks or which cannot be used for energy production. The building blocks will be used for anabolism, which involves building new molecules for growth and replacement of dead cells. The energy produced, stored as ATP, will be used to build new molecules, to run processes inside the body and for activity.

10.6 Main components of the feed

10.6.1 Protein

Protein is the main ingredient for the growth of salmon. In general, the protein requirement is higher for fish than for humans and animals. Protein is also used in a number of other areas in salmon, such as in hormones and enzymes (Figure 10.6). A limited amount will also be used for energy production, and the waste products of the oxidation of protein for energy production will be ammonia, carbon dioxide and water. The proteins in the muscle are composed of a number of various amino acids that are linked together with peptide bonds. Salmon do not take up protein directly; the protein supplied with the feed is broken down in the intestine into amino acids and smaller peptides of a molecular size that can be absorbed (Figure 10.8). These are taken up and rebuilt (synthesized, termed protein synthesis) inside the fish cells, unless they are deaminated and used for energy production. Amino acids can be divided into essential or indispensable

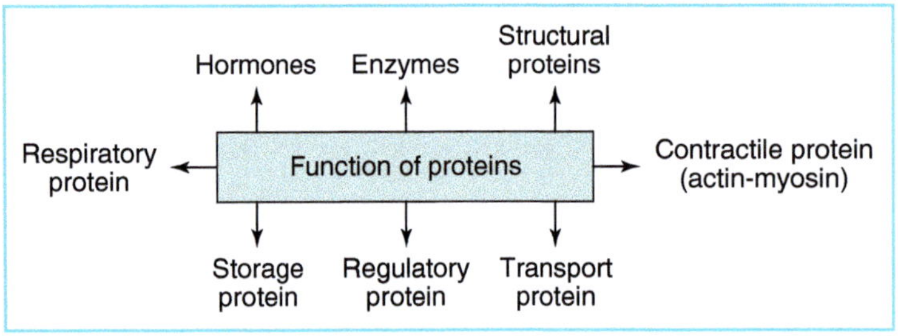

Figure 10.6 Protein is used for many biological functions in the body of the salmon.

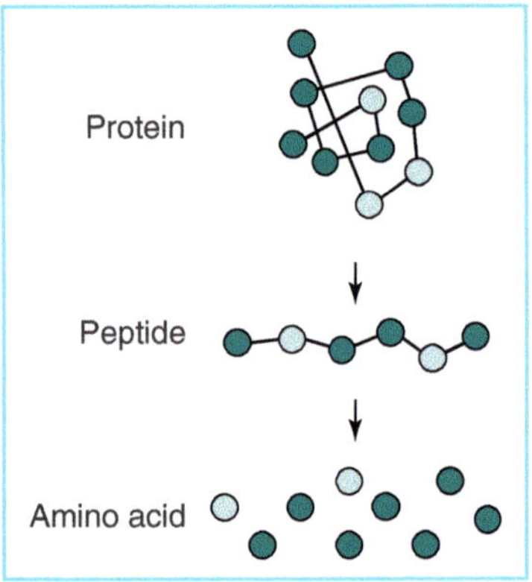

Figure 10.7 The structure of amino acids, peptides and protein. First protein is broken down to amino acids then these are reconstructed to form muscles.

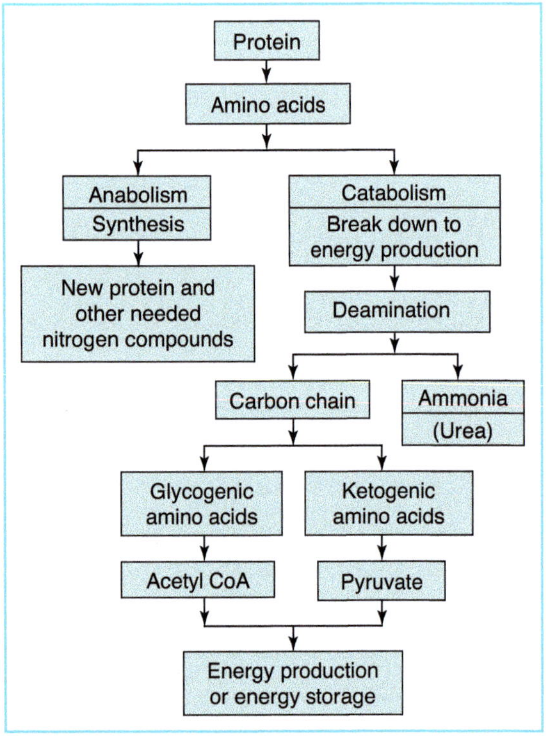

Figure 10.8 Pathways for protein/amino acid metabolism.

amino acids (IAA) and non-essential or dispensable amino acids (DAA). Essential amino acids cannot be synthesized, so must be included in the feed; 10 amino acids are essential for fish [7]. It is important that the IAA to DAA ratio is adequate: a ratio of 50–60 : 40–50 seems satisfactory [9]. The IAA group includes the following amino acids: methionine, lysine (often first limiting), arginine, threonine, tryptophan, histidine, isoleucine, leucine, valine and phenylalanine. If the protein in the feed does not contain these essential amino acids, the feed will be suboptimal. The fish also need a number of non-essential amino acids, but these can be synthesized from other sources and include alanine, asparagine, aspartic acid, glutamic acid, glutamine, glycine, proline, serine and tyrosine [8]. No growth will be achieved if the food source does not include protein, because protein synthesis will stop. This can be illustrated with a dam constructed with vertical staves linked together. If the length of one of the staves is shorter than the rest, the water in the dam will drain at that point and it will not be possible to fill up the dam: no IAA means an incomplete dam. Lack of essential amino acids may also result in disease such as cataract, which is caused by deficiencies of methionine, tryptophan or histidine.

Salmon, which are carnivorous fish, have higher requirements for protein than herbivorous (plant eating) species. Fry and juveniles have a higher protein requirement than ongrowing salmon because the growth rate is higher and relatively more muscle has to be built; therefore, the composition of the feed differs with fish size. The need for protein is also dependent on the water temperature, and for ongrowing salmon the protein requirement drops, one example shows a drop of 10–15% [153]. There seems to be a linear relationship between protein content of the feed and growth until a certain protein level is reached, which means more protein in the feed will result in increased growth (Figure 10.9). As well as being necessary for growth, the amino acids taken up by the fish are also needed for general maintenance and replacement of dead cells, and a limited amount of the protein is used for energy production. Through evolution, salmon have adapted to high protein diets as a carnivorous species, eating smaller fish, so are not adapted to only use fat or carbohydrates for energy production. Amino acids are used as precursors for glucose digestion, so some protein will always be used for energy production. The ratio between digestible protein (DP) and digestible energy (DE) is an important value (DP : DE). If too much protein is used in relation to the total energy content of the feed, more protein will be used for energy with higher ammonia release over the gills as a result. Production of energy

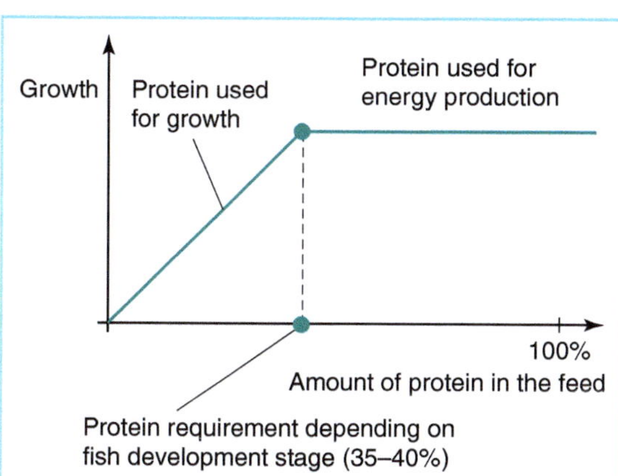

Figure 10.9 Increased growth is seen with increased protein levels until a given point (partly based on [19]).

based on protein is also a less effective energy-converting process, using more oxygen. The ratio must neither be too low because this will result in the fish obtaining sufficient energy without getting enough protein, as the fish will stop eating when they have taken in enough energy. This will result in growth depression. Ratios of digestible protein to digestible energy of between 23 g/MJ for smolt to 16–17 g/MJ for late ongrowing (2.5–5 kg) are suggested values [18]. Salmon diets typically include 40–55% crude protein, mostly from small fish. When protein is used for production of energy (catabolism), it must first be deaminated; this means the amino group is removed, and ammonia is released. The remaining carbon skeleton can be broken down to produce energy, carbon dioxide and water, and the energy can be stored temporarily as ATP (through glycolysis, citric acid cycle and oxidative phosphorylation, described later).

Protein is the most expensive (>50% more) of the three macro components in the feed and therefore it is important that the protein supplied is used for muscle building, not for energy production. The typical protein composition includes carbon (50%), nitrogen (16%), oxygen (21.5%) and hydrogen (6.5%) [1]. Limited amounts of sulphur and phosphorus will also be included.

The digestibility of the protein is of major importance; if the salmon does not digest the protein sources contained in the feed, the protein supplied is of no use. Therefore, a good protein source in salmon feed has high digestibility. Typical digestibility (bioavailability, described later) of natural protein sources used in the feed is around 90% but varies with the source.

It is important that the protein in the feed contains the correct amino acids, and the protein must be balanced. This sets requirements for the protein ingredients that can be used in the feed. Fishmeal is a good raw material for production of fish feed, as it contains the essential amino acids and is correctly balanced. Plant material does not have the optimal amino acid profile, so if too high an amount of plant material is used then greater quantities are needed to satisfy demands, and the fish will take in too much carbohydrate which is also suboptimal. A deficit of one of the essential amino acids will impede the overall protein digestion and suboptimal growth will be the result. Therefore, even if enough total protein is available but there is a lack of essential amino acids, the growth will be inhibited. Examples show that up to 6% growth (SGR) reduction can result for each percentage drop below the needs of the salmon [153]. Health problems may also occur if wrongly balanced or insufficient protein is supplied. The production cost will also increase if the dietary protein is not correctly balanced.

Nitrogen excretion from salmon is mainly across the gills as ammonium (ion dissolved in water) produced from metabolism, but some will also be released as faecal waste (10–15%), mainly from indigestible feed ingredients containing nitrogen. Salmon only excrete urea to a limited extent, when the fish are in the first life stages. If there is too much protein in the feed, more will be used for energy production and there will an increased loss of nitrogen to the environment. Even though the nitrogen is excreted as ammonium, there is an equilibrium with ammonia and the form that occurs in the water is dependent on the pH level.

Protein yield refers to how much of the total protein supplied is deposited as muscle; this has been measured at 28% [20] but depends on the size of the fish. In relation to this, protein utilization can be divided into anabolic, meaning it is used in the body (building new protein, protein synthesis), and catabolic, used in production of energy.

As stated, the protein requirement will vary with fish size. During start feeding and the juvenile stage the requirement is higher and the recommended protein content is around 50%; as the salmon increase in size the recommended protein content drops below 50%, and for ongrowing salmon in the sea cages a content of around 45% is recommended [21].

The amount of nitrogen in the feed is related to the protein content: 16% of the protein is nitrogen. Dividing the protein value by 6.25 will therefore give the nitrogen value of the feed. The nitrogen value is important in relation to the discharge from salmon farming, because nitrogen acts as a nutrient or fertilizer when released to the environment, both in fresh and seawater.

10.6.2 Lipids

Lipids can be divided into fats and oils, waxes, phospholipids and sterols (Figure 10.10). Fats and oils are within the group called triglycerides (TAGs) and most of the lipids in salmon feed are in this form. Typically, they represent 98–99% of the lipid group and therefore lipids are often simply named fat. TAGs are hydrophobic and will not dissolve in water, which increases the requirement to digest them. In salmon the various lipids have different purposes; fat is used as the main energy source for the metabolism, steroids provide building blocks for hormones, and phospholipids are included in the cell membrane (Figure 10.11). Fat also serves as a transporter

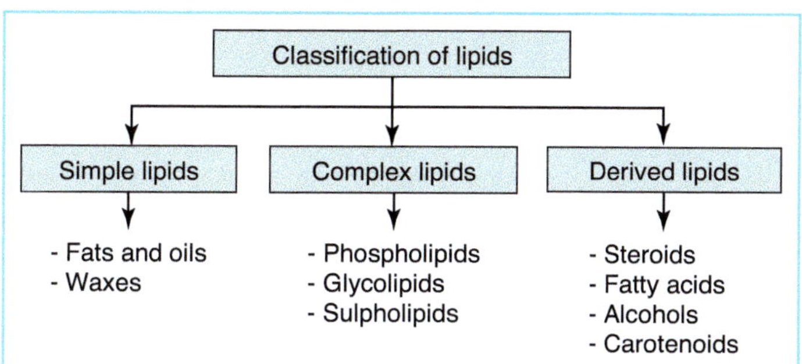

Figure 10.10 Classification of lipids.

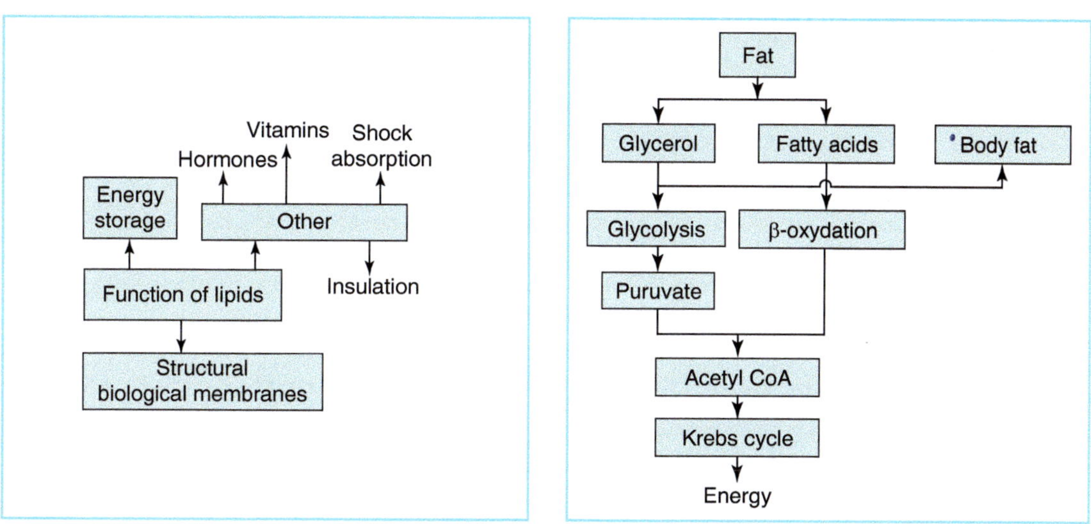

Figure 10.11 Function of lipids and pathways for lipid metabolism.

for fat-soluble vitamins, and some fat is deposited in the muscle. TAGs are a good energy source due to their high-energy density; they will not bind water. Much energy is released on oxidation of TAGs because they are large molecules with several energy-rich bonds. The main components of lipids include carbon, hydrogen and oxygen.

A triglyceride contains three fatty acids and one glycerol unit (Figure 10.12). In phospholipids a phosphate group is linked to the glycerol unit, but it also includes the fatty acids group. The special property of phospholipids is that they are water soluble, unlike triglycerides. A wax contains a fatty acid and a fatty alcohol connected with an ester bond.

The fatty acids are recognized and categorized by their carbon chain and the bonds within them, and are divided into saturated (SFAs) and unsaturated fatty acids (UFAs). Unsaturated fatty acids contain one or more double bonds between the carbon atoms in the carbon chain, which differs from saturated fatty acids that do not have double bonds. Unsaturated fatty acids can be further divided into monounsaturated fatty acids (MUFAs) and polyunsaturated fatty acids (PUFAs). Among the polyunsaturated fatty acids a group named highly unsaturated fatty acids (HUFAs) can be included (>20 carbons in the chain); the name long-chain PUFA (LC-PUFA) may also be used for this group, which is recognized by long carbon chain lengths and more than four double bonds. Unsaturated fatty acids can also be divided into omega-3, -6 and-9 fatty acids (n3, n6 and n9) (Figure 10.13). Which group they belong to is determined by the location of the first double bond in the carbon chain starting from the methyl end of the molecule. In omega-3 fatty acids this is found between carbon atoms 3 and 4 in the chain and in omega-6, between carbon atoms 6 and 7. The fatty acids can be described in a certain way, called n-nomenclature. To illustrate this, an example is alpha-linolenic acid which can be listed as 18:3:n3. Here the first number gives the number of carbon atoms (the length of the carbon chain), the second number gives the number of double bonds and the third shows where the first double bond is located in relation to the methyl end of the carbon chain.

SFAs and UFAs are different in many ways; an important difference is related to the melting point. SFAs are in solid form at room temperature, while UFAs can be in liquid form at room temperature, but can also be in liquid form at lower temperatures, including below zero degrees. The length of the carbon chain gives the degree of unsaturation, with longer chains being more unsaturated; HUFAs therefore have very low melting points. For example, the HUFAs EPA and docosahexaenoic acid (DHA) have melting points of –54°C and –44°C, respectively. As for

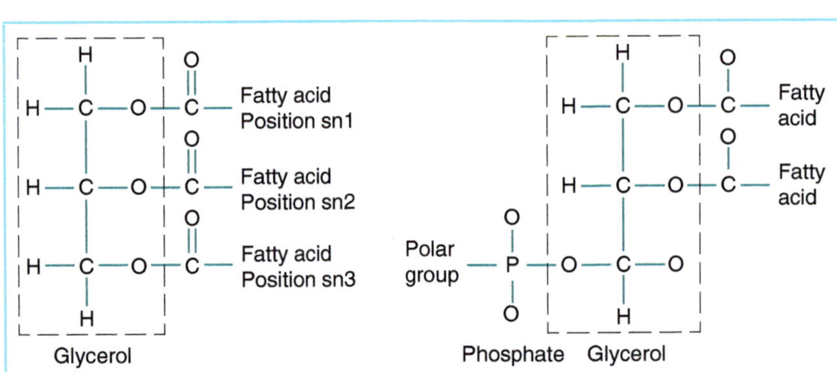

Figure 10.12 Structure of a triglyceride and a phospholipid.

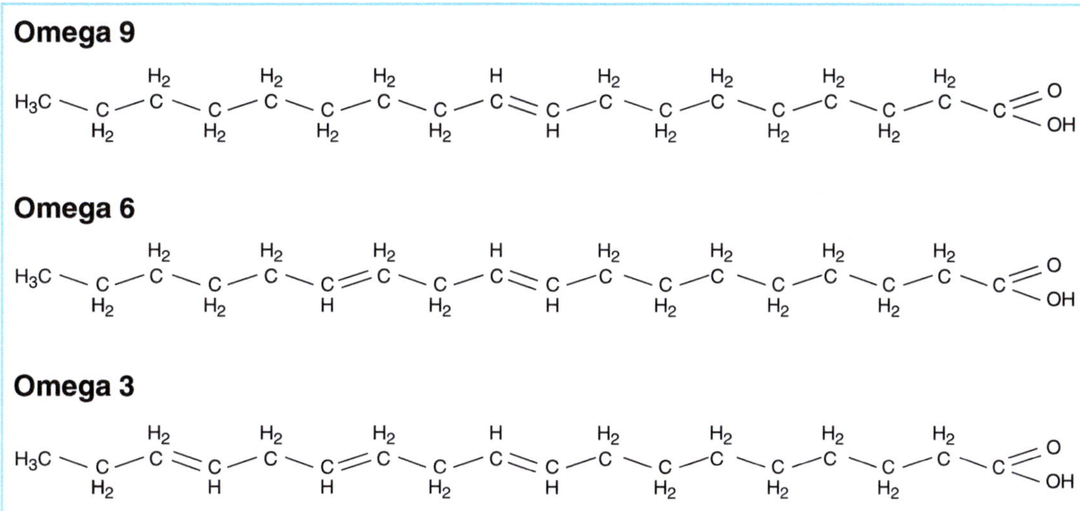

Figure 10.13 Structure of omega-3, -6 and -9 fatty acids.

protein, it is the digestibility of fat that is important when used as an ingredient in the feed, and UFAs have a better digestibility than SFAs. The digestibility is, based on this, also dependent on the water temperature, and at low temperatures long-chain UFAs are the best. This also shows why feed composition is dependent on species, and why salmon need a specially adapted feed because they are adapted to low water temperatures.

Essential fatty acids (EFAs) are fatty acids that the salmon need, but cannot synthesize themselves. Therefore, EFAs have to be supplied with the feed; these include both omega-3 and omega-6 fatty acids. If they are not supplied at the correct level and ratio, growth will drop and health problems may occur. The requirement depends on the size of the fish and the water temperature. The most expensive EFAs, and the most difficult to obtain, are the HUFAs, which include the fatty acids EPA (20:5:n3) and DHA (22:6:n3). Freshwater fish can synthesize these fatty acids themselves by elongating shorter ones by adding carbon atoms to the chain and increasing the number of double bonds (desaturation), but salmon do not have this capability and therefore HUFAs must be supplied with the feed. If EFAs are not supplied at satisfactory levels the weight gain will drop, but not immediately because at first the fish will try to obtain them from the endogenous tissue lipids [8]. Salmon need a low amount of HUFAs (EPA and DHA) for metabolic functions and cellular membrane components; to ensure welfare they must comprise 1% of the diet [22], which is why they are named essential fatty acids. It is for the same reason that humans need to include these fatty acids in their diets.

Salmon is a fatty fish species that deposits fat in the muscle in contrast to lean fish, such as cod, which deposit fat in the liver. The fat represents an energy store for the fish that can be used in periods where the dietary supply is limited. This can be seen when the fish are starved and the fat content in the fish subsequently drops.

Digestibility of fats is generally high, within the range 80–90% [21], and the fats with the highest digestibility are the unsaturated fats, especially the long-chain UFAs. The drawback here is that these are also highly susceptible to oxidation, causing rancidity, with spoilage of the feed as a result. To avoid this, antioxidants are added to the feed but it still sets limitations on the storage conditions and storage time for the feed. To mitigate the oxygenation risk the feed should be stored in an environment with as low an amount of air as possible, and not stored open in windy conditions nor directly exposed to sunlight.

The fat content of feed has increased a lot compared to the first formulated feed made for salmon, where the protein content was much higher (later described as high-energy feed). However, too high a lipid content will result in more deposits of lipids in the intestine. The amount of lipids also affects the protein utilization, which is improved up to levels of 24% [21], but salmon feed today typically contains more than this.

The minimum required content of lipids varies with size; generally larger salmon have higher requirements, 24–30% for ongrowing in sea cages, while the requirement for fry is 16–18% [21]. If there is too much dietary fat, fat will be stored in the muscle tissue and viscera. The amount and composition of fat in the salmon is dictated by the feed, so to have a healthy salmon it is necessary to provide healthy fat in the diet.

10.6.3 Carbohydrates

Salmon contain a very limited amount of carbohydrates. Some are stored as glycogen in the muscle and the liver. This is an energy store that can rapidly be transferred into glucose, used when the fish need energy, for instance, for short periods of fast swimming. Some carbohydrates may also be converted to lipids [8]. The conversion of lipids to energy takes more time.

Carbohydrates can be classified as simple sugars (mono- and disaccharides, consisting of one and two sugar units, respectively) and more complex sugars (polysaccharides, containing many sugar units) such as starch, cellulose and glycogen (Figure 10.14). The latter group have a high molecular weight and are insoluble in water. They may also be classified as starch and non-starch polysaccharides. Carbohydrates are formed from carbon dioxide and water by photosynthesis, where the energy for synthesizing (building larger molecules) comes from the sun. The main elements in carbohydrates include carbon, hydrogen and oxygen, and when fully oxidized the waste products will be water and carbon dioxide in addition to energy. Salmon can digest simple carbohydrates such as simple sugars but for forms that are more complex, digestion rapidly decreases. Starch can be digested to a very limited degree, up to 15% [18], even if it is cooked (gelatinized), and salmon cannot digest cellulose, which is an even more complex sugar form. Cellulose is a fibre found in plant cells used in wood and stalks to ensure strength. Low digestibility of complex

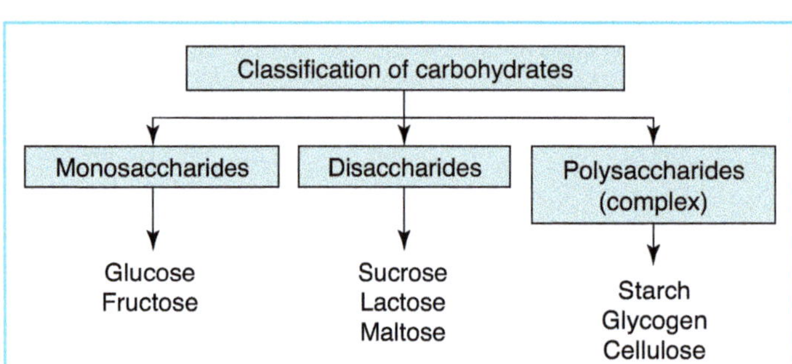

Figure 10.14 Classification of carbohydrates.

carbohydrates is recognized in monogastric animals, and fish lack the enzyme cellulase that is necessary for breaking down cellulose. The typical digestibility of glucose (a monosaccharide) is almost 100% while the digestibility of raw starch can be as low as 5% [19]. Salmon have a slow turnover of glucose; blood glucose will rise quickly but will not drop for a long time, and a large amount will not undergo glycolysis (described later). A high carbohydrate content therefore leads to poor regulation of glucose [23] and salmon can be compared with humans with diabetes, as carbohydrates at high levels create hyperglycaemic conditions. Similar to other carnivorous species, salmon utilize (metabolize) carbohydrates to a very limited degree. This is based on evolution, where they have never adapted to eat carbohydrates. The carbohydrate level, including simple sugars and starch, is not recommended to be above 12% in the diet and the crude fibre fraction should be less than 3% [19]. Increasing levels of starch will reduce the overall digestibility, not only for the starch but also for the fat [18]. The addition of indigestible carbohydrates to the feed will also increase the amount of faeces released.

Adding carbohydrates in large ratios to salmon feed may therefore result in negative growth, and may also inhibit the uptake of the other main nutrients. Intestinal problems causing reduced fish health can also be observed when giving diets containing too much carbohydrate. The addition of, for instance, cellulose to the feed in large amounts will result in growth depression in salmonids. The major reason for including as much carbohydrate in the feed as possible is because it is a cheap raw material/feed ingredient and large volumes are available on the market. The aim is that the salmon use the carbohydrates for energy production, rather than the more expensive lipids. Some carbohydrates also have a good binding capacity, meaning they are well suited to bind the feed particles together when producing the pellets. They are therefore good feed binders but have, as discussed, zero or even negative nutritional value.

Digestion of starch can be increased by pre-cooking the starch during the feed manufacturing process (gelatinization) or pre-treating with enzymes. Cellulose may be used as an ingredient, but

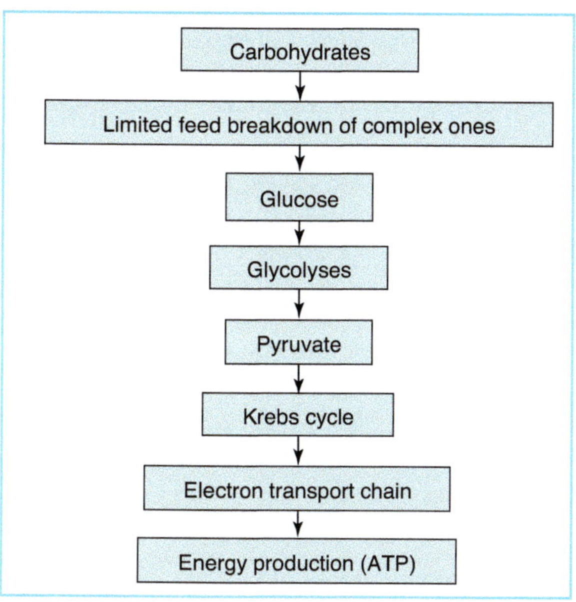

Figure 10.15 Pathways for carbohydrates in metabolism.

the structure has to be broken down, for instance, with enzymes, before adding to the feed. As stated, the carbohydrate digestibility will vary a lot, from 0% for cellulose to up to almost 100% for pure glucose, and the reason for not including too much simple sugar in the diet is due to problems with regulation of blood sugar.

10.6.4 Vitamins

Vitamins are included in the group of micronutrients. Vitamins are complex organic elements necessary for the growth and health of salmon (see [18] for an overview). There are 15 essential vitamins which must be supplied in food because the fish can either not synthesize them or cannot synthesize them fast enough. Vitamins can be divided into water-soluble or fat-soluble vitamins; 11 water soluble and 4 fat-soluble are essential. Water-soluble vitamins include the vitamin B group which includes a total of eight vitamins, for example riboflavin (B2), pyridoxine (B6) and biotin (B7), and are typically involved in catabolism. Vitamin C, or ascorbic acid, is another water-soluble vitamin that salmon cannot synthesize, and is important for growth and a number of body functions. Ascorbic acid is also a powerful antioxidant important for the immune system in fish. The fat-soluble vitamins include A vitamins, such as retinols (for example, responsible for vision and mucus secretion), D vitamins, such as cholecalciferol (D3) (required for bone integrity), E vitamins, such as the tocopherols (antioxidants in the cell membrane) and K vitamins (required for blood clotting, skin integrity). The lack of specific vitamins will lead to specific symptoms, but reduced growth is the most common symptom of any vitamin deficiency. An example of a disease caused by a vitamin deficiency is scoliosis (curvature of the backbone), caused by a deficiency of vitamin A, and dark colouration may result from a deficiency of folic acid. Vitamin supplements in the form of premixes are typically added to the feed to ensure correct doses. However, as with all other substances, vitamins can reach toxic levels if the volume included is too high; even though this is not is common, it is important for levels to be balanced.

Incorrect storage of feed resulting in oxidation may result in vitamin degeneration; vitamin C is susceptible, as are the lipid-soluble vitamins A, D and E. Since signs of lipid oxidation produce a specific odour, it is easy to test for incorrect storage by smelling the feed.

10.6.5 Minerals

Minerals are inorganic elements needed for normal body functions, for example to maintain osmotic balance, for digestion and for respiration [18,19,24,25]. In addition, minerals are necessary for formation of the skeleton. There are 13 essential minerals that fish have to obtain from food [25]. Minerals are grouped into macro-minerals and micro-minerals or trace minerals, based on the quantity that the salmon require and therefore the amount supplied in the feed [26,27]. Common essential macro-minerals are sodium, chloride, potassium and phosphorus. These minerals are important for regulation of the osmotic balance and for bone formation. The micro-minerals or trace minerals are required in small amounts and are needed for well-functioning enzyme and hormone systems, for example. Common essential trace minerals include copper, chromium, iodine, zinc and selenium. Fish have the ability to absorb minerals directly from the water through both the gills and the skin. If the mineral content of the feed is suboptimal, salmon can compensate to some degree if those minerals are available in the water. Minerals are, as for vitamins, typically added to the feed as premixes in the production process, but care should be taken if the raw materials used are changed, to ensure levels are still adequate. An example here is phosphorus, also described later, where bioavailability is of major importance; this is much reduced in plant proteins. Ratios between the minerals are also important, for example, the best balance between calcium and phosphorus should be 2 : 1 or less. Based on their origin, minerals may either be organic or inorganic.

A problem with the addition of minerals to the feed, both as a premix and with those included in the feedstuff, is that the faeces from the fish may contain levels of minerals that create problems for later utilization of the collected faeces. The levels may be above the limits for use as fertilizer on agricultural land, particularly in relation to zinc and cadmium. Similar to vitamins, if levels of minerals, for example, iron and zinc, are too high this may be toxic for the fish.

10.6.6 Pigments – carotenoids

Salmon need carotenoids (pigment) in the feed to colour the meat [28–30]. The market requires salmon that has red meat. If pigment is not included in the feed, salmon meat will be grey. There are no absolute requirements for adding pigments in the feed to colour the meat, but pigment is necessary for viable eggs. When salmon are mature, this can be observed because the eggs become red and the colour of the salmon muscle tissue becomes less red. Pigment also plays a part in sexual dimorphism: the males become red to attract females, and females use the pigment to colour the eggs. Carotenoids are also powerful antioxidants.

The carotenoids found in salmon belong to a group called xanthophylls, a class of oxygen-containing carotenoid pigments. Astaxanthin, a lipid-soluble pigment, is part of this group and is the major contributor to the red colour in salmon. This is the same both in wild and farmed salmon. Typically, synthetic astaxanthin, identical to that found in nature, is included in the feed because salmon are unable to synthesize it. Synthetic astaxanthin has the same biological function as the natural form. Natural products containing astaxanthin or products with similar properties in high concentrations may also be used, such as krill meal, shrimp shells, yeast, or marine algae that contains high levels of astaxanthin [31,32]. In the early days of the salmon farming industry natural carotenoids were used, but they were gradually changed to the synthetic form. The problem with the natural substance is that a large volume is necessary to fulfil the demands of the salmon industry. Canthaxanthin was also used for a while, but this gives a more orange colour to the flesh and today astaxanthin totally dominates.

Even though it is not an absolute requirement to include carotenoids in the feed, a positive effect on growth rate and survival of juvenile fish has been observed. Astaxanthin is also a good antioxidant and protects against oxidative damage in the salmon and improves immunity. It is also a pro-vitamin A source, another antioxidant. Astaxanthin will protect the fat in the salmon meat from oxidation after slaughter, and thereby increase the shelf life.

The amount of carotenoids in the feed should be above 6–10 mg/kg [33]; recommended levels are around 50 mg/kg. When levels above 50–60 mg/kg are reached, any additional effect decreases [34]. EU regulations set a maximum allowed content of 100 mg/kg of astaxanthin [35]. The digestibility of astaxanthin is low and the muscle retention is less than 10%, but a higher fat content seems to improve the absorption (as astaxanthin is absorbed with fat). There are different strategies for colouring the meat; one is to start with a level of 25–35 mg astaxanthin per

Figure 10.16 Colour scales are used to determine the colour of the salmon meat, either manually or digitally.

kilogram of feed when the smolt are transferred to the sea and then to continue with 50–60 mg when the salmon pass 1 kg [18]. The measured amount of astaxanthin in farmed salmon flesh is 6–8 mg/kg [32].

The colour of the salmon meat can be measured with several methods. One much-used industrial standard is the SalmoFan™ colour scale (DSM). Based on the colour of the salmon meat, it ranges from 20 (pinkish) to 34 (dark red) (Figure 10.16). A digital version of this is also under development. Values above 25–26 on the colour scale are recommended in order to have the correct colour to satisfy the market. However, a colour above 23 will be satisfactory for most consumers, and once above 26 the willingness to pay for increased colour is limited [36]. Computer vision in general may also be used for measuring of colour [37].

10.6.7 Other substances

The feed ought to contain some fibre. This seems to increase growth and improve protein utilization [38]. Fibre is a complex carbohydrate, such as cellulose and lignin. However, since fibre has no or low digestibility, the level should be low, below 8%. The feed also needs to contain some binders, to make a good-quality pellet. These may include complex carbohydrates, alginate and guar gum. The binders will also have no or low digestibility, but it is important to understand what they do and how they react in different parts of the gut. A high content of UFAs also requires the inclusion of antioxidants in the feed. It is also possible to use what is termed functional food, which is food that includes ingredients with particular properties, which may relate to health (described in Chapter 11). A range of functional ingredients can be included, depending on the goal. Typically, such feed is used in special situations, such as before the smolt are transferred to the sea to strengthen the immune system or to strengthen the fish against sea lice attacks. Prebiotics or probiotics are other ingredients to include. Antibiotics or other medications may also be included in the pellet, which is then termed medical food (described in disease chapter). This is, however, only used when infection is diagnosed by a veterinarian. Special appetizers may also be included in the diet.

10.7 Energy requirements – bioenergetics

As already discussed, salmon need energy for several functions, such as activity, growth, excretion of waste products and reproduction. The study of how energy flows within the body is called bioenergetics. The basis for bioenergetics are the laws of thermodynamics, the first of which states that energy cannot be created nor destroyed, but can only change forms. The second states that when energy changes form, some will always be lost, meaning it is impossible to retain the same amount and the energy quality is degraded. Typically, some of the energy is transferred to non-useful heat.

The energy used by salmon is supplied by the feed and transformed to ATP, the fuel that salmon use. The energy is created by catabolism of the feed, meaning its breakdown/oxidation. The result is the release of chemical energy from breaking the bonds in the macronutrients supplied. During periods of starvation, energy for basal metabolism is created by breaking down the salmon's own muscle, resulting in weight reduction.

The fish eat to satisfy their need for energy, not for nutrients for metabolism and growth, which illustrates the importance of having a feed with the right balance between the energy and the required nutrients, otherwise the fish may not get the correct amount and balance of nutrients.

The salmon need a supply of energy to support basic body functions, for growth and for activity. If not enough energy is supplied, there will be growth reduction and reduced activity.

10.7.1 Salmon use less energy than mammals

Compared to mammals, salmon are much more energy efficient. This is due to three major reasons: they are cold blooded, they do not need energy to stand up (they are weightless in water) and their waste excretion is more effective. Since salmon are cold blooded (poikilothermic), they do not need any energy to heat their bodies. There will always be some energy produced and released as heat during the processes occurring within the body, but very little compared to warm-blooded animals (described further with calculations later).

Salmon are weightless in water, regulating buoyancy with the swim bladder. They do not need energy and nutrients to form a heavy skeleton that can hold them in an upright position, as is the case for mammals. Their motion is also much more energy efficient compared to moving a body with a heavy skeleton.

Catabolism of protein results in production of ammonia, which is toxic for the salmon if the concentrations get too high; therefore, the fish have to remove the ammonia from the body. Salmon continuously excrete nitrogen as ammonia over the gills. This is far more energy effective than transforming it into urea or uric acid and storing it in the body before releasing it in batches. In mammals, the kidney turns ammonia into urea (CH_4N_2O) as it is a less toxic form for the body. Birds transform urea further to uric acid ($C_5H_4N_4O_3$), an even more concentrated form than mammals, in order to release it in a drier form with the faeces and thereby avoid carrying water as urine. These processes require more energy compared to releasing nitrogen as ammonia directly over the gills, therefore salmon need less energy to breakdown protein (catabolism) and excrete nitrogenous waste compared to mammals and birds. Approximately 5% of the salmon's energy is used for the process, while the same energy requirement for mammals is 10–20% higher [39].

10.7.2 Energy requirement: metabolic rate

The energy requirement of salmon directly depends on the metabolic rate, which describes the velocity or kinetics of the metabolism. It can be compared with the production rate in a factory. Overall, the metabolic rate can be divided in three: standard metabolic rate, routine metabolic rate and active metabolic rate. The first includes maintenance, with no growth and no activity. The second includes growth and activity. The third can be described as maximum metabolic rate, and this is represented by maximum activity that is still aerobic, and can be compared with human training. There are a number of factors affecting the metabolic rate and therefore the salmon's energy requirements, and of course requirements will increase when going from standard metabolic rate to active metabolic rate.

The most important factor is the water temperature. Increased temperature will increase the metabolic rate and also the growth rate within the temperature tolerance of the salmon. However, if the temperature gets too high the bodily functions will not adapt and will start to fail. Other factors that affect the metabolic rate will therefore also affect the energy requirement. Under farming conditions, increased metabolic rate requires an increase in the supply of feed and oxygen for the oxidation process. High water temperatures can be a problem because the available oxygen in the water reduces with temperature, so it can be necessary to add pure oxygen gas to the water to reach maximum metabolism and growth potential at higher temperatures.

The size of the fish is also important for the energy requirements because smaller fish have faster growth and body development (internal organs, muscle, skeleton, and so on) which also dictate the energy requirements, both regarding age and development stage. Brood fish will also need additional energy due to the development of gonads.

More activity will increase the energy consumption, as will a suboptimal environment resulting in the fish becoming stressed. This also results in higher oxygen consumption, as oxygen is required for energy production.

10.7.3 How does digestion release energy?

When salmon digest food (metabolism), some is transformed to energy by oxidative reactions requiring access to oxygen (breathing), an aerobic reaction (described later). Simplified, the process converts the carbon chains in protein, fat and carbohydrates in the feed to carbon dioxide and water, and during this reaction energy is released. The approximate amount of energy released in digestion of protein, fat and carbohydrate in salmon is documented. Fat will release around 39 MJ/kg while protein releases 23.5 MJ/kg and carbohydrates release around 17 MJ/kg [40]. When protein is digested, oxides of nitrogen and some sulphur and phosphates will also be released.

The energy requirement for production of 1 kg salmon meat is at least 15 MJ, of which at least 6 MJ should be lipid based [41], but this value varies with the fat content among other factors; this is described in more detail later, along with models for estimation.

To obtain energy, the major nutrients (protein, fat and carbohydrates) are, as discussed, broken down. This includes several biochemical processes described briefly below, where the end product is adenosine triphosphate (ATP). ATP is the energy form that the fish utilize for processes that require energy, so can be considered as the fuel for energy-requiring processes in the fish, like gasoline is in car engines. Protein, fat and carbohydrates can all be broken down for production of ATP, but the biochemical breakdown process varies.

10.7.4 ATP production – fuel production

As stated above, the energy salmon use for metabolic processes is in the form of adenosine triphosphate (ATP), and therefore the salmon need to produce ATP. To release the energy in ATP, it is broken down to adenosine diphosphate (ADP). The energy is released due to the breakage of the high-energy chemical bond (one phosphoanhydride bond) which links ADP to another phosphate group to create ATP. The energy release is around 30 kJ per mol ATP converted to ADP. The ADP can then be 'recharged' to ATP by using new energy from the digested macronutrients to add a phosphate group (phosphorylation), also called cellular respiration (Figure 10.17).

Production of ATP involves complex biochemical processes based on catabolism of the macronutrients (protein, fat and carbohydrates). The energy that is used to produce ATP from ADP is found in the chemical bonds of the macronutrients that are broken down to the simple ones. The major end products of the catabolism are carbon dioxide and water in addition to the produced energy. The biochemical breakdown processes are based on a number of redox reactions where electrons are transferred to a lower energy level, oxidation, and thereby release energy. The ATP is produced within the cells, the majority inside the mitochondria, which are also named the energy factories of the cells. In ATP the energy is also stored in the chemical bonds, and by breaking those bonds energy is released.

When salmon eat the large molecules in the macronutrients and they are broken down through metabolism, energy is released and temporarily stored as ATP (for a very short time, milliseconds) before being used. The pathways for breakdown are different for protein, fat and carbohydrates but all produce energy when broken down; it is the start of the process that differs, with the last steps being similar. The main steps in breakdown for energy production start with production of pyruvate by glycolysis; this is further converted to acetyl-CoA that enters the citric

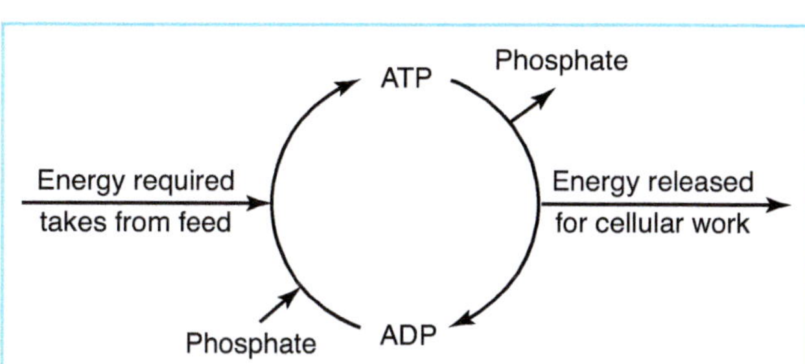

Figure 10.17 ATP–ADP cycle: ATP is transformed to ADP with release of energy, and from ADP to ATP energy is absorbed.

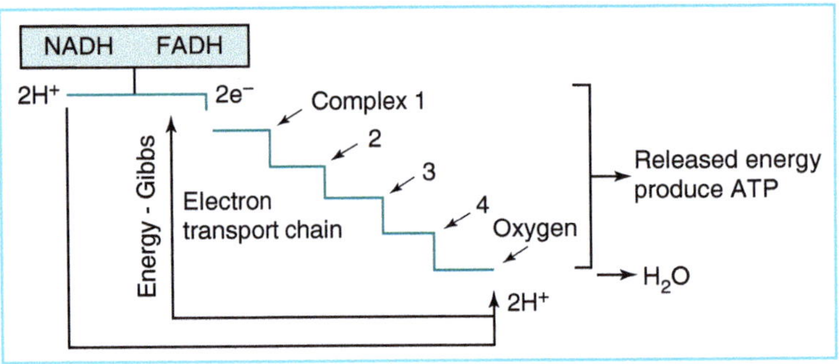

Figure 10.18 In the electron transport chain energy is gained through several redox reactions.

acid cycle, and from this electrons are transferred for oxidative phosphorylation (Figure 10.18). The overall process needs oxygen and produces carbon dioxide and water in addition to energy.

Let us start by looking at carbohydrates. The process begins with the breakdown of polysaccharides and disaccharides to monosaccharides (mainly containing six carbon atoms), such as glucose. The monosaccharide undergoes a process named glycolysis (glycol = glucose and lysis = breaking down) to form pyruvate (a smaller molecule with three carbon atoms); this process has ten steps catalysed by ten different enzymes. During glycolysis, first two ATP molecules are used and then four are produced for each mol of glucose, resulting in an overall net production of two ATP, and therefore phosphate groups are also supplied. In addition, two nicotinamide adenine dinucleotide (NADH) molecules are produced. This is a coenzyme (necessary for function of an enzyme). NADH is a very effective electron carrier, produced by reduction of NAD^+ (a reduced form of NADH).

$$NAD^+ + 2e^- + 2H^+ \rightarrow NADH + H^+$$

$$FAD + 2\,e^- + 2H^+ \rightarrow FADH_2$$

Flavin adenine dinucleotide (FAD) is a similar redox-active coenzyme that also functions as an electron carrier, active in later processes. The energy in both NADH and $FADH_2$ is stored in the electrons and released after transportation to the mitochondria.

The overall chemical reaction for glycolysis is:

$$C_6H_{12}O_6 \text{ (glucose)} + 2\text{ ADP} + 2\text{ Pi (phosphates)} + 2\text{ NAD}^+ \rightarrow 2\,C_3H_3O_3^- \text{ (pyruvate)} + 2\text{ ATP} + 2\text{ NADH} + 4\text{ H}^+$$

The pyruvate is then converted; first 1 CO_2 molecule is formed, based on the release of 2 oxygen and 1 carbon molecules from pyruvate, and one molecule of NADH is produced by reduction of NAD^+, then acetyl-CoA is produced by transferring an acetyl group to the coenzyme A.

The chemical equation it follows is:

$$2\text{ pyruvate} + 2\text{ CoA} + 2\text{ NAD}^+ \rightarrow 2\text{ acetyl-CoA} + 2\text{ H}^+ + 2\text{ CO}_2 + 2\text{NADH}$$

The acetyl-CoA is further converted in the next process, the citric acid cycle (also named Krebs cycle or tricarboxylic [TCA] cycle), while the NADH is sent further to the electron transport chain; carbon dioxide is released from the cell as waste, while the H^+ is kept within the cell. The citric acid cycle includes eight steps with chemical reactions (Figure 10.18). In the cycle, the acetyl group in acetyl-CoA is linked to an oxaloacetate molecule (containing four carbon atoms) to form a citrate molecule (six carbon atoms). The citrate is then oxidized and carbon dioxide is released. Two carbon dioxide molecules are released for each acetyl molecule entering the cycle, and NADH and FADH are produced in addition to ATP. This results in the following overall chemical reaction for the citric acid cycle:

$$2\ \text{acetyl-CoA} + 6\ NAD^+ + 2FAD + 2\ ADP + 2\ Pi \rightarrow 6\ NADH + 2\ FADH_2 + 2\ ATP + 6\ H^+ + 4\ CO_2$$

The NADH and $FADH_2$ then deliver electrons to a linked process named oxidative phosphorylation. The main aim of the citric acid cycle is therefore to produce not ATP but the electron carriers NADH and $FADH_2$ that can be used for energy in a later related step (oxidative phosphorylation).

Through oxidative phosphorylation, ADP is recharged to ATP by adding a phosphate group. The energy for this is taken from the energy carried with the electrons from the citric acid cycle, therefore it is sometimes also called the electron transport chain.

A mitochondrion has an internal membrane which surrounds an inner matrix. When the NADH releases its two electrons and protons (H^+), the electron is released into the membrane where a number of redox reactions occur in protein complexes (four in total) inside the membrane (Figure 10.19). The complexes are special proteins embedded in the membrane. Through the redox reactions inside complex 1, the electrons pass from redox centre to redox centre, and each time an electron is passed energy is released. The released energy is then used to 'pump' protons into the inner matrix of the mitochondrion, setting up a proton gradient across the membrane. In complex 2 a similar process happens, but here it is FADH that donates two electrons and two protons, and the energy released is also used to pump protons. Next, coenzyme Q (a mobile electron carrier) takes the electrons released from complexes 1 and 2 and transports them into complex 3, where a similar process to that in complexes 1 and 2 occurs. Another mobile electron carrier, coenzyme C, transports electrons further into complex 4. This is the last complex, and here four electrons are used to convert an oxygen molecule into two water molecules. The oxygen is taken in over the gills and the water is released in the body. The energy to pump protons into the inner matrix during this process is supported by reduction of the oxygen. An increasing proton gradient develops across the membrane. The next step in the process, ATP synthase action, involves a special protein complex embedded in the membrane where ATP is created. For this, ADP is needed in addition to energy. The energy is gained by letting the protons leak out through the membrane through the special protein complex, which can be visualized as a turbine which produces energy when it rotates. The exiting protons link to oxygen, producing water. The four complexes comprise the electron chain while the last process is oxidative phosphorylation, the step in the overall oxidation process where the majority of the ATP is produced.

The equation that describes the electron chain and oxidative phosphorylation is as follows:

$$10\ NADH + 2\ FADH_2 + 6\ O_2 \rightarrow 10\ NAD^+ + 2\ FAD + 12\ H_2O$$

For each NADH that becomes NAD+, 3 ATP are produced: in total 30

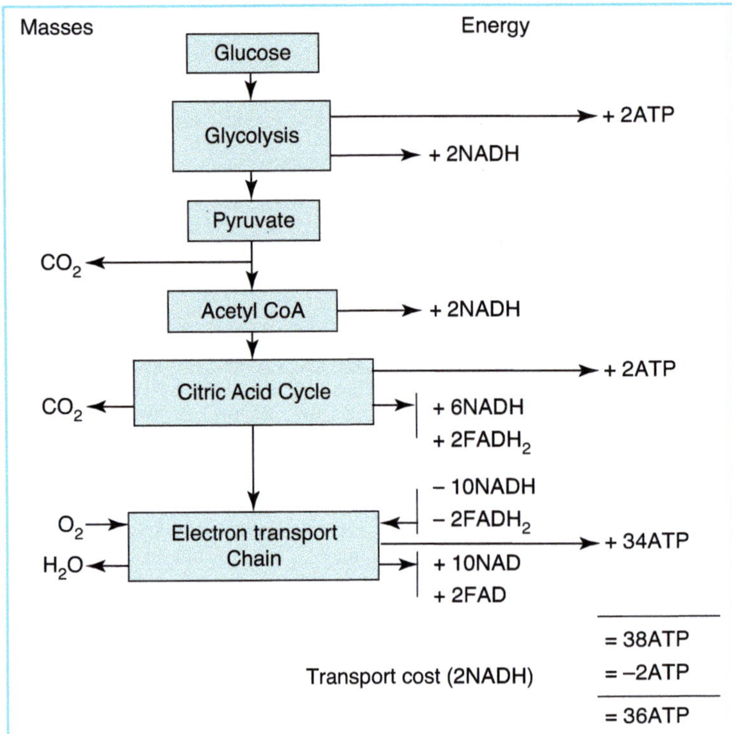

Figure 10.19 The entire production chain for ATP including glycolysis, pyruvate to acetyl-CoA, citric acid cycle, electron transport chain and oxidative phosphorylation.

For each $FADH_2$ that becomes FAD, 2 ATP are produced: in total 4
Total ATP production: 34

The majority of ATP production occurs, as discussed, in the mitochondria; glycolysis occurs in the cytoplasm of the cells where two ATP are produced, the next step of pyruvate conversion occurs in the mitochondrial matrix, two ATP are consumed in the process of transporting NADH into the mitochondrion, and the citric acid cycle occurs in the matrix producing two ATP. Oxidative phosphorylation occurs in the mitochondrion, and here 34 ATP are produced. The total net ATP production will therefore end up as 36.

This is how energy is produced from carbohydrates but the other main nutrients, lipids and proteins, can also be used for energy production; they are broken down and enter into the same system (Figure 10.20).

Moving on to lipids or fat, triglycerides are first split into fatty acids and glycerol by a process named lipolysis (described later). The glycerol enters the glycolysis process as dihydroxyacetone phosphate, together with the monosaccharides, while the fatty acids will be directly broken down to acetyl-CoA through fatty acid oxidation or beta (β)-oxidation. The process then follows the citric acid cycle and oxidative phosphorylation as for carbohydrates. Much energy is stored in the triglycerides because they each contain three fatty acids which typically have long carbon chains.

Breakdown of glycerol first involves conversion to glycerol-3-phosphate by adding a phosphate group using ATP. The glycerol-3-phosphate is then converted to DHAP and NADH is produced. DHAP enters the process of glycolysis (not the first step).

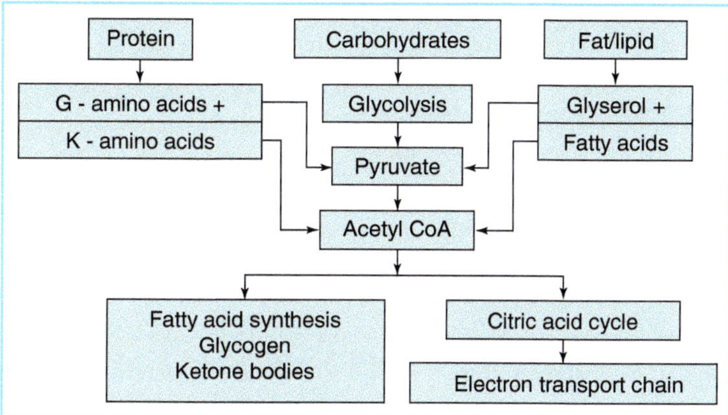

Figure 10.20 All major nutrients may be oxidized to acetyl-CoA through different pathways.

When the fatty acids are broken down by β-oxidation (β due to the second carbon molecule in the fatty acid chain), this involves a circular oxidation process with several steps where a two carbon unit is detached from the fatty acid for each step. In more detail, first the fatty acid is broken down to fatty acyl CoA that can be transported over the outer mitochondrial membrane. This is then linked with carnitine to form acyl carnitine that can pass through the inner mitochondrial membrane; it then reforms to fatty acyl CoA and is further converted into acetyl-CoA that enters the citric acid cycle. During this process energy is released which is stored in FADH2 and NADH; one of each is produced per cycle.

Protein is first broken down to amino acids that are used for protein synthesis, but if there are more than needed they will be used for energy production. Values for energy needed for protein synthesis vary, but 0.06–0.075 mol ATP per gram protein synthesized have been reported [42]. Transferring protein to energy is also a multi-step process. The first is to remove the amine group from the amino acid, a process called oxidative deamination. The amino group will end up as ammonia released across the gills of the salmon. The carbon chain now remains, which will be used for energy production. Depending on the type of amino acid, they can be categorized as either glucogenic amino acids that end up as pyruvate or as ketogenic amino acids that end up as acetyl-CoA, both processes involving a number of steps. Both then enter the Krebs cycle and ultimately produce energy.

If too much acetyl-CoA is released relative to requirements, ketone bodies (water dissolvable molecules) may be produced by ketogenesis. Ketone bodies are produced in the liver where the majority of fatty acid oxidation occurs. Ketone bodies can be looked at as energy packages from not fully broken down fatty acids that can be transported in the body and used when needed. Too high a concentration of ketone bodies is not good, but they have an important function when too little glucose is supplied, such as when starving the fish. When used in the cells, the ketone bodies are converted back to acetyl-CoA before entering the citric acid cycle.

There are therefore three ways to produce acetyl-CoA: from oxidation of pyruvate (the result of glycolysis); from oxidation of fatty acids; and from oxidation of amino acids. This shows how energy can be produced from all macronutrients.

If there is plenty of glucose, glycerol and acetyl-CoA, new triglycerides can be built by lipogenesis. This is also a way of establishing fat deposits when there is a surplus of glucose, but the other factors must also be included.

The energy used by the fish is ATP, but energy is stored as fat or glycogen. Fat is stored in the muscle tissue and in the body cavity/viscera, and this energy reserve will be used if the fish do not get enough energy through the feed. The process of breakdown occurs as for the fat in the food. Glycogen is a complex water dissolvable molecule made of a number of glucose molecules. It can be broken down to glucose through glycolysis and therefore functions as an energy storage. Glycogen is stored in the liver and in the muscle tissue, and is quickly accessible, much faster than fat. When salmon need additional energy, for example, when swimming fast for short periods using light muscle, glycogen will be used.

There is an important connection between the main nutrients and the energy content of the feed. There is a requirement for both energy and the main nutrients, and this must be adapted to the needs of the salmon. For instance, if using too much fat and a reduced protein content in the diet, there will be a reduction in growth (due to insufficient protein for building new muscle) because the food intake will be suppressed. The salmon stop eating because they are satiated (described later). If using too much protein in relation to fat, the result will be that more protein is used as a source for energy production, which is more expensive, and the waste discharge will increase. The salmon will also need to eat more for the same weight gain. The energy to protein ratio is therefore an important factor when developing feed recipes.

10.7.5 Measuring feed energy content: bomb calorimeter

The energy content of the supplied feed can be measured in a bomb calorimeter (Figure 10.21). This can also be used to measure the single ingredients and the energy content in the fish, which makes it possible to perform energy balances and calculate the retained energy.

In a bomb calorimeter the raw material is combusted, and the amount of energy released is measured. A known weight of the feed or raw material (the sample) is loaded in a cup standing in water inside an insulated container. An ignition coil surrounds the cup and pure oxygen is added to the container above the sample so that the atmosphere is pure oxygen. This produces optimal conditions for starting a fire, combustion, which occurs when a current is sent to the ignition coil. The sample will totally combust, and carbon dioxide, water and ash remain (for proteins there will also be oxides of nitrogen and sulphur). This is called complete combustion and the energy released is named gross energy (GE). The energy released in the combustion within the bomb calorimeter is absorbed by the surrounding water, increasing its temperature. Since the amount of water is equal for every combustion, the temperature increase will document the amount of energy released from the sample. The energy needed for heating water one degree is known (specific heat capacity of water), so by monitoring the temperature increase the amount of energy released from the sample can be calculated. This value is equal to the amount of energy released during the metabolism of the feed, if fully oxidized. However, there will always be indigestible inorganic parts in the feed. If such feed is combusted, ash will be the result, as this represents the indigestible inorganic part. Minerals are examples of materials that are not combusted, as well as other inorganic elements.

The total energy content in a feed sample may be estimated if the amount of protein, fat and carbohydrate is known, since the energy released from those components is documented but differs somewhat from mammals. There are also some differences between the values given for fish; below is one example based on proposed values [40].

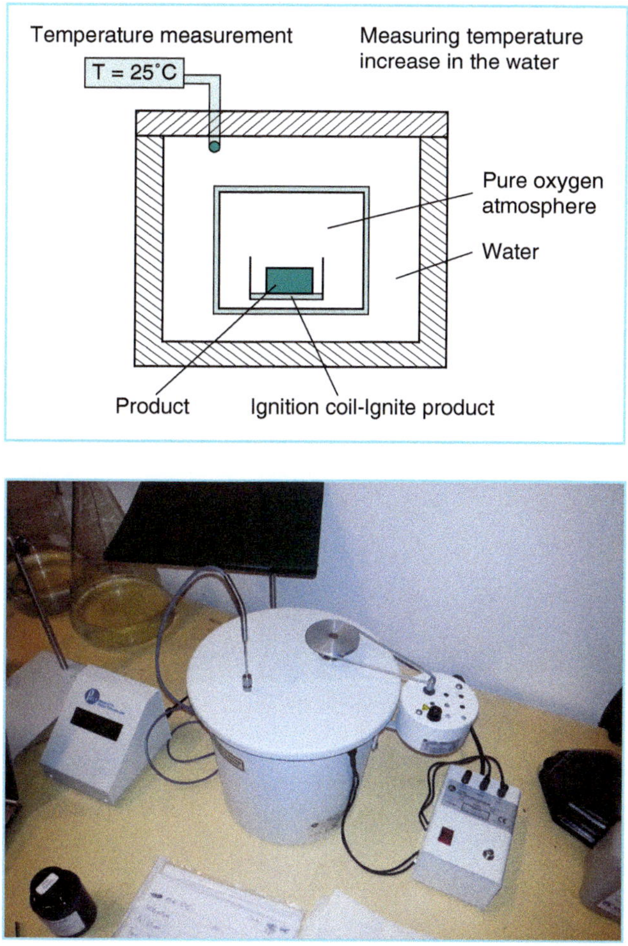

Figure 10.21 A bomb calorimeter and the basic design.

Estimated energy value (MJ/kg) = 23.5 (MJ/kg) × Protein content (kg) + 39 (MJ/kg) × Fat content (kg) + 17 (MJ/kg) × Carbohydrate content (kg)

These are overall values and will vary with composition, for instance with fatty acid chain length. In salmon this energy will be in the form of ATP.

10.7.6 Utilization of energy

The energy provided to the fish through the feed is distributed and used as listed below (Figures 10.22 and 10.23):

- energy lost in faeces (F)
- energy lost in excretion across gills and with urine (E)
- energy used for heating of the body (H)
- energy for basal metabolism, maintenance or fasting (M)
- energy retained as growth and gonad development (G).

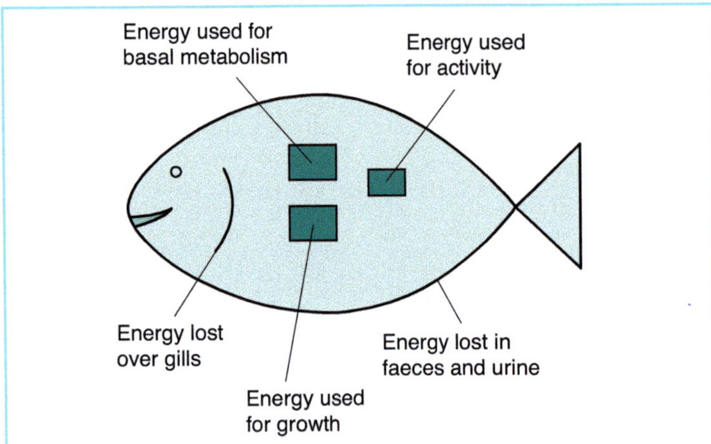

Figure 10.22 Nutrient flow through the salmon.

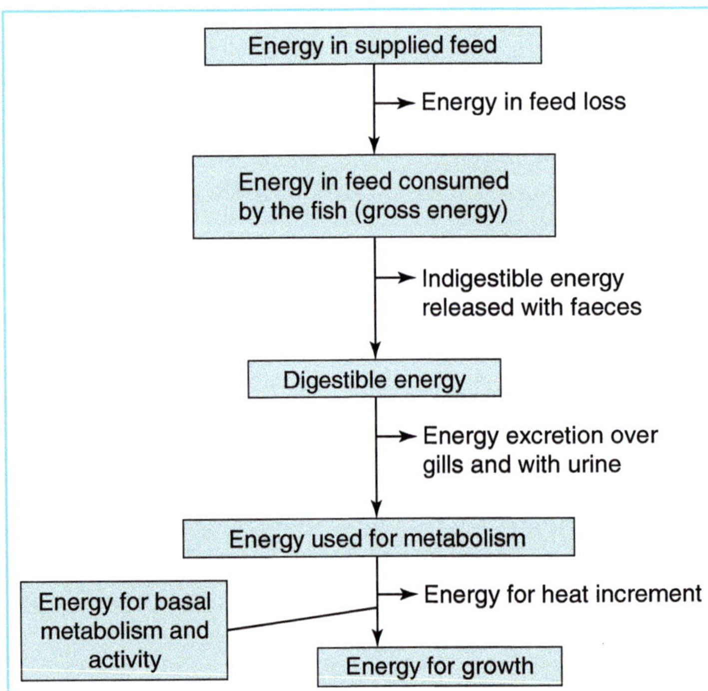

Figure 10.23 Energy flow diagram [46].

Some feed particles supplied will not be consumed by the salmon, and so some energy will be lost. Some minor feed loss seems to be impossible to avoid in practice if optimal growth is to be achieved, but almost all of the feed will be consumed. This energy is the gross energy (GE) and is the energy measured in the bomb calorimeter. However, some parts of the feed will be indigestible for the salmon, such as some feed binders and some carbohydrates, because the feed is composed of the ingredients and raw materials that are available, rather than the ideal chemical composition. The indigestible material goes directly through the digestive system and is released as faeces, but it may contain combustible parts even though the salmon cannot utilize it. As it would have been combusted in the bomb calorimeter, it therefore represents energy losses. Typically, this value is around 25% of the GE but of course is highly dependent on the digestible parts of the feedstuff [19,43]. The remainder of the energy is termed digestible energy (DE) which is taken up in the intestine and utilized by the salmon. Minor amounts, 3–6%, will be lost in energy-requiring processes such as releasing urine, gases across the gills and branchial energy [44]. These are limited values for salmon compared to mammals, see equation below. The remaining energy is used for metabolism, metabolic energy (ME), and the difference between DE and ME is therefore small. Some of the DE is used for heat production inside the fish's body (15%). Even though salmon are cold blooded, some energy is still used for this. It can be looked on as energy to run the body. On RAS farms with a high degree of water reuse, however, this will be noticeable and will lead to a temperature increase, but in flow through farms this will be negligible because so much energy is needed to heat up the large amount of water in such a farm. The low amount of energy used for heat production compared to warm-blooded mammals is, as mentioned earlier, one of the major reasons for the advantages of fish compared to mammals regarding feed used compared to weight gain. The maintenance energy requirement for fish is in general less than for mammals because the energy needed to heat the body is reduced, as is the energy for release of ammonia across the gills instead of converting to urea or uric acid [45]. The following estimated values listed for metabolic energy illustrate the difference between fish and mammals [45]:

$$\text{Subadult rainbow trout: ME} = 8.85\ \text{BW}^{0.82}\ \text{kcal per kg per day (BW = body weight)}$$

$$\text{Mammals: ME} = 70\ \text{BW}^{0.73}\ \text{kcal per kg per day}$$

$$\text{Birds: ME} = 83\ \text{BW}^{0.75}\ \text{kcal per kg per day.}$$

Another equation shows the differences between terrestrial animals and fish but in the form of average maintenance energy expressed as the basal metabolic rate, with corresponding values of 0.7 and 0.01–0.07 (MJ/ kg $^{0.75}$/day) respectively for animals and fish [43].

The remaining energy is then used for basal metabolism, activity and growth, and is named net energy (NE); this is around 55% of the gross energy, and of this around 30% is used for growth [19].

The energy requirement will vary. It is lowest for fasting, for which values from 30–80 kJ/ kg body weight/day have been reported [43]. Maintenance energy requirements of 75–100 kJ/ kg body weight/day have been reported, with corresponding values for salmon in full growth of 15–18 MJ DE/kg live weight gain [43]. Those values are, however, only rough values because they depend on several factors including water quality and temperature, body size and growth rate (Figure 10.24). More detailed models to calculate energy requirements, both for the entire requirement and for different parts, are described later. Energy needs increase exponentially when going above the temperature for optimal growth (Figure 10.25).

Growth is, of course, the most important factor when farming salmon for the consumer market. However, when the salmon start to mature energy is also used for development of gonads, and this is a reason for slaughter before maturation. The energy used for activity will vary, and if the aim is only to keep the fish alive with no growth the energy requirement drops; this is called standard

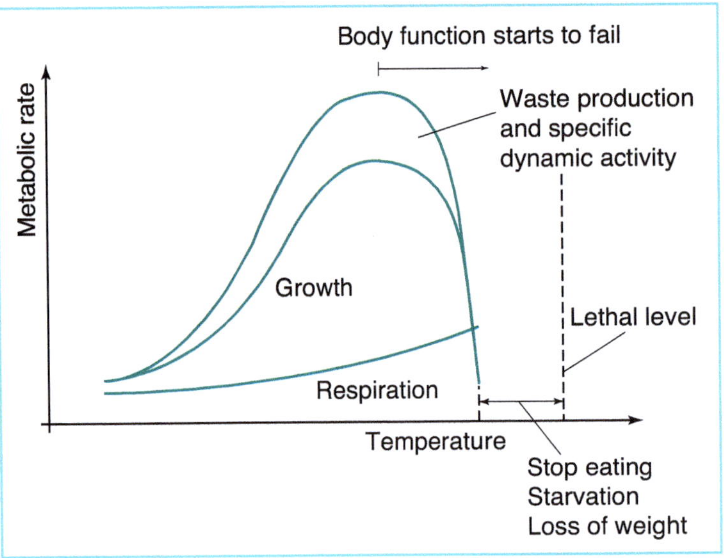

Figure 10.24 The need for feed increases with temperature, and as much as possible must be used for growth (based on [47]).

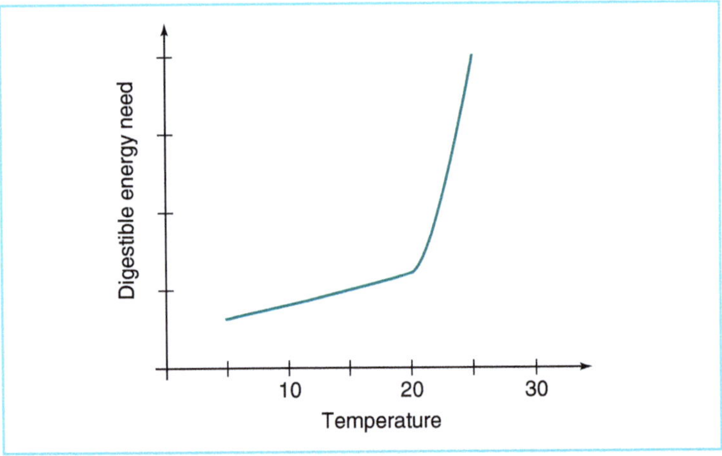

Figure 10.25 The energy requirement increases exponentially when exceeding the temperature for optimal growth (based on [48,49]).

metabolic rate. This also shows that when not feeding, the oxygen consumption will drop along with the energy requirements, which is why feeding must be stopped when there are problems maintaining satisfactory oxygen levels in production units. Metabolism for normal growth and activity is called routine metabolic rate. Active metabolic rate is when the salmon are forced to increase activity and the aerobic metabolism increases. This requires an increased uptake of oxygen, so the oxygen consumption will increase above normal levels. The amount of feed to get 1 kg weight gain will also increase, because more energy is used for purposes other than growth. The increased activity may, however, improve both growth rate and salmon welfare.

The digestible energy of the feed is not so easy to monitor for salmon as for animal livestock. The major reasons for this is because it is difficult to measure gases released across gills, substances released with the urine that will be diluted in the water and the soft faeces that dissolve easily in the water and are therefore difficult to collect. A respirometer can be used for this purpose, but generally the apparent digestibility coefficient is measured (apparent digestibility coefficient [ADC], described more later).

10.7.7 High-energy feed

In recent decades a change in the composition of salmon feed has taken place; the amount of protein has reduced and the amount of fat has increased (Figure 10.26). This has resulted in increased growth [50,51]. The ratio change is a benefit not only due to protein being expensive but also because the available energy per kilogram feed supplied is higher when the amount of fat is increased. The result is that the number of feed particles that the fish must consume is reduced, as there is more energy in each feed particle. But it is, of course, important to have the correct energy to protein ratio [52]. Using feed with a high-energy content may increase the fat deposition. This can be seen lying in the belly cavity and around the intestines, but the deposits in the muscle tissue will also increase, meaning a higher fat content in the fillet [53]. Use of higher quantities of lipids in particular will also require increased numbers of antioxidants in the feed to avoid oxidation, and this represents an increased cost.

10.8 How salmon take up nutrients: digestion

Digestion breaks down nutrients so they can be transported from the intestine to the blood and then to the cells for utilization. Digestion, or catabolism as it is also called, of the main nutrients

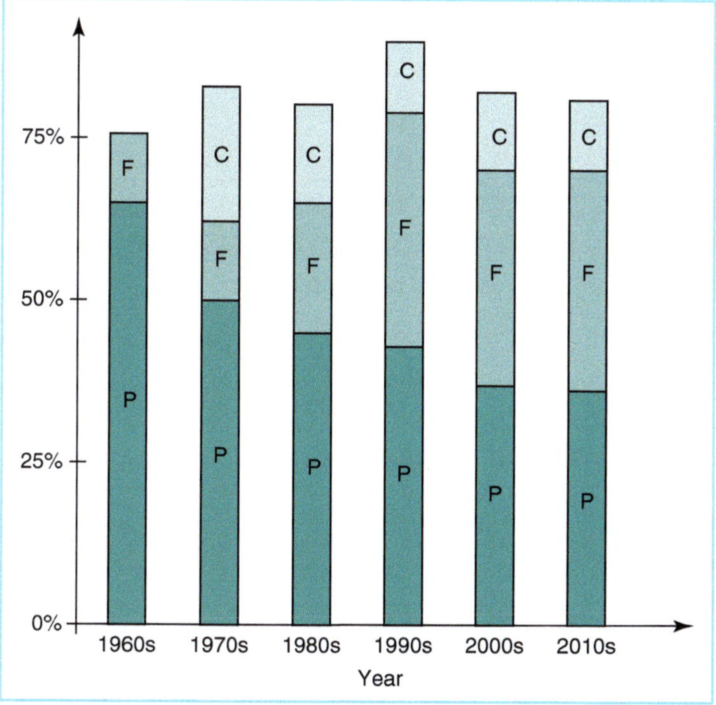

Figure 10.26 Development of the content of the macronutrients in the salmon feed, based on [54–56] (P = protein, F = fat, C = carbohydrates).

occurs by a combination of pH changes, enzyme activity and microbiota in the different parts of the digestive system, all controlled by hormones. The digestive tract, or the gastrointestinal tract, goes from the mouth to the anus, and can be compared with a hollow fibre of various diameters. It can be likened to a reversed assembly line where the feed is taken apart to produce smaller, more basic components that can be absorbed [57]. The feed is continuously broken down and becomes liquidized (hydrolysed) when it goes through the digestive tract. The feed is taken in through the oral cavity (mouth/teeth) and travels through the oesophagus to the stomach, then passes through the pyloric caeca and the intestine; undigested feed is then released through the anus (Figure 10.27). Two other organs are also of major importance when digesting the feed: the pancreas and the gallbladder, which release enzymes and bile salts, respectively, into the small intestine/pyloric caeca. The pyloric caeca, located in the small intestine, comprise a number of blind sacs, creating an increased volume for digestion and absorption. Salmon differ from mammals because in mammals digestion starts in the mouth with chewing and secretion of saliva rich in enzymes, a process called mastication; here enzymes such as amylase start the digestion of starch. As discussed, the digestibility of the feed is of major importance, and can be calculated by comparing the contents of the feed with the undigested components exiting as faeces (described later).

The uptake or absorption of nutrients occurs in the intestine. However, before they can be taken up the feed has to be broken down into smaller elements (molecules) of a size that can be absorbed. This is done through chemical, mechanical and minor biological breakdown processes. The breakdown is mainly chemical (hydrolysis), but there is also some mechanical breakdown during the transport of the feed through the digestive system. Biological breakdown, through enzymes produced by bacteria, is little developed in salmon compared to herbivorous (plant eating) fish species. Chemical breakdown is mainly performed by enzymes working under different pH conditions in the stomach and intestine. After the breakdown process the molecules will be of a size that can be absorbed from the intestine and transferred over membranes and into the blood for further transport to the cells. This occurs by osmosis (concentration gradient) or by active transport (which requires energy).

Salmon have a stomach but their gut length is short, as for other carnivorous fish species, and is typically illustrated by relative gut length (RGL) (gut/body length). Carnivorous species in general have a typical ratio between gut length and body length of 0.2:2.5 (in salmon this is around 0.5 [58]), while the same value for herbivorous fish is in the area of 0.8:15 [21]. To compare, the value for humans lies in the area of 3:5. This results in a reduced time for digestion of

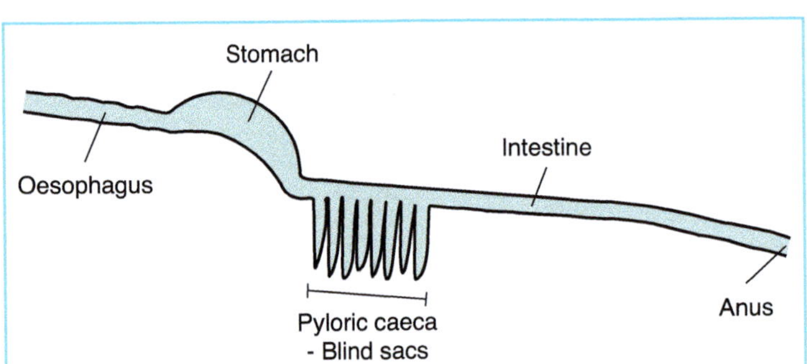

Figure 10.27 The digestive system of salmon.

feed compared to other fish species, such as tilapia, which is a reason why salmon cannot digest high amounts of carbohydrates, such as plants. If plant material is added to the salmon's diet, pre-treatment of the raw material is necessary, for instance enzymatic treatment, so that it is more easily digestible for the fish.

10.8.1 The journey of a feed particle

The feed is first taken into the mouth and is then pushed down through the oesophagus to the elastic S-shaped stomach. In salmon the oesophagus is short and goes directly into the stomach. The walls are elastic so quite large particles can pass. This is not important in aquaculture because here the particle size is optimized for the mouth opening, but it is important under wild conditions where the prey varies in size. Mucus is produced in the mouth and throughout the entire digestive tract to enable feed to pass smoothly and to protect the tract from damage. In the stomach the feed meets an acidic (low pH) environment, due to the release of gastric acid (hydrochloric acid, HCl). This will also destroy some microorganisms that might have accompanied the feed. Breakdown of the large molecules in the feed is initiated here by the enzymes released in the stomach, such as pepsin, a protease. The stomach is surrounded by muscles that can change the volume and thereby apply mechanical forces to the feed. The degree of filling of the stomach controls the appetite and food intake; if the stomach is empty the fish will eat, while a full stomach inhibits the fish from eating. After the stomach, the now-liquid mix continues into the intestine; this is relatively short, shorter than the length of the salmon. It first enters the small intestine, and here the environment changes from acidic to neutral/alkaline and new enzymes are released into the mix of partly broken-down feed: the pancreas releases enzymes, and bile (produced in the liver) is released from the gallbladder. At the start of the intestine the pyloric caeca are located, which are a number of blind sacs that increase the surface area of the first part of the intestine. Absorption of the digested materials begins here and continues throughout the intestine. Water is absorbed in the last part of the intestine, making the faeces a bit denser. The pH level is an important part of digestion: the stomach is acidic (pH 3–4), then pH increases to slightly alkaline (pH 7–9) at the start of the intestine and drops to neutral in the hindgut. The pH is higher than in mammals because more bicarbonate is secreted. Mucus is secreted in the stomach and intestine protecting the surfaces against the pH conditions; this secretion is continuous and the excess is excreted with the faeces, so there needs to be a balance between the secretion and the excretion to avoid problems. The health of the gut is important both to ensure optimal uptake of nutrients but also to protect against pathogens. How the actual digestion process occurs depends on the nutrient, and differs between protein, fat and carbohydrates (described later).

10.8.2 Gastric evacuation time

Several factors affect how long the digestion process (gastric evacuation time [GET]) takes [59–61]. These factors include water temperature (faster with increasing temperature), fish size (longer with increased size), energy content of the feed particle size (increases with size), particle consistency (increases with looser pellet) and amount of feed (increases with consumption) [62]. In one experiment using salmon in seawater (1131 g), the small intestine emptied after 48 hours, but some contents were still present in the hindgut [61]. In the same experiment 50% of the small intestine was emptied after 6–12 hours. For rainbow trout, values from 60 hours to 100 hours have been observed at temperatures of 11°C for fish with approximate sizes of 200 g and 1500 g, respectively [62]. A number of models have been developed for estimation of the gastric evacuation time (for example [63]); this includes a simple approximate one for rainbow trout [62]:

$$\text{GET} = 1995 \times (12{,}809\text{–}0.387 \times \text{T}) \times \text{W}^{0.1192}$$

where GET = gastric evacuation time, T = temperature, W = fish weight in grams.

It is important to remember that this relates to total emptying of the stomach. The GET is an important value because it provides information about feeding frequency: salmon will start to eat again when the stomach is empty, as this is the appetite trigger. The appetite will gradually increase as the stomach empties, and at a certain degree of stomach filling fish start to eat again (Figure 10.28).

10.8.3 Protein digestion

Protein must be broken down into smaller molecules, either to amino acids or to smaller peptides, before it can be absorbed. To illustrate protein digestion in a simple way is difficult because the pathways are so complex, but a simplified description follows (Figure 10.29). In salmon the breakdown of protein starts when the food enters the stomach where it meets an acidic environment (HCl) suited to hydrolytic breakdown of the peptide bonds in protein. This is carried out

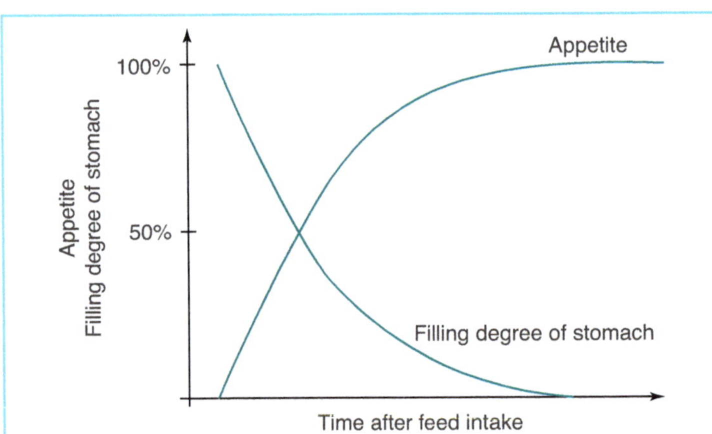

Figure 10.28 Appetite of fish in relation to the filling degree of the stomach (based on [62]).

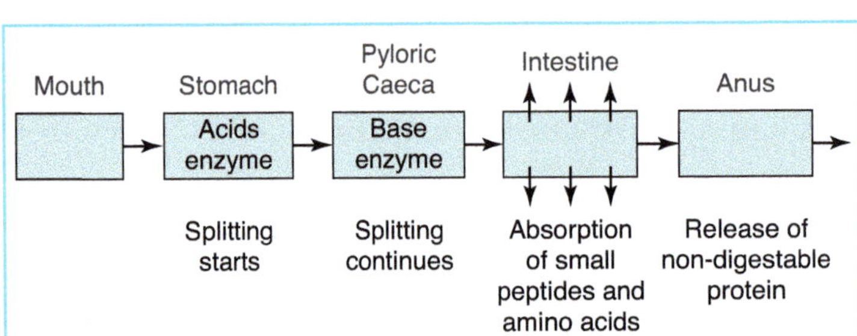

Figure 10.29 Protein digestion.

by enzymes released in the stomach (proteases) which are adapted to work under acidic environments. The protein continues to the pyloric caeca in the small intestine, where the pH changes to neutral/slightly alkaline, based on release of calcium carbonate. Here enzymes suited to this environment, for example trypsin released from the pancreas into the small intestine, work on the breakdown of the remaining protein and peptides. The broken-down protein now consists of small peptides and amino acids, which continue through the intestine and are now small enough to be absorbed and carried in the blood to the organs and muscle. Here they are used for building new proteins (protein synthesis) and other molecules and as a substrate for energy (as part of gluconeogenesis and lipogenesis) [4]. One experiment shows that more than 50% of the protein is absorbed in the pyloric region [64].

10.8.4 Fat digestion

Fat or lipid breakdown occurs in stages and is a more difficult process to describe than breakdown of protein [65,66] (Figure 10.30). Fat is not dissolvable in water, regardless of pH, and the enzymes that break it down do not act directly as for protein. Fat flows through the stomach as separate fat droplets in the digestive liquid (chyme). In the first stage, the fat droplets are broken down into smaller globules (drops) with the help of bile salts and lecithin (phospholipids) released from the gall bladder. The bile salts have the same function as a detergent (soap) when washing; they emulsify the fat (emulsification). The bile salt molecule has a part that dissolves in water (hydrophilic) and a part that dissolves in fat (hydrophobic). This takes place in the small intestine, so there is no digestion of fat in the stomach. In the pyloric caeca, the surface area and retention time are increased, and there are high levels of digestive enzymes. Through emulsification the fat is split into several smaller globules in the digestive liquids and becomes more water soluble. The bile salts reduce the forces that initially keep the fat droplets intact. Emulsification increases the surface area of the fat, so creating a larger area on which the different lipase enzymes can work. The lipase enzymes that are released from the pancreas split the fat fraction through hydrolysis, again with the help of bile salts. The triglycerides are now separated into the free fatty acids and a glycerol molecule. This is therefore a two-step process: first globules, very small fat bubbles, are formed and then these are divided again to create what are named micelles. These are tiny globules, diameter typically around 3–7 nm, with a lipid core and surrounded by bile salts. These dissolve in the digestive liquids; the fat core with fatty acids, monoglyceride and glycerol are absorbed by diffusion, while the bile salts are used again. The lipid absorption mainly occurs in the pyloric caeca and the anterior intestine (duodenum) [64]. The free fatty acids are then transported to the liver where lipid metabolism occurs (described later). Wax esters, for instance

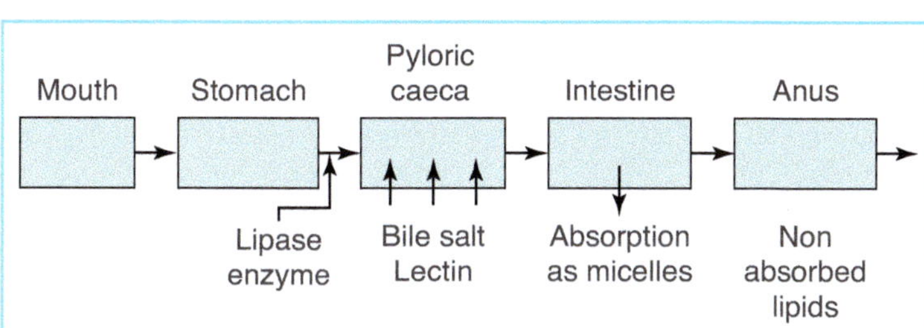

Figure 10.30 Fat digestion.

from copepods, may also be included in the diet as a part of the lipid fraction for salmonids [65]. They are hydrolysed slower than triglycerides, so to include more than 30% of the total lipid fraction as wax esters may result in growth depression and reduced digestibility [67]. More fat is deposited in the viscera when high levels are included in the feed [50,53].

10.8.5 Carbohydrate digestion

Digestion of carbohydrates starts with the enzyme amylase working under acidic conditions in the stomach (Figure 10.31). This creates a mix of partly digested feed called chyme, an acidic semi-fluid mass, which continues into the small intestine where more amylase is released from the pancreas and where the pH is increased by bicarbonate. This results in breakdown of some polysaccharides to disaccharides. In the small intestine several other enzymes are also released (lactase, maltase, sucrose) and the disaccharides are further broken down to monosaccharides (simple sugars) such as glucose, which can be absorbed in the intestine through diffusion. The problem with salmon is that this process is not very well developed; salmon, a carnivorous species, are adapted to eat a diet containing protein and fat. Adding a high amount of carbohydrates to salmon feed can result in not only a negative nutrient value, but also a negative environment in the gut, and high amounts of carbohydrates may in addition inhibit uptake of other nutrients. Salmon are therefore suited for what is referred to as a low-carb diet for humans. Starch, a polysaccharide much used in feed production as a binder, has low digestibility in salmon. Fibres (cellulose), also used for the same purpose, are indigestible and will pass through the fish undigested. Even though starch and cellulose in large amounts are unwanted, they may be present in the ingredients or raw materials that are used when composing the feed. Non-starch polysaccharides (NSP) may, however, be broken down with enzymes during feed manufacturing [68].

10.8.6 Digestibility or bioavailability – key words

It is not the exact value of the nutrient content in the raw material that is important, nor the gross energy, but what is called the bioavailability. This is the biologically utilizable fraction: the digestible part that salmon can take up and utilize in relation either to nutrients or to energy. This can be illustrated with raw starch or cellulose, which have an energy content when combusted in a bomb calorimeter, but almost none can actually be digested by salmon; therefore the useable energy will be low, so it has a low bioavailability or digestibility. Another example here is

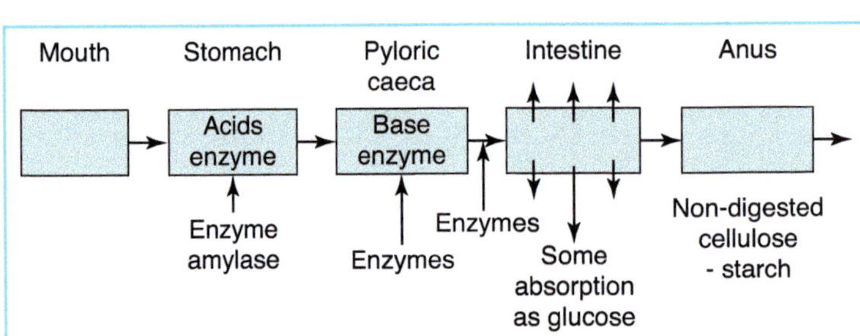

Figure 10.31 Carbohydrate digestion.

phosphorus, a mineral that is of major importance for bone formation in salmon. In addition to bone formation, which accounts for approximately 75% of phosphorous use, small amounts are used for ADP/ATP in muscle metabolism, for phospholipids in cell membranes and in DNA. The ingredients typically used in fish feed contain enough phosphorus if measured directly, but only 30–40% is available for the fish and the rest is in indigestible forms. This means that additional phosphorus must be added to the feed. The indigestible phosphorus will be released into the environment creating potential problems of eutrophication. Bioavailability is also a major consideration when changing the ingredients in the feed. For instance, in plants the major phosphorus component is in the form of phytate which has low bioavailability because salmon lack the digestive enzymes required to release the phosphorus from phytate [69]. Therefore, monocalcium phosphate or an alternative inorganic phosphorus source [70] are typically included in the diet. The enzyme phytase that increases the bioavailability of phosphorus may also be included in the feed.

Crude protein, crude lipid or crude carbohydrate are therefore not the same as digestible or bioavailable protein, lipid or carbohydrate. The difference between crude and digestible values are important for waste discharge from the salmon farms. Larger differences will increase the amount of faeces released. To quantify the digestibility, digestibility coefficients (DC) can be calculated. The true digestibility coefficient (TDC) shows how much of the supplied nutrients are digested and can be calculated in the following way:

$$\text{TDC} = \frac{\text{Nutrients consumed} - (\text{Nutrients excreted in faeces} - \text{Endogenous losses})}{\text{Nutrients consumed}}$$

Compared to mammals the metabolism of fish is much more difficult to measure because collection of urinary excretion is difficult in water, the faeces are soft and dissolve easily and the gases released across the gills dissolve in water. Due to the fact that digestibility is tricky to measure in fish in general, a typical method for measurement is represented by the apparent digestibility coefficient (ADC).

The following equation is used for calculation of the APC

$$\text{ADC}(\%) = 100 - 100\left[\left(\frac{I_D}{I_F}\right) \times \left(\frac{N_F}{N_D}\right)\right]$$

where I_D = inert marker in the diet, I_F = inert marker in the faeces, N_D = actual nutrient in the diet as protein and N_F = actual nutrient in faeces.

The equation may also be used for energy and dry matter.

Using this method, it is possible to measure protein digestibility, which is typically in the area of 80–95% (Table 10.1). For lipid digestion this method gives variable results, but the digestibility in general increases with the degree of unsaturation and carbon chain length. The ADC of lipids is generally high, above 80%, and values of up to 97% for fishmeal lipids have been reported ([71,72]. For energy, the same values are attained.

As shown, an inert marker must be included in the feed. Various markers can be used [73,74], one of which is yttrium oxide (a rare earth metal oxide) [75,76].

Table 10.1 Apparent digestibility for some protein sources for Atlantic salmon (based on [21,77–79]).

Source	Apparent digestibility (%)
Fish meal	91–96
Meat, bone, blood meal	70–94
Soybean meal dehulled	83
Soybean concentrate	93–97
Wheat gluten	98–99
Alga meal	83–85
Brewer yeast	87

The retention of the nutrients will also give some input to overall digestibility. This can be found by data from total feed consumed and total fish produced, and knowledge of the compositions of both, described more later.

10.8.7 Theoretically optimized feed

Fish feed is, as shown, not a theoretically optimized feed, but it is interesting to consider this in relation to the understanding of the digestive system and feed composition. This is therefore only a theoretical exercise because it is impossible in practice with today's knowledge and feed production methods based on the use of available raw materials as ingredients. There is a lack of knowledge about precisely how fish digest feed, and the feed is not produced optimally based on separate and single molecules: it is not fully chemically composed or synthetically produced from single molecules. It is not yet known how to produce a fully synthetic feed optimally in relation to growth and all the other processes in the fish that the molecules in the feed support. Through evolution, salmon have adapted to a certain type of feed.

The closest comparison to a fully synthetic food is intravenous feeding in humans, but this is not optimized for maximum growth. Use of a theoretically fully optimized feed would mean that all of it is consumed and digested, only releasing carbon dioxide, water and ammonia over the gills with no release of faeces from the fish. This is, however, only a hypothetical thought because it is impossible to do. Among other reasons, the digestive system needs something to 'work on'.

10.9 Practical composition of salmon feed

When salmon feed is produced today, traditional raw materials are used in a mix in the best possible way with additional necessary micronutrients, vitamins and minerals (Table 10.2.). Typically, some type of dried raw materials/ingredients, meal, is used with addition of oil to include the lipid fraction. When selecting the ingredients/raw materials to produce feed several factors must be taken into consideration. First of all, the mix of ingredients in the feed must meet the nutritional requirements of the salmon, and it must of course be digestible, or bioavailable, for the salmon. Typically, there is a min–max range that is used when deciding upon the

Table 10.2 Example of composition of salmon feed (values from [54] from Norway in 2016).

Content	%
Marine proteins	14.5
Marine oils	10.4
Plant proteins	40.3
Plant oils	20.2
Carbohydrates	10.6
Micro-ingredients	4.0

proportions of the feed components. The price is also of major importance, and the produced feed must satisfy the required technical quality.

The main ingredients in salmon feed have traditionally been fish meal and fish oil produced from small pelagic fish species, such as anchovies, sardines and menhaden, from the south Pacific, and also from species in the north Atlantic, such as blue whiting and capelin. The fishing of these species is called industrial fishery or reduction fishery and the fish harvested may also be called forage fish. A large part of this fishing occurs in south America, off the coasts of Peru and Chile, where anchovy is the major species. Asia and Norway are the main consumers of fish meal and fish oil. An increasing proportion of the fish meal and fish oil is based on cut offs/trimmings from fish processing, but in total it is still a limited tonnage compared to industrial fishery.

The production of fish meal and fish oil has been quite stable for the last 25 years, and there is very limited potential for any increase. Today's fishing of the natural fish stocks is at a responsible level and stocks cannot be exploited further. The aquaculture industry uses the majority of the available fish meal, up to 70%, and also uses up to 90% of the available fish oil. The need for ingredients for producing fish feed, salmon included, is increasing worldwide due to the increase in the aquaculture industry. A growth of over 6% per year makes feed for fish in the aquaculture sector the fastest growing sector among animal feeds. The increased demand is also reflected in the prices of fish meal and fish oil on the world market, which have multiplied during the last 20 years. The high prices also highlight the possible risks of overexploiting the natural stocks of forage fish, so control of this fishery is of major importance.

Salmon feed composition has changed considerably over the years. The main trend is a gradual replacement of the marine ingredients, fish meal and fish oil, with land-produced ingredients. In 1990, for example, almost 90% of the feed for salmon in Norway was based on marine ingredients (fish meal 65.4% and fish oil 24%), while in 2016 marine ingredients accounted for only around 25% (fish meal 14.5% and fish oil 10.4%) [54]. The changes include new protein and lipid sources, and different types of pre-treatment of carbohydrates to achieve higher digestibility and enable the use of cheap ingredients for energy production in the fish.

Protein comes from two sources: animal protein and plant protein. Animal protein will have a better amino acid balance for use in salmon feed compared to plant protein. The traditional animal protein source, fish meal, has a high protein content and an amino acid balance directly adapted to the needs of the salmon, and also has high palatability and digestibility. Unlike other possible new protein sources, it does not contain any antinutrients that must be specifically handled. More and more plant protein is, however, included in salmon feed, due to the increased demand for fish meal and the resulting increase in price. Plant protein has a different amino acid profile than animal protein (fish meal) which may result in a lack of essential amino acids that must then be supplied additionally. Typically, these include the amino acids lysine, methionine and arginine [21], and histidine in the early growing stages. Too high a concentration of some amino acids, such as leucine, may result in toxicity. Feeding salmon on plant-based diets can, as an extreme, be compared to feeding carnivorous animals such as wolves with plants. When using diets for salmonids that contain high amounts of plant protein care should therefore be taken. The plants also contain carbohydrates that are of low digestibility for salmon, and high amounts can again have negative effects. The amount of starch can be higher than that which is healthy for salmon. It may also be necessary to add other vitamins and minerals [80].

So far, soybean meal has been the main replacement for protein but other plants are also used, such as wheat, maize and other beans.

When starting to change the diet from fish meal to an increased amount of vegetable meal, gut problems were identified and the gut health of salmon deteriorated, which illustrates the problems that can occur when changing ingredients in the diet. This was due to the addition of soybean meal directly into the feed, and the solution was to change the soy characteristics by processing it (by removing some water-soluble carbohydrates) into soy protein concentrate (SPC); this solved the problems. Of the protein, SPC from soybeans was the most important, representing 19% of the feed in 2016 even though it dropped from the same measurements in 2012. Wheat and wheat gluten are also major ingredients in salmon feed (17.9% in Norway) [54].

Animal protein may also be sourced from land animal by-products (LAP). Europe, in general, uses no or small amounts of animal by-products while Chile and Canada include them to a high degree, with over 20% dietary inclusion being reported [81]. The reason for the lack of use in Europe is due to the BSE crisis of the mid-1990s (bovine spongiform encephalopathy, or mad cow disease). Even though it is legal to use non-ruminant processed animal protein (PAP) today and LAP is supported as a sustainable resource in salmon feed, it is still not used, probably due to unknown consumer perception.

The lipid content in salmon feed has, as discussed, typically been based on marine fish oil which has all the necessary EFAs, including HUFAs. During recent years there has also been a change to include more vegetable oil due to the increasing prices and lower availability of traditional fish oil. No negative effects on growth and feed utilization have been observed by changing from fish oil to plant oil. The predominant plant oils are based on rapeseed and camelina. Most of the lipids are used for energy production but some HUFAs not found in vegetable oil must be included in the lipid fraction. When changing the lipid fraction to other lipid sources, attention must be paid to ensure salmon are provided with the necessary EFAs. Therefore, even though vegetable oil is used to an increasingly higher degree, fish oil is still included in the feed at a lower ratio to ensure the supply of HUFAs.

The composition of the feed will dictate the composition of the salmon that enters the consumer market. The salmon fillets bought in the shops are a result of the feed that the salmon have eaten. The easiest observation here is, of course, whether astaxanthin has not been added at high enough levels, resulting in less colour. However, the lipid fraction will also be affected and if feed containing low levels of HUFAs is used, this will affect the fatty acid profile of the fillet. Since the salmon cannot decide which lipids to use for energy production and which to use for muscle deposits, it is impossible for the salmon to only deposit the HUFAs and use the shorter carbon chain and cheaper lipids for energy production. When including high levels of fish oil, HUFAs will therefore be oxidized for energy production. The fatty acid profile in the supplied feed has been changed by decreasing the traditional marine fish oil content and increasing the content of plant oil, and this is reflected in the composition of the salmon fillet where the level of the healthy EPA-DHA fatty acids has reduced (Figure 10.32). The development of average EPA-DHA content in farmed salmon was compared to wild salmon in one survey [82,84]. An example from Scottish salmon farming showed the quantity almost halved from 2006 to 2015 [82]. The result is that the consumer portion needed to obtain the required EPA-DHA intake levels recommended by the health advisory organizations must be doubled [82]. Values from Norwegian salmon showed a drop in concentrations of EPA-DHA from 3% in 2010 to 1.2% in 2016 [54].

It is important to remember that when changing the main ingredients in salmon feed, the requirements for the addition of micronutrients also change. Fish meal and fish oil have the great advantage of being well suited to the needs of salmon, but are geographically and seasonally dependent. In Norway, for example, micronutrients were added at a level of 4% in 2016 compared to only 1% in 1990 [54]. Micronutrients that need to be added include, for example, the mineral phosphorus, which is present in plants but in a non-available form. The composition of the added premix of vitamins and minerals must therefore be adapted when changing the main feed ingredients.

The chemical composition of salmon feed varies to some degree between the salmon producing countries based on the locally available ingredients. To illustrate this, the composition of Scottish salmon feed has been listed as 37% lipid, 36% protein, 11% carbohydrate/fibre, 9% ash and 7% moisture, with an energy content not below 22 MJ/kg [81], while in another example from Norway in 2016 the average fish feed composition included 93.4% dry matter, 35.6% crude protein, 33.5% crude lipid and 11% carbohydrates, with an energy content per kg of 23.7 MJ [54]. The ash content is typically in the area of 8–9% [83]. The composition will, as previously mentioned, vary with the size of the fish.

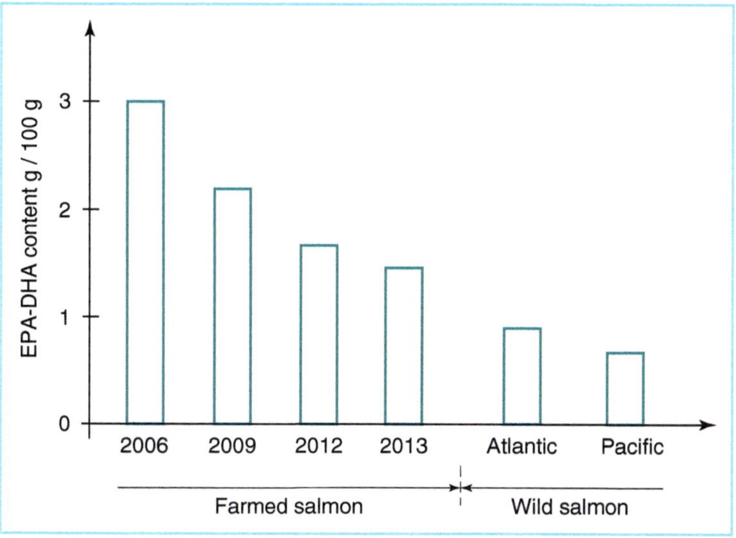

Figure 10.32 Development of average EPA-DHA content in farmed salmon compared to wild salmon in one survey [82,84].

10.10 The race for new fish feed resources

The general demand for biomass in the world, of either marine or vegetable origin, is increasing. This includes feed ingredients for all types of animals including fish, but a new important contender has entered the arena: biomass is needed for biofuel and bioplastic production. Biomass for the new areas of biofuel and bioplastics, plus the growth in the world aquaculture industry, is stretching biomass sources. The traditional marine resources used for production of fish meal and fish oil are finite resources. The availability is therefore limited, and prices increase. However, fish meal and fish oil are not essential ingredients in salmon feed, as the required nutrients can be taken from other sources. An important factor to take into consideration here is that it is difficult to find ingredients that fulfil salmon's nutritional requirements as well as fish meal and fish oil do. The prices that are demanded today for fish meal and fish oil are high and to find new resources that are competitive in price and quantity is not easy. When searching for new raw materials for salmon feed it is, of course, also an important consideration that they can produce a healthy salmon; trends have already been observed of a reduction in HUFA content when changing to a more plant-based diet.

As discussed, plant material has taken over from fish meal and fish oil in the salmon's diet, and today salmon diets typically contain more than 70% plant-based materials. Compared to the overall world production of protein and edible oil, fish meal and fish oil account for a very limited share. A 4% increase in the world soybean production will nearly equal the world fish meal supply [85]. Today plants provide the majority of both protein and edible oil in the world (94% and 86%, respectively).

New production methods to optimize the use of limited resources is one way to handle the deficits, for example the utilization of the valuable HUFAs (EPA and DPA). Trials to provide higher contents of these fatty acids for a limited period just before slaughter to boost the content in the salmon fillet have been carried out with good results. Again, fish oil is used but the total amount needed for this limited period is, of course, reduced compared to using fish oil through the entire production period; however, if early slaughter is required this strategy is problematic.

A range of factors must be taken into consideration when discussing which sources are to be used for feed production in the future. These may include the wish to avoid using agricultural land for production of raw materials for fish feed. It is also of interest to find alternatives that do not compete with food sources that can be directly used for human consumption. Regarding energy and nutrient utilization, it is of course best if food can be used directly for humans, rather than first being used in fish which then become feed for humans. The plants used in salmon feed today are to a large degree the same that can be used for direct human consumption, such as soy, wheat, corn, beans and sunflower. Also, regarding sustainability of the salmon industry, the focus on finding new raw materials to replace fish meal and fish oil is a matter of discussion. The reduction, industrial or forage fish species used for production of fish meal and fish oil may be classified as low trophic level species (LTLS) and occur in great numbers; they are important species for larger fish and marine mammals. Instead of utilizing them for reduction to fish meal and fish oil, some claim that it is better to let them remain as prey for species at a higher trophic level, or instead of using them for feed production, to use them for direct human consumption (however, this is debated [86], IFFO position statement, [87]).

Of course, it is of major importance to have good control of the fishery for these species because as they are at a lower trophic level, overfishing will affect the upper part of the food chain. Today the control of this fishery is generally good, but high prices may challenge this in the future. The main species used for production of fish meal and fish oil is anchoveta caught in the south Pacific, off the coast of Peru and Chile, and this species may also be used for direct human consumption. However, although efforts have been made to increase their use for direct human consumption the results have been negative. It is important to remember that these are small fish with many bones. But this raises questions regarding the use of fish for production of salmon, and the ratio of fish in to salmon out, or the 'fish in fish out factor' (described more chapter 10.11.3). In the early days of salmon farming the industrial fish requirement for the production of 1 kg salmon was in the range of 3–4 kg, which means 3–4 kg industrial fish was needed to produce 1 kg salmon. This value has, however, been highly reduced by the utilization of plants in the salmon diet and today the ratio is around 1 kg fish for production of 1 kg salmon (described more precisely later). But even though the aquaculture industry is the major consumer of HUFAs, it is also the major supplier of EPA-DHA to the human population [84], and makes it available in a preferable form for humans.

10.10.1 Potential new resources

The industry is continuously searching for new raw materials for production of feed, either new resources or by-products from other industries. By-products or residuals from other industries are, of course, of great interest because these are resources that are already there and easy to collect. The problem is that the quantities required in order to evaluate whether they are of use is quite high. One newly successful introduced material is guar meal, a by-product from processing guar seed, which is used as a protein concentrate.

Potential new biomass to include in salmon feed, whether protein or fat, may be grown/farmed or harvested from natural stocks of other sources than those utilized today. The latter is typically harvested from the ocean, while biomass production through farming can be carried out either in water, mainly marine, or on land. A distinction can also be made between organisms that need to be fed and those that do not, where an example of the latter includes biomass harvested from the ocean, but may also include organisms that are 'planted' in the sea and use naturally occurring nutrients for growth, including, for example mussels, seaweed and tunicates. The interest in biomass production is high today, driven by the biofuel industry that needs large quantities of biomass. However, to grow biomass, as for growing salmon, there is a requirement for nutrients for both metabolism and growth. The organisms or plants that are grown either use photosynthesis, which takes energy from the sun, or take energy from oxidation of organic matter. Both will, however, require a supply of nutrients for growth, if not using naturally occurring nutrients in the environment (non-fed). As soon as nutrients must be supplied it presents a challenge because it affects the price of the produced biomass. Utilization of different waste streams as a nutrient source is therefore of major interest.

10.10.1.1 Resources from marine sector

Resources from the sea have always been an important component of salmon feed, and in the future the potential is good for the inclusion of other marine organisms [88–91]. Harvesting of non-traditional sources from the ocean is under development; these include, for example, mesopelagic fish and sources from lower trophic levels such as krill, zooplankton (copepods) and seaweed (macroalgae). Krill oil and krill meal have gradually become commercially available and included in the salmon feed, even though the price is quite high. The composition is well suited for use as an ingredient in aquaculture feed, and the content of HUFA is particularly interesting. There is interest in utilizing zooplankton such as copepods (a group of small crustaceans) in feed for fish, and several trials are ongoing. These organisms may also contain HUFAs. One copepod whose commercial harvesting is under development is *Calanus* (*Calanus finmarchicus*). The total biomass of copepods in the oceans is very large, indicating their potential as ingredients in feed.

Mesopelagic fish stocks have not previously been used commercially. These are fish living at great depths, from 200 m to 1000 m, and represent large biomasses; estimates say between 2 and 19 gigatons or up to 100 times the total amount of traditional fishing stocks [92–94]. A number of small-sized species, a few centimetres long, are included in this group of fish, where the lantern fish is the most common. There are also mesopelagic species containing high amounts of lipids (≥50% of dry weight), and all species are good sources of HUFA (≤20%). The protein content is also good, slightly below 20% of wet weight, and therefore mesopelagic fish are a real source of interest for further feed production.

However, there are challenges with deep water fishing [95]. Effective fishing gear is not available and profitability has not yet been achieved. After catching, the fish are rapidly broken down by enzymes, as for krill, meaning that early onboard processing is important. Much of the lipids are in wax form, so are less easy to digest. Harvesting earlier in the nutrient chain also sets much higher requirements for control because taking out too high a number here could have catastrophic consequences. The foundation of the nutritional pyramid may be taken away, resulting in full collapse. Therefore, it is extremely important not to overexploit. In addition, it is very difficult to estimate the mesopelagic fish biomass with today's methods, so to find the maximum sustainable yield is almost impossible [97]. More knowledge is therefore required before taking mesopelagic fishery into a full-scale industry [92].

Filter feeders utilize suspended solids naturally occurring in the water as nutrients and no additional feed needs to be supplied. Filter feeders include scallops, clams and mussels. Because there is no need for additional feed, there is interest in utilizing these as ingredients in salmon feed. Mussel meal is used to a degree today but the production cost is still high compared to the price for the final product. Tunicates, also called sea squirts (*Ciona intestinalis*), are a potential raw material that is being tested now [98]. They are small filter-feeding animals that trap nutrients (particles) by pumping water through their oral siphon. They are an excellent source of protein and may contain 50–60% protein (DM basis), and this is the reason for the great interest in their use. In addition, they contain cellulose and also some lipids. Trials in commercial farming are ongoing using technology based on scallop/mussel farming, and only utilizing naturally occurring nutrients in the water. Estimates of production of 100–200 kg/m^2 are given [99]. With an expected water content of 95%, 4.5 kg protein can be taken out per square metre. Taking into account the large coastal areas available there is great potential.

Plants in the ocean, also called macroalgae or seaweed, have been produced for centuries in Asia, and the world production is large [100]. Around 29 million tonnes are farmed and an additional 1 million tons are harvested from the wild stocks [101]. Today a very limited amount of this, about 1%, is utilized in the total production of animal feed, fish included. However, the

Figure 10.33 Seaweed is a possible new feed ingredient.

potential for increased growth is huge with large areas available in the ocean (Figure 10.33). As a biomass resource the interest in utilization is therefore increasing, including as an ingredient for feed production [100,102]. Compared to using land areas for growing feed ingredients, an advantage in utilizing the ocean is that much larger areas are available throughout the globe (71% versus 29%). Another great advantage is that in the ocean three-dimensional growing is possible (or at least more easy). Seaweed production is an example of an organism that grows into deeper layers. It is also an interesting organism because it includes both amino- and fatty acids, including both omega-3 and omega-6. The seaweed composition will depend on the species. It can therefore be an excellent source of marine protein [103], and some seaweed species may be a good source of lipids including HUFAs [104,105]. However, the biomass cannot be used directly in salmon feed because the nutrients are mainly in indigestible forms, resulting in low digestibility. Possible solutions to convert the nutrients to digestible ingredients for salmon feed are enzymatic treatment for breakdown or use of the seaweed as feed for microorganisms that could then be used as feed ingredients. Using seaweed compared to land plants is currently a more expensive production process, the drying cost representing a major cost [106]. However, by increasing and optimizing the production the potentials are large, and high yields can be achieved.

10.10.1.2 Land production

New feed resources grown on land will, as for the marine sector, include organisms from lower trophic levels. Insects are one of the interesting resources from a lower trophic level, and a number of insects can be grown under farming conditions [107–111]. One example where growing insects for industrial production is under development is larvae from the black soldier fly [112]. These are processed to insect meal and have come quite far in development.

Microalgae, yeast, filamentous fungi and bacteria are all microorganisms in lower trophic levels that contain nutrients that could be used in feed (review [113]). EPA-DHA are naturally produced in the marine environment by algae, and as the organisms higher up the nutrient pyramid eat the algae there is accumulation in higher trophic levels, such as in zooplankton and fish. Since HUFAs are based on marine algae, the first suggestion will, of course, be to start farming such organisms. The other possibilities are to transfer genes from algae into other organisms and then grow these organisms, resulting in GM products (later described).

Growing microalgae for use in fish feed is therefore of great interest, in particular to obtain HUFAs but also for general biomass production (Figure 10.34). Algae are either produced heterotrophically by fermentation (organic carbon, produced in fermenters) or phototropically (produced by photosynthesis in bioreactors, requiring access to light and carbon dioxide). Fermentation with sugar as the organic carbon source is easy, but the use of waste products will be of interest, for example cellulose or other products indigestible for salmon.

Algal biomass is produced today and has been included in fish feed on a commercial scale, but is based on non-HUFA algae [114]. There is also commercial production of microalgae containing astaxanthin [115–117]. Algal meal, independent of species, will contain both protein and lipid but to varying degrees, and only some algal species contain HUFA. Typically, DHA has been produced in fermenters, but other species have now been taken into consideration which produce both EPA and DHA. The interest in producing algae containing EPA-DHA is increasing and a number of companies are starting production [84]. This also includes production of GM yeast with genes from algae that makes the yeast able to produce PUFA, and oils based on this have been included in salmon feed [118]. However, producing HUFA based on algae is still costly so the price is high, as for algal production in general, but commercial products that contain both EPA and DHA are entering the market [119]. One of the suppliers is setting up large factories [120] and have based their production on sugar fermentation. Fungi, both in yeast form and as filamentous fungi, is also of interest. In both algae and fungi, the content of protein and lipid can be varied by both chosen species and growth conditions. Both categories may therefore include PUFAs, but not necessary HUFAs.

An important factor in lower trophic levels and when growing microorganisms for biomass is the growth rate of the organism, or generation time as it is typically named for these sizes of organism. The algae, protozoa and fungi that are used have a generation time or generation interval usually in the range of 2–15 hours at temperatures from 20°C to 30°C. Bacteria will typically have an even shorter interval, 1–3 hours, but several are <1 hour. As an example, the generation interval of *E. coli* is 20 minutes, but typically the temperature requirement is higher, at 37–40°C.

Unicellular or single celled microorganisms (SCM) can be further divided into single cell protein (SCP) and single cell oil (SCO) microorganisms. Single cell protein can be defined as protein extracted from the produced microorganisms, while single cell oil can be defined as

Figure 10.34 Growing phototropic microalgae requires light.

edible oil extracted from the produced microorganisms [121,122]. Several organisms contain both protein and oil, but this is dependent on the type of microorganism, species and growing conditions.

The single cell organisms need nutrients for growth; some utilize photosynthesis while others use the breakdown of organic substances for energy production. The first category may utilize carbon dioxide and therefore can be part of carbon dioxide fixation. The other uses organic carbon chains that are broken down releasing energy. However, both will need compounds, such as nitrogen and phosphorus, for growth, in addition to the carbon. This will represent an important cost for such production and today the production cost of the single cell organisms is still rather too high for use as ingredients in salmon feed. For fermentation, sugar is a good basic substance, but the problem is that relatively large quantities are necessary to produce, for instance, omega-3-rich algae.

It will therefore be a great advantage to utilize available waste streams and residuals. This can, for instance, be waste and waste water from the food industry. Single cell protein can then be produced from inexpensive carbon sources. By-products from fermentation industries, such as breweries, could also be of interest as a feedstuff. Seaweed and spruce wood have also been tested as material for growth [123].

Residuals from biofuel (bioethanol) and the bioplastic industry will also be attractive as ingredients for fish feed. These production systems primarily utilize the carbohydrate part, and protein and lipids included in the residual by-product could be used as ingredients for salmon feed. When producing biodiesel, the lipid fraction is used but protein is available in the residuals. A combined production of biomass for biofuel and fish feed ingredients may be a solution, and then the valuable PUFA parts can be taken out for fish feed production [85].

SCP may also be grown on natural gas [122,124,125]. This can be based on *Methylococcus* fermentation where methane is the carbon source and inorganic nitrogen is the amino acid source. Today there is a commercial product available on the market with similar characteristics to fish meal, >60% protein and 10% fat, which is based on algae produced in this way [126]. In the future, growth on biogas may be a possibility for production of SCPs.

Other sources than those traditionally used today may also be included, for instance, forestry and forestry waste [127]. This provides carbohydrate-rich materials containing cellulose, hemicellulose and lignin, all components that cannot be utilized directly for salmon. They may, however, be broken down biologically, chemically or physically, so parts of the nutrients can be utilized by the salmon. An example here is enzymatic treatment of cellulose. However, a better way where higher utilization of the overall products can be achieved is to utilize it as nutrients for microorganisms such as yeast or filamentous fungi that can later be harvested and used as ingredients in fish feed.

10.10.1.3 Genetically modified ingredients

Including plants in the diet of salmon also raises the question of utilizing genetically modified (GM) crops, also named transgenic plants. In genetically modified crops, the genome is changed by the use of gene technology, and transgenic crops are such examples. By using GM crops more plants could be included in the salmon's diet (Figure 10.35). Today most of the soy on the market is GM, and it is becoming difficult to avoid non-GM soy. The reception of GM ingredients varies around the world: Europe is restrictive while America and Asia are more tolerant. Salmon outside Europe is therefore fed on feed that typically contains GM ingredients such as soy. The reason for the use of GM crops has been to modify them to tolerate herbicides and/or insects. However, a new question related to the use of GM crops/organisms is raised when considering the highly limited sources for HUFA. A new golden age for GM crops/organisms may occur. As an example, a GM yeast strain has been utilized in feed for salmon in Chile for production of a niche product [81,128], but it is still expensive and produced in small volumes. Large volumes containing PUFA will, however, become available if GM oilseed crops containing PUFA are grown where areas and infrastructure are well established for non-GM products. It is possible today to produce vegetable oil containing PUFA by utilizing genetically modified oilseed crops such as soy or canola [84,129–131]. Including genes from algae may be utilized to produce crops containing EPA and DHA. So far there is limited scientific publication within the area, but this is increasing. No negative effects of using GM canola were observed in one trial, but this contained PUFA, not EPA-DHA [132]. More studies have been done on oil produced from GM camelina and no negative effects were observed in these either [84]. A recent study showed that both fish oil and rapeseed oil could be replaced with camelina oil that contains high levels of EPA-DHA [133]. However, GM ingredients that contain EPA and DHA are not available on the commercial market in significant quantities. In the future these may, however, be a major ingredient for salmon feed.

Figure 10.35 Genetically modified plants containing EPA-DHA may be a major feed ingredient in the future.

10.11 Feed and metabolic calculations

10.11.1 Feed conversion ratio

The amount of feed needed for a 1 kg increase in fish weight is named feed conversion ratio (FCR). This is also described as feed factor (FF). To calculate the FCR the salmon has to grow. If growth is limited, much lower than expected, the FCR will be high. This is because the supplied feed is used for running the basal metabolism, not for growth. If using a lot of feed with high feed loss, the FCR will be high because the fish are not taking up and utilizing all the supplied feed. In a normal growth situation, around half of the supplied feed will be used for basal metabolism, or maintenance.

Fish use feed very effectively compared to other farmed animals. In general, fish use slightly above 1 kg dry feed per kilogram weight gain, depending on feed composition, fish size and any

Box 10.1

The weight gain of the fish is 5 kg and 5.2 kg feed is used. Calculate the feed conversion rate:

$$FCR = 5.2 / 5 = 1.04$$

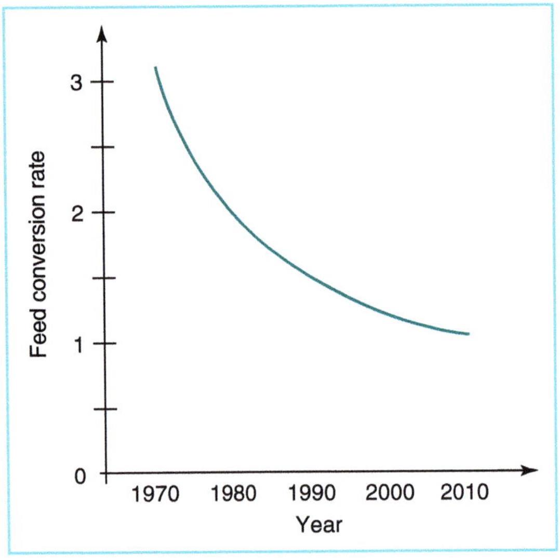

Figure 10.36 Development of the feed conversion rate in salmon farming (based on [55,56,134]).

loss of feed during feeding. Compared to other farmed animals, broiler production has the best results, utilizing slightly below 2 kg feed per kilogram growth, followed by pigs with just below 3 kg feed per kilogram growth, while for cattle over 6 kg dry feed is needed per kilogram growth. The reason for the high effectiveness is, as described earlier, based on three factors: salmon are cold blooded, weightless in water and continuous ammonia release.

Compared to other fish species the FCR achieved in commercial salmon farming is very low. FCR below 1 has been observed on smolt farms. The FCR in salmon farming has been significantly reduced since the industry began (Figure 10.36).

In practical fish farming, the terms economical (EFCR) and biological feed conversion rate (BFCR) are used. BFCR is the weight of delivered feed divided by the total biological production growth, where the biological growth also includes dead fish collected over the same period. EFCR is delivered feed divided by growth of live fish from beginning to end (so only uses harvested fish).

Instead of FCR the term feed efficiency (FE) can be used. FE refers to how effectively the fish utilize the feed: how much of the supplied food is used for growth. It can be calculated by the following equation:

$$FE = \text{growth} / \text{supplied feed} = 1 / FCR$$

Box 10.2

The weight gain of the fish is 5 kg and 5.2 kg feed is used. Calculate the feed efficiency:

$$FE = \text{growth} / \text{supplied feed} = 5\text{ kg} / 5.2\text{ kg} = 0.96$$

$$FE = 1 / FCR = 1 / 1.04 = 0.96$$

10.11.2 FCR does not provide the full picture

The FCR for farmed salmon depends on a range of factors including feed type, species, size, genetics, activity levels, environmental parameters and the culture system: therefore, it does not provide the full picture. It is not possible to compare FCR for different feeds with different energy and nutrient contents, and where the bioavailability of the ingredients is different. Using high-energy feed, as for today's salmon farming, will typically reduce the FCR, given that the nutrient requirement is covered. In smolt farming an FCR down to 0.8 has been observed.

But how is this possible when the fish need to use some energy for metabolism, and not all is used for growth? The main reason is due to the water content: the DM in 1 kg feed is typically 92–93% while in salmon meat it is around 35%.

Box 10.3

The FCR is set to 1, which means that 1 kg feed results in 1 kg fish growth. The DM in the feed is set to 0.92, meaning 0.08 kg water, while for the salmon these values are set to 0.35, meaning 0.65 kg water. Considering DM, 0.92 kg is transferred to 0.35 kg. If calculating FCR on a DM basis (FCR-DM) it will then be 2.63, which means the production of 1 kg salmon measured as DM needs 2.63 kg feed.

10.11.3 Other equations regarding feed and feed utilization

The fish in fish out ratio (FIFO) is an interesting value showing how much fish is used for production of salmon:

$$\text{FIFO} = \frac{\text{Fish used (kg)}}{\text{Salmon produced (kg)}}$$

The FIFO has, as mentioned earlier, been reduced greatly from the start of the salmon farming industry until today. In Norway, for example, in 2016 the FIFO was 0.84 and 1.45 for meal and oil, respectively [54].

The forage fish dependency ratio (FFDR) is a similar value but only shows how much forage fish or industrial fish is used for production of 1 kg salmon. The difference from FIFO is that trimmings/cut offs from the fish processing industry are not included.

$$\text{FFDR} = \frac{\text{Forage fish used (kg)}}{\text{Salmon produced (kg)}}$$

The FFDR has also been greatly reduced, and in Norway in 2016 it was 0.63 and 1.09 for meal and oil, respectively [54].

This value can also be described as the marine nutrient dependency ratio (MNDR) [135]. This shows the dependence on the marine nutrients in the feed and can be calculated as marine protein dependency (MPDR) and marine oil dependency (MODR):

$$\text{MPDR} = \frac{\text{Marine protein used (kg)} \times \text{\% protein in protein source}}{\text{Salmon produced (kg)} \times \text{\% protein in the salmon}}$$

$$\text{MODR} = \frac{\text{Marine oil used (kg)} \times \text{\% oil in lipid source}}{\text{Salmon produced (kg)} \times \text{\% lipid in the salmon}}$$

The MPDR and MODR have also been greatly reduced; in Norway in 2016 they were 0.6 and 0.5 respectively [54].

Nutrient or energy efficiency can be calculated by dividing the quantity of the nutrients or energy incorporated in the salmon by the total values supplied with the feed, and this, as stated earlier, also provides information about bioavailability.

Protein utilization ratio (PER) gives a value showing how the supplied protein is utilized, and the same ratio can be calculated for the lipids (LER) and for the energy (EER):

$$\text{PER} = \frac{\text{Biomass produced (kg)}}{\text{Protein used (kg)}}$$

$$\text{LER} = \frac{\text{Biomass produced (kg)}}{\text{Lipid used (kg)}}$$

$$\text{EER} = \frac{\text{Biomass produced (kg)}}{\text{Energy used (kg)}}$$

The values for Norwegian salmon farming in 2016, measured on the whole salmon body, were protein retention 37%, crude lipid 49% and energy 41% [54]. The corresponding value for DM was 34% and looking at EPA-DHA in particular, the value was 37%. Carbohydrates are difficult to measure because of the conversion to lipids.

10.11.4 How much feed to supply?

How much feed to supply to a production unit per day depends on several factors, including feed composition, water quality and temperature, size of salmon, season and production system. Typically, a mix of restricted feeding and ad libitum feeding is used. It is attempted to feed the fish to satiation even though it can be difficult to fully attain this in practice.

The way this is done is by predicting how much feed the fish need the next day. This amount is then fed, and the appetite of the fish is assessed by additional hand feeding or by observing feed loss. Feed loss can be observed by particles on the bottom or in the outlet of closed units or by having video cameras in sea cages. Special sensors or dynamic feeding systems may also be utilized. By monitoring the appetite, the feeding can be stopped when the appetite is low and feed loss is observed, while additional feed can be added if the appetite is greater; the prediction gives a valuable input for the amount.

10.11.4.1 Prediction of growth

There are different models developed for prediction of growth [42]. Two commonly used models are the specific growth rate (SGR) model [136] and the thermal unit growth coefficient (TGC)

Table 10.3 An example of a feeding table for ongrowing salmon in seawater, the numbers giving the percentage body weight gain per day (based on [21,138–140]).

Weight	Temperature °C							
	2	4	6	8	10	12	14	16
30–150 g	0.35	0.6	0.7	1	1.3	1.7	2	2.1
150–600 g	0.3	0.5	0.7	0.9	1.2	1.4	1.6	1.7
600–2000 g	0.2	0.4	0.5	0.6	0.8	1	1	1
> 2000 g	0.15	0.3	0.35	0.5	0.5	0.6	0.6	0.6

model [41,137]. The SGR model depends on water temperature and fish size and estimations are given through tables. The TGC values depend on the temperature to a slight degree, and to some degree on fish weight. From the predicted growth for a given period, for instance 1 day, the amount of feed to supply is decided by the chosen feed conversion rate. Typically, this is set to 1, meaning for 1 kg fish growth 1 kg feed is needed. Therefore, the amount of feed supplied will be equal to the expected growth. The estimated growth is typically described in growth tables (SGR or TGC) and if the FCR is set to 1 the growth table can also be named a feeding table (Table 10.3).

As stated earlier, the feed supplies both nutrients and energy, and the ratio here is important. Since the major part of salmon is protein, the DP : DE ratio is, of course, of major importance. When performing predictions of the required amount of feed to supply it is assumed that the feed has the correct composition. Calculations in relation to optimal feed composition based on available feed stuffs are not included here.

10.11.4.2 Growth tables – specific growth rate

The SGR gives a prediction of how much 1 kg fish biomass will grow during a given period. The values are based on experiments and the calculation is based on the natural logarithm of the weight:

$$SGR = 100 \times \frac{\ln W_n - \ln W_0}{n}$$

where W_n = final weight at the end of the period, after n days, W_0 = initial weight at the start of the period and n = number of days.

From the SGR, the weight after n days can be calculated:

$$W_n = W_0 \times (1 + SGR / 100)^n$$

where W_n = final weight at the end of the period, W_0 = initial weight at the start of the period, SGR = daily specific growth rate (SGR) from table and n = number of days.

In the 1980s data was collected from a range of growth experiments to find the specific growth rates for salmon. Size of the salmon, water temperatures and whether the fish were in freshwater or seawater were important for specific growth rate and were therefore used as categorizing units. However, time of the year, day length, water quality and other factors will also affect the growth rate, although to a minor degree. In more advanced models such factors may also be included.

When using the SGR model for prediction of growth, it will overestimate when long periods are used because it predicts exponential growth in relation to the number of days. Since the SGR is dependent on both fish size and water temperature (and other factors) it is time consuming to develop the growth tables. It is also difficult to compare with other farms, because they may have other temperature regimes. Higher temperatures for periods of time mean an increased SGR.

For established farms, growth data in relation to fish size and any other factors can be recorded and used to develop their own SGR values. By monitoring feed consumption and fish production the FCR can also be calculated. Having both SGR and FCR values can form the basis for development of site-adapted feeding tables.

The growth will, of course, also be dependent on the composition of the supplied feed; therefore several feeding companies have developed their own growth tables adapted to their feed, and some have also included more factors in the tables.

Box 10.4

Salmon in seawater is 1 kg, the daily weight gain is to be calculated based on SGR tables. The water temperature is 12°C.

$$W_n = W_0 \times (1 + d / 100)^t$$

where W_n = final weight at the end of the period, W_0 = initial weight at the start of the period, SGR = daily specific growth rate (SGR) from table and n = number of days.

From the table, specific daily growth is found to be 1 at a temperature of 12°C and a weight of 1 kg. The calculations are for 1 day:

$$Wn = 1\ \text{kg} \times (1 + 1 / 100)1$$

$$W_n = 1.01$$

$$W_g = 0.1\ \text{kg}$$

A mathematical model based on the growth tables for salmon [139] has also been developed [141] limited to seawater and a size range from 50 g to 3000 g.

$$SGR = 0.9\ T^{-0.97} \times W^{-0.34}$$

where T = temperature and W = weight.

10.11.4.3 Thermal unit growth coefficient (TGC)

The TGC or growth factor (GF) may also be used for estimation of growth. This is the predicted growth in relation to water temperature, time and fish size. These values have also been found from experiments. The following equation can be used for finding the TGC:

$$TGC = 1000\ ((W_n^{1/3} - W_0^{1/3})\ /\ DGR)$$

where W_0 = start weight (gram), W_n = end weight after n days and DGR = the number of day degrees in the period = $T_{avg} \times D$.

By reversing the equation, it can be used to estimate the end weight when knowing the TGC:

$$(TGC \times DGR)\ /\ 1000 + W_0^{1/3} = W_n^{1/3}$$

$$W_n^{1/3} = W_0^{1/3} + (TGC \times DGR)\ /\ 1000$$

$$W_n = (W_0^{1/3} + (TGC \times DGR)\ /\ 1000)^3$$

As seen, the growth is predicted here by an exponent of 1/3 of the weight and this prediction seems to be more accurate than the one utilized in the SGR model, especially when calculating over long periods [142]. While the SGR model predicts growth for a certain period for a given fish size and water temperature, the TGC is affected by fish size to a lesser degree (Figure 10.37). It also includes both water temperature and days in the equation. By calculating the TGC it is also possible to see how much of the growth potential is achieved. It also gives the possibility of comparing growth at different sites to some degree, unlike the SGR that only permits an overall view of the production at one specific site. Even though comparing sites using TGC will have some value, it is important to remember that other factors also affect growth rate. The calculated weight gain based on TGC can also be used to establish feeding tables, like for SGR.

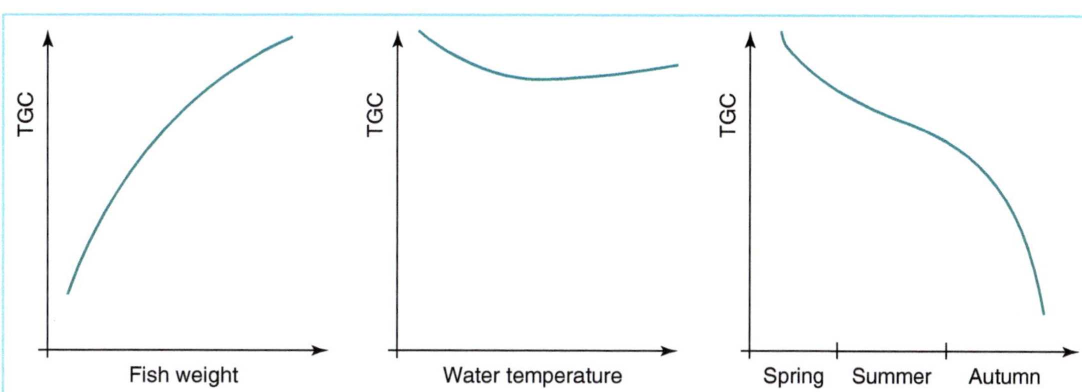

Figure 10.37 The connection between TGC and weight and water temperature (based on [62]).

Note: y axis not to scale, only to visualize effects.

Table 10.4 Some examples of achieved TGC for salmonids of different sizes [143,144], but this varies in practice.

Size	TGC
Start feeding (0.15–1 g)	1.0
Fry (1–0 g)	1.5
Juvenile – smolt (>10 g)	2.0
Sea water – ongrowing	2.7–3.5

10.11.4.4 Energy requirement

Another way to calculate how much feed to supply is based on the energy requirement of the fish [145]. The model is based on two parts: prediction of daily weight gain and daily energy use. This method, however, also requires the feed to be of optimal composition.

Using this model requires knowledge of the theoretical energy requirement (TER) for all fish in the production unit. This can be calculated based on the digestible energy needs (DEN) for production of 1 kg fish body weight (wet body weight, BW). The DEN value will vary with species, strain and weight of the fish (described later). The DEN value will also depend on the fat content and fat deposition in the fish because more energy is needed for weight gain as fat compared to weight gain as protein. The DEN value represents the sum of the different processes that require energy (energy pathways, see next chapter) and it is difficult to measure with good accuracy. A number of models for the energy-requiring processes have been developed, some shown later, but some have shown low accuracy when comparing to real aquaculture conditions.

One simplified model suggested to calculate the DEN value for salmonids is as follows [49]:

$$\text{DEN} = 9.22 \times 1.12 \,\text{Ln}\,(\text{BW})$$

where BW is in grams and DEN in kJ/g.

The DEN value also seems quite stable within temperatures from 5°C to 17°C.

The predicted weight gain (W_g) can, for instance, be calculated based on TGC. Now both the energy requirement and the estimated weight gain can be found and the theoretical energy requirement (TER value) can be calculated:

$$\text{TER} = \text{DEN} \times W_g$$

Based on feed supplier information, the DE in the feed can be identified. The feed allowance (FA) can then be calculated by the following equation:

$$\text{FA} = \text{TER} / \text{DE}$$

where TER = theoretical energy requirement (MJ digestible per kg fish) and DE = digestible energy per kg (MJ/kg).

Box 10.5

Salmon in seawater is 1 kg as the start value, the daily weight gain is to be calculated based on energy requirement. The digestible energy of the feed used is set to 20 MJ/Kg. The water temperature is 12°C.

TGC from table is set to 3.5.

First calculate the expected weight after 1 day:

$$Wn = (W_0^{1/3} + (TGC \times DGR) / 1000)^3$$

$$Wn = ((1\ kg)^{1/3} + (3.5 \times 12 \times 1) / 1000)^3$$

$$Wn = 1.131\ kg$$

$$Wg = 0.131\ kg$$

The weight gain is then 0.131 kg

Then the energy required for 1 kg wet weight gain can be calculated, the DEN value:

$$DEN = 9.22 + 1.12\ Ln\ (BW)$$

$$DEN = 9.22 + 1.12\ Ln\ (1000\ g)$$

$$DEN = 17.0\ kJ/g = 17\ MJ/kg$$

Then the theoretical energy requirement can be calculated:

$$TER = DEN \times W_g$$

$$TER = 17\ kJ/g \times 131\ g$$

$$TER = 2227\ kJ = 2.227\ MJ$$

Then the amount of feed to supply can be found:

$$FR = TER / DE$$

$$FR = 2.227\ MJ / 20\ MJ/kg$$

$$FR = 0.111\ kg = 111\ g$$

10.11.4.5 Full bioenergetic model

Instead of using the DEN value directly it can be based on the individual factors that make up the energy requirement. The total energy budget of the salmon may be separated as follows:

$$I = F + E + M + H + G$$

where I = energy content in the feed, F = energy lost in faeces, E = energy lost in excretion, H = energy used for heating, M = energy for basal metabolism, maintenance or fastening and G = energy retained for growth and gonad development.

By not accounting for digestibility in the calculation by proposing 100% digestibility (not the actual value), the equation will be as follows where I_d represents the DEN from the last paragraph.

$$I_d = E + M + H + G$$

It is important to remember that this is a model, not exact values, and it is dependent on a number of factors. Several models have been developed and the theoretical models are getting more and more complex, and include more details that make them more accurate [146]. Here only some simple calculations are described including some of the possible values, with the aim of demonstrating and aiding an understanding of the principles of calculating bioenergetics. Software has also been developed for modelling of fish bioenergetics [147]. The same basic principles used in aquaculture will of course also be used for fish in wild conditions, the main difference being that wild fish have to hunt for prey.

The total energy retained in the body (G) can be measured on the whole wet body in a bomb calorimeter. The body can also be dried first and the total energy content based on dry weight; then the percentage DM of the overall wet body weight must also be measured. Measurement of the whole body is in the area of 11–13 MJ/Kg but of course is highly dependent on the fat content. Energy content on a DM basis will be in the area of 25–28 MJ/kg. This value is also called retained energy (RE) and represents the energy needed to build body weight when none is used for gonad development.

$$RE = W_g \times DM(\%) \times \text{Energy in dry matter kJ/g DM}$$

The energy required for maintenance or fasting, Rf, is dependent on the body weight and the water temperature. One model for calculating this is as follows [142]:

$$Rf = (-1.04 + 3.26\,T - 0.05T^2)\,(\text{kg } BW^{0.824})\ \text{kJ/day}$$

As shown from the equation, the energy requirement for maintenance increases up to a temperature of around 15°C before declining. One way to calculate this indirectly is to base it on oxygen consumption: for production of 13.6 kJ energy, 1 g of oxygen is needed [44]. Maintenance energy required for rainbow trout production measured as oxygen consumption is 3.5–4 g O_2 per kilogram fish per day [43].

There will also be minor energy losses by heat increment (Eh). One model for estimation of this can be based on 60% of the maintenance energy [142]:

$$Eh = Rf \times 0.6$$

However, literature indicates that the heat increment is related to digestible nitrogen intake, 27–30 kJ/g digestible nitrogen in rainbow trout [148].

There will also be energy loss in the non-faecal loss, proposed to be from 3–6% of the digestible energy [44]. This depends on all energy used and the following model is proposed to be used for calculation (from rainbow trout) [142]:

$$Enf = (RE + Rf + Eh) \times 0.06$$

Based on this the total digestible energy requirement (TDER) can be estimated:

$$TDER = Re + RF + Eh + Enf$$

This value represents the digestible requirement, meaning no energy loss in faeces. Based on this the feed requirement can be calculated if the digestible energy in the feed is known.

$$FA = TDER / DE$$

where TDER = digestible energy requirement for production of 1 kg fish biomass and DE = digestible energy per kilogram feed.

10.11.5 Digestible energy

The digestible energy in the feed depends on the composition and the digestibility of the feedstuff included in the feed. It is possible to calculate the overall digestibility of the feed by summing the digestibility of the separate ingredients with the following equation:

$$DE = \sum (ADC \times \%\ \text{ingredient in the diet})$$

By knowing the proportions and energy values of the major ingredients (protein, fat and carbohydrates), the overall energy digestibility can be calculated (Table 10.5).

Table 10.5 The overall energy content of the feed can be calculated from the percentage and digestibility of the macronutrients.

Major ingredient	% per kg feed	Digestibility (%)	Energy content (MJ/kg)	Total digestible energy content (MJ/kg)
Protein	45	90	23.5	9.52
Lipid	30	90	37	9.99
Carbohydrates	10	50	17	0.85
			Total	20.37

Box 10.6

Calculate the energy requirement for a weight gain of 1 kg for a salmon weighing 1 kg. The energy content is set to 12 MJ/kg (wet body weight).

Retained energy in the body:

$$RE = 12\ MJ/kg \times 1\ kg = 12\ MJ$$

Then the energy needed for maintenance or fasting, Ef, can be calculated:

$$Rf = (-1.04 + 3.26\ T - 0.05\ T^2)\ (kg\ BW^{0.824})\ kJ/day$$

$$Rf = (-0.0104 + 3.26 \times 12 - 0.05 \times 12^2(\times 1\ kg^{0.824})$$

$$Ef = 0.040\ MJ$$

Then the energy loss for heat increment (Eh) can be calculated:

$$Eh = (RE + Ef) \times 0.17$$

$$Eh = (12 + 0.040) \times 0.17$$

$$Eh = 2.05\ MJ$$

Energy loss in non-faecal loss:

$$Enf = (RE + Ef + Eh) \times 0.09$$

$$Enf = (12 + 0.04 + 2.05) \times 0.09$$

$$Enf = 1.29\ MJ$$

Total digestible energy requirement can be calculated:

$$TDE = Re + RF + Eh + Enf$$

$$TDE = 12\ MJ + 0.04 + 2.05\ MJ + 1.29\ MJ$$

$$TDE = 15.38\ MJ$$

Not all the supplied feed will be digested and therefore the digestibility must be included.

10.12 Practical feeding

First it is, of course, essential to have feed of an optimal nutritional quality, then handling and storage of the feed must be done correctly. The feed must not be stored in a way that can degrade the nutritional quality or the technical quality. If the feed contains too many broken particles and dust it will be suboptimal for feeding and the feed loss (and FCR) will increase.

10.12.1 Frequency of feeding

Feeding frequency depends on the size and age of the salmon. In general, one to four meals per day is enough to achieve optimal growth. Young fish, fry, need to be fed more regularly than larger ongrowing fish, typically from once per 15 minutes to several meals per day. It is, however, important to supply enough feed particles each time to avoid development of dominant individuals. The salmon search for feed by sight and therefore feeding must be done in light [149,150]. How often the salmon eat is, as described earlier, strongly correlated to the filling degree of the stomach.

10.12.2 Seasonal variations

The feed intake of the salmon varies with season, increasing during springtime and reducing during autumn. This is probably due to their fat deposits; in nature they build up fat deposits in the summer and autumn when nutrient access is good in order to survive the winter season when there is lower access to feed, and when the deposits have built up they will reduce their feed intake.

10.12.3 Feeding rate

The feeding rate or how much to feed per day depends on fish size, water quality and water temperature. If the water quality is kept optimal and there is enough oxygen in the water and enough light, the feeding rate will only depend on water temperature for a given fish size. Typically, it will be in the range of 1% to 4% of the body weight per day.

A practical example from the Faroe Islands showed an average growth when transferred to sea from 500 g to 7.5 kg over a 12-month period [151]. The SGR was 4.13 (RGI 119). This illustrates the possible growth potential of salmon when factors are optimal.

10.12.4 Feed particle size

The particle size depends on fish size. A rough rule of thumb says that it should be 20–30% of the fish mouth gape [14]. Another suggestion is a particle diameter of 2.5% of the fork length for juvenile salmon [152]. The large feed companies have developed tables listing recommended feed particle sizes for different fish sizes (Table 10.6). It is important to use neither too small nor too large a particle size. Too small a size will increase the number of particles that the fish need to eat, and the salmon will use more energy to search for and eat the increased number of feed particles. Too large a particle will be difficult for the fish to take up.

Table 10.6 Possible particle sizes for different fish sizes in fresh- and seawater (from one feed supplier [153]).

Water	Fish size (g)	Particle size (mm)
Freshwater	0.12–0.40	0.5
	0.18–1.00	0.75
	0.9–3.5	1
	2–7	1.2
	5–20	1.5
	15–60	2
	40–200	3
	100–500	4
Seawater	75–150	3
	150–600	4.5
	600–1200	7
	1200–	9
	2500–	12

10.12.5 Feed storage

Correct storage is important when using dry feed. The feed ought to be stored under dry conditions and ideally in cool storage or at least air conditioned storage. The nutritional value of the feed can be reduced, rancidity of lipids may occur when storage conditions are too hot or when exposing the feed to sunlight, moulds can result if the temperature and humidity are high, and vitamins can be degraded under suboptimal storage conditions.

High mechanical forces on the feed particles must also be avoided, as this may lead to breakage. Loading the feed in high stacks is not recommended due to the weight on the lower sacks; the feed should not be stacked above ten small sacks or three large sacks in height [12]. Proper

Figure 10.38 Feed can be stored in big sacks or silos at the farm.

ventilation in the feed storage area is also a requirement. Today it is common to use silos for bulk storage of the feed (Figure 10.38).

10.12.6 Feeding equipment

Today some form of automatic feeding is used on all salmon farms, both smolt farms and ongrowing farms. This is not described further in this book, but an overview is given in [154] in addition to some information in the production chapters. But in general, more and more automation is becoming included in feeding systems, taking the feed from the bulk storage to the fish. Feedback systems controlling the feeding systems are also included. In the near future, artificial intelligence will also be included in the feeding systems, meaning that a computer will take over the entire feeding process based on input from a number of sensors, for example from all types of environmental factors.

References

1. Abowei, J.F.N. and Ekubo, A.T. Some principles and requirements in fish nutrition. *British Journal of Pharmacology and Toxicology*. 2011. 2(4): 163–179.
2. Lovell, T. *Nutrition and Feeding of Fish*. 2012. Springer: New York.
3. Halver, J. *Fish Nutrition*. 2013. Elsevier Science: Amsterdam.
4. Halver, J.E. and Hardy, R.W. *Fish Nutrition*. 2002. Academic Press: New York.
5. Webster, C.D. and Lim, C. *Nutrient Requirements and Feeding of Finfish for Aquaculture*. 2002. CABI: Wallingford.
6. Jobling, M. Fish bioenergetics. *Oceanographic Literature Review*. 1995. 9(42): 785.
7. National Research. *Nutrient Requirements of Fish and Shrimp*. 2011. The National Academies Press: Washington, DC.
8. Gatlin, D.M. Principles of Fish Nutrition. 2010. SRAC Publication No. 5003.
9. Oliva-Teles, A. Nutrition and health of aquaculture fish. *Journal of Fish Diseases*. 2012. 35(2): 83–108.
10. Jobling, M. Fish nutrition research: past, present and future. *Aquaculture International*. 2016. 24(3): 767–786.
11. PetMD. How Fish Metabolism Works. 2016. https://www.petmd.com/fish/care/evr_fi_fish_metabolism.
12. Kaushik, S.J. Feed management and on-farm feeding practices of temperate fish with special reference to salmonids. *FAO Fisheries and Aquaculture Technical Paper*. 2013. 583: 519–551.
13. Sørensen, M. and Denstadli, V. Alkaline preserved herring by-products in feed for Atlantic salmon (*Salmo salar* L.). *Animal Feed Science and Technology*. 2008. 144(3–4): 327–334.
14. Craig, S. et al. Understanding Fish Nutrition, Feeds, and Feeding. 2017. https://fisheries.tamu.edu/files/2019/01/FST-269.pdf.
15. Knights, B. Energetics and fish farming, in *Fish Energetics*. 1985. Springer: Dordrecht, pp. 309–340.
16. Bureau, D.P., Kaushik, S.J. and Cho, C.Y. Bioenergetics, in *Fish Nutrition*. 2003. Elsevier: Amsterdam, pp. 1–59.
17. Krogdahl, Å. et al. Important antinutrients in plant feedstuffs for aquaculture: an update on recent findings regarding responses in salmonids. *Aquaculture Research*. 2010. 41(3): 333–344.

18. Storebakken, T. Nutrient requirements and feeding of finfish for aquaculture, in *Atlantic Salmon, Salmo salar*, C.D Webster and C. Lim (eds). 2002. CABI: Wallingford, pp. 79–102.
19. Tacon, A.G.J. *The Nutrition and Feeding of Farmed Fish and Shrimp: A Training Manual.* 1987. FAO: Rome.
20. Fry, J.P. et al. Feed conversion efficiency in aquaculture: do we measure it correctly? *Environmental Research Letters*. 2018. 13(2): 024017.
21. Lall, S. *Atlantic salmon, Salmo Salar*. 2019. http://www.fao.org/fishery/affris/species-profiles/atlantic-salmon/atlantic-salmon-home/en/.
22. Ruyter, B. et al. Essential fatty acids in Atlantic salmon: effects of increasing dietary doses of n-6 and n-3 fatty acids on growth, survival and fatty acid composition of liver, blood and carcass. *Aquaculture Nutrition*. 2000. 6(2): 119–127.
23. Hemre, G.I. et al. Effect of gelatinized wheat and maize in diets for large Atlantic salmon (*Salmo salar* L.) on glycogen retention, plasma glucose and fish health. *Aquaculture Nutrition*. 1996. 2(1): 33–39.
24. Antony Jesu Prabhu, P. Minerals in fish: does the source matter? 2015. Wageningen University and Research Center, Wageningen University.
25. Hardy, R. FAO/UNDP Training Course in Fish Feed Technology. 1980. ACDP/REP.
26. Antony Jesu Prabhu, P., Schrama, J.W. and Kaushik, S.J. Mineral requirements of fish: a systematic review. *Reviews in Aquaculture*. 2016. 8(2): 172–219.
27. Antony Jesu Prabhu, P. et al. Recommendations for dietary level of micro-minerals and vitamin D(3) to Atlantic salmon (Salmo salar) parr and post-smolt when fed low fish meal diets. *PeerJ*. 2019. 7: e6996
28. Daniel, N. et al. Application of carotenoids on coloration of aquatic animals. *International Journal of Fisheries and Aquatic Studies*. 2017. 2: 1–7.
29. Bjerkeng, B. Natural functions, in *Carotenoids in Aquaculture: Fish and Crustaceans*,, G. Britton, S. Liaaen-Jensen and H. Pfander (eds). 2008. Birkhäuser Basel: Basel, pp. 237–254.
30. García-Chavarría, M. and Lara-Flores, M. The use of carotenoid in aquaculture. *Research Journal of Fisheries and Hydrobiology*. 2013. 8(2): 38–49.
31. Gupta, S.K. et al. Use of natural carotenoids for pigmentation in fishes. *Natural Product Radiance*. 2007. 6: 46–49.
32. Ambati, R.R. et al. Astaxanthin: sources, extraction, stability, biological activities and its commercial applications—a review. *Marine Drugs*. 2014. 12(1): 128–152.
33. Torrissen, O.J. and Christiansen, R. Requirements for carotenoids in fish diets. *Journal of Applied Ichthyology*. 1995. 11(3-4): 225–230.
34. March, B.E. and Macmillan, C. Muscle pigmentation and plasma concentrations of astaxanthin in rainbow trout, chinook salmon, and Atlantic salmon in response to different dietary levels of astaxanthin. *The Progressive Fish-Culturist*. 1996. 58(3): 178–186.
35. Efsa. Scientific Opinion on the safety and efficacy of astaxanthin (CAROPHYLL® Pink 10% CWS) for salmonids and ornamental fish. *EFSA Journal*. 2014. 12(6): 3725.
36. Steine, G., Alfnes, F. and Rørå, M.B. The effect of color on consumer WTP for farmed salmon. *Marine Resource Economics*. 2005. 20(2): 211–219.
37. Quevedo, R.A., Aguilera, J.M. and Pedreschi, F. Color of salmon fillets by computer vision and sensory panel. *Food and Bioprocess Technology*. 2010. 3(5): 637–643.
38. Buhler, D.R. and Halver, J.E. Nutrition of salmonoid fishes: IX. Carbohydrate requirements of chinook salmon. *The Journal of Nutrition*. 1961. 74(3): 307–318.
39. Brett, J.R. and Groves, T.D.D. Physiological energetics. *Fish Physiology*. 1979. 8(6): 280–352.
40. Jobling, M., Energetic models | bioenergetics in aquaculture settings, in *Encyclopedia of Fish Physiology*, A.P. Farrell (ed.). 2011. Academic Press: San Diego, CA, pp. 1664–1674.

41. Cho, C.Y. Fish nutrition, feeds, and feeding: with special emphasis on salmonid aquaculture. *Food Reviews International.* 1990. 6(3): 333–357.
42. Dumas, A., France, J. and Bureau, D. Modelling growth and body composition in fish nutrition: where have we been and where are we going? *Aquaculture Research.* 2010. 41(2): 161–181.
43. Kaushik, S.J. and Médale, F. Energy requirements, utilization and dietary supply to salmonids. *Aquaculture.* 1994. 124(1–4): 81–97.
44. Kaushik, S.J. Nutritional bioenergetics and estimation of waste production in non-salmonids. *Aquatic Living Resources.* 1998. 11(4): 211–217.
45. Lovell, R.T., Bush, L.J. and Getz, W.R. Energy requirements of fish. *The Professional Animal Scientist.* 1994. 10(4): 145–149.
46. Lekang, O.-I. *Aquaculture Engineering.* 2020. John Wiley and Sons: Chichester.
47. Hanson, P. et al. Fish Bioenergetics 3.0. 1997. University of Wisconsin Sea Grant Institute, Madison, WI.
48. Alanärä, A. and Strand, Å. The energy requirements of percid fish in culture, in *Biology and Culture of Percid Fishes: Principles and Practices*, P. Kestemont, K. Dabrowski and R.C. Summerfelt (eds). 2015. Springer: Dordrecht, pp. 353–368.
49. Bailey, J. and Alanärä, A. Digestible energy need (DEN) of selected farmed fish species. *Aquaculture.* 2006. 251(2–4): 438–455.
50. Hillestad, M. and Johnsen, F. High-energy/low-protein diets for Atlantic salmon: effects on growth, nutrient retention and slaughter quality. *Aquaculture.* 1994. 124(1–4): 109–116.
51. Hillestad, M. et al. Long-term effects of dietary fat level and feeding rate on growth, feed utilization and carcass quality of Atlantic salmon. *Aquaculture Nutrition.* 1998. 4(2): 89–98.
52. Weihe, R. et al. Improving production efficiency of farmed Atlantic salmon (*Salmo salar* L.) by isoenergetic diets with increased dietary protein-to-lipid ratio. *Aquaculture Research.* 2018. 49(4): 1441–1453.
53. Weihe, R. et al. Increased protein-to-lipid ratio in energy dense diets improves slaughter yields and muscle thickness of different weight classes of farmed Atlantic salmon (*Salmo salar* L.). *Aquaculture Reports.* 2019. 13: 100173.
54. Aas, T.S., Ytrestøyl, T. and Åsgård, T. Utilization of feed resources in the production of Atlantic salmon (*Salmo salar*) in Norway: An update for 2016. *Aquaculture Reports.* 2019. 15: 100216.
55. Åsgård, T., *Aquaculture Development 40 Years Lesson.* 2018. https://documen.site/download/lysbilde-1-5ae3ab3dc7695_pdf.
56. Hardy, R. Fish Nutrition Research Differences and similarities with livestock nutrition and what the future holds. http://www.pitt.edu/~super4/33011-34001/33021.ppt.
57. Prabu, E. et al. An overview on significance of fish nutrition in aquaculture industry. *International Journal of Fisheries and Aquatic Studies.* 2017. 5(6): 349–355.
58. Peruzzi, S., Hagen, Ø. and Jobling, M. Gut morphology of diploid and triploid Atlantic salmon (*Salmo salar* L.). *Aquaculture International.* 2015. 23(4): 1105–1108.
59. Handeland, S.O., Imsland, A.K. and Stefansson, S.O. The effect of temperature and fish size on growth, feed intake, food conversion efficiency and stomach evacuation rate of Atlantic salmon post-smolts. *Aquaculture.* 2008. 283(1–4): 36–42.
60. Sveier, H., Wathne, E. and Lied, E. Growth, feed and nutrient utilization and gastrointestinal evacuation time in Atlantic salmon (*Salmo salar* L.): the effect of dietary fish meal particle size and protein concentration. *Aquaculture.* 1999. 180(3–4): 265–282.
61. Aas, T.S. et al. Measurement of gastrointestinal passage rate in Atlantic salmon (*Salmo salar*) fed dry or soaked feed. *Aquaculture Reports.* 2017. 8: 49–57.
62. Alanärä, A. *Pelletologi – läran om fiskens foder.* Vol. rapport 34. 2001: Sveriges lantbruksuniv.

63. Ruohonen, K. Gastrointestinal responses of rainbow trout (*Oncorhynchus mykiss walbaum*) to dry pellet and low-fat herring diets: consequences for growth, food utilization and nutrient load in the water. 1994. University of Wales, Bangor.
64. Krogdahl, Å. et al. Effects of diet composition on apparent nutrient absorption along the intestinal tract and of subsequent fasting on mucosal disaccharidase activities and plasma nutrient concentration in Atlantic salmon *Salmo salar* L. *Aquaculture Nutrition*. 1999. 5(2): 121–133.
65. Olsen, R.E. and Ringø, E. Lipid digestibility in fish: a review. *Recent Research Developments in Lipid Research*. 1997. 1: 199–265.
66. Bogevik, A.S. Marine wax ester digestion in salmonid fish: a review. *Aquaculture Research*. 2011. 42(11): 1577–1593.
67. Bogevik, A.S. Marine wax ester digestion in salmonid fish: a review. *Aquaculture Research*. 2011. 42(11): 1577–1593.
68. Sinha, A.K. et al. Non-starch polysaccharides and their role in fish nutrition–A review. *Food Chemistry*. 2011. 127(4): 1409–1426.
69. Davis, D.A. and Gatlin, D.M. Dietary mineral requirements of fish and marine crustaceans. *Reviews in Fisheries Science*. 1996. 4(1): 75–99.
70. Morales, G.A. et al. Effect of different inorganic phosphorus sources on growth performance, digestibility, retention efficiency and discharge of nutrients in rainbow trout (*Oncorhynchus mykiss*). *Aquaculture*. 2018. 495: 568–574.
71. Cho, C. Apparent digestibility measurement in feedstuffs for rainbow trout, in *Proc. World Symp. on Finfish Nutrition and Fishfeed Technology Vol. II*. 1979.
72. Cho, C. Apparent digestibility measurement in feedstuffs for rainbow trout, in *Proc. World Symp. on Finfish Nutrition and Fishfeed Technology Vol. II*. 1979.
73. Krontveit, R.I., Bendiksen, E.Å. and Aunsmo, A. Field monitoring of feed digestibility in Atlantic salmon farming using crude fiber as an inert marker. *Aquaculture*. 2014. 426: 249–255.
74. Witkowski, S.J. et al. Development and assessment of novel endogenous markers in commercial aquafeeds to measure apparent digestibility in large Atlantic salmon (*Salmo salar*) on salmon farms. *Aquaculture*. 2019. 508: 192–198.
75. Austreng, E. et al. Evaluation of selected trivalent metal oxides as inert markers used to estimate apparent digestibility in salmonids. *Aquaculture*. 2000. 188(1–2): 65–78.
76. Hatlen, B. et al. Addition of yttrium oxide as digestibility marker by vacuum coating on finished pellets—A method for assessing digestibility in commercial salmon feeds? *Aquaculture*. 2015. 435: 301–305.
77. Carter, C., Foale, M. and Hauler, R. Feed characteristics and apparent digestibility of ingredients with potential for inclusion in Atlantic salmon feeds, in *Aquaculture Feed Development for Atlantic Salmon (Salmo salar)*, C.G. Carter, R.C. Hauler and C. Foster (eds). 2002. Fisheries Research and Development Corporation: Deakin.
78. Kim, S.W. et al. Meeting global feed protein demand: challenge, opportunity, and strategy. *Annual Review of Animal Biosciences*. 2019. 7: 221–243.
79. Burr, G. et al. Apparent digestibility of macro-nutrients and phosphorus in plant-derived ingredients for Atlantic salmon, *Salmo salar* and Arctic charr, *Salvelinus alpinus*. *Aquaculture Nutrition*. 2011. 17(5): 570–577.
80. Hemre, G.I. et al. Aquaculture nutrient requirements: understanding vitamins, minerals and other nutrtients in fish feed diets based on plant derived ingredients. *ARRAINA Technical*, 2016. 16: 2016b.
81. Shepherd, C.J., Monroig, O. and Tocher, D.R. Future availability of raw materials for salmon feeds and supply chain implications: The case of Scottish farmed salmon. *Aquaculture*. 2017. 467: 49–62.

82. Sprague, M., Dick, J.R. and Tocher, D.R. Impact of sustainable feeds on omega-3 long-chain fatty acid levels in farmed Atlantic salmon, 2006–2015. *Scientific Reports*. 2016. 6: 21892.
83. Dessen, J.-E. et al. Different growth performance, lipid deposition, and nutrient utilization in in-season (S1) Atlantic salmon post-smolt fed isoenergetic diets differing in protein-to-lipid ratio. *Aquaculture*. 2017. 473: 345–354.
84. Tocher, D.R. et al. Omega-3 long-chain polyunsaturated fatty acids, EPA and DHA: bridging the gap between supply and demand. *Nutrients*. 2019. 11(1): 89.
85. Rust, M. et al. The future of aquafeeds NOAA/USDA alternative feeds initiative. 2011.
86. Pikitch, E. et al. *Little fish, big impact: managing a crucial link in ocean food webs*. 2012. Lenfest Ocean Program, Washington, DC.
87. Huntington, T.C. and Hasan, M.R. Fish as feed inputs for aquaculture–practices, sustainability and implications: a global synthesis. Fish as feed inputs for aquaculture: practices, sustainability and implications, 2009: 1–61.
88. SAPEA. *Food from the Oceans: How Can More Food and Biomass Be Obtained from the Oceans in a Way That Does Not Deprive Future Generations of Their Benefits?* 2017. SAPEA: Berlin.
89. Costello, C., Cao, L. and Gelcich, S. *The Future of Food from the Sea*. 2019. World Resources Institute: Washington, DC.
90. Duarte, C.M. et al. Will the oceans help feed humanity? *BioScience*. 2009. 59(11): 967–976.
91. Turchini, G.M., Ng, W.-K.and Tocher, D.R. *Fish Oil Replacement and Alternative Lipid Sources in Aquaculture Feeds*. 2010. CRC Press, Baton Rouge, FL.
92. Hidalgo, M. and Browman, H.I. Developing the knowledge base needed to sustainably manage mesopelagic resources. *ICES Journal of Marine Science*. 2019. 76(3): 609–615.
93. Súni Lamhauge et al. *Fishery and Utilization of Mesopelagic Fishes and Krill in the North Atlantic*. 2008. Nordic Council of Ministers.
94. Gjøsaeter, J. et al. *A Review of the World Resources of Mesopelagic Fish*. 1980. FAO: Rome.
95. Remesan, M.P. et al. A review on techniques and challenges in the harvest of mesopelagics. 2019. https://www.researchgate.net/publication/337075459_A_Review_on_Techniques_and_Challenges_in_the_Harvest_of_Mesopelagics.
96. Prellezo, R. Exploring the economic viability of a mesopelagic fishery in the Bay of Biscay. *ICES Journal of Marine Science*. 2018. 76(3): 771–779.
97. St John, M.A. et al. A dark hole in our understanding of marine ecosystems and their services: perspectives from the mesopelagic community. *Frontiers in Marine Science*. 2016. 3: 31.
98. Oceanbergen. *A new marine biomass resource*. 2017. https://www.oceanbergen.no/news/.
99. Berthelsen, K. Ny marin råvare kan dekke fôringrediens-behovet ni ganger. 2018. https://www.kyst.no/article/ny-marin-raavare-kan-dekke-foringrediens-behovet-ni-ganger/.
100. Buschmann, A.H. et al. Seaweed production: overview of the global state of exploitation, farming and emerging research activity. *European Journal of Phycology*. 2017. 52(4): 391–406.
101. FAO. *The Global Status of Seaweed Production, Trade and Utilization*. 2018. FAO: Rome.
102. Tiwari, B.K. and Troy, D. *Seaweed Sustainability: Food and Non-Food Applications*. 2015. Elsevier Science: Amsterdam.
103. Angell, A.R. et al. Seaweed as a protein source for mono-gastric livestock. *Trends in Food Science & Technology*. 2016. 54: 74–84.
104. van Ginneken, V.J.T. et al. Polyunsaturated fatty acids in various macroalgal species from North Atlantic and tropical seas. *Lipids in Health and Disease*. 2011. 10: 104–104.
105. Rajauria, G. Seaweeds: a sustainable feed source for livestock and aquaculture, in *Seaweed Sustainability*, B.K. Tiwari and D.J. Troy (eds). 2015. Academic Press: San Diego, CA, pp. 389–420.

106. Forster, J. and Radulovich, R. Seaweed and food security, in *Seaweed Sustainability*. 2015. Elsevier: Amsterdam, pp. 289–313.
107. Makkar, H.P.S. et al. State-of-the-art on use of insects as animal feed. *Animal Feed Science and Technology*, 2014. 197: 1–33.
108. Gasco, L. et al. Fishmeal alternative protein sources for aquaculture feeds, in *Feeds for the Aquaculture Sector*. 2018. Springer: Dordrecht, pp. 1–28.
109. Henry, M. et al. Review on the use of insects in the diet of farmed fish: past and future. *Animal Feed Science and Technology*. 2015. 203: 1–22.
110. Belghit, I. et al. Potential of insect-based diets for Atlantic salmon (*Salmo salar*). *Aquaculture*. 2018. 491: 72–81.
111. Halloran, A. et al. *Edible Insects in Sustainable Food Systems*. 2018. Springer: Dordrecht.
112. Biancarosa, I. et al. Replacing fish meal with insect meal in the diet of Atlantic salmon (*Salmo salar*) does not impact the amount of contaminants in the feed and it lowers accumulation of arsenic in the fillet. *Food Additives & Contaminants: Part A*. 2019. 36(8): 1191–1205.
113. Linder, T. Making the case for edible microorganisms as an integral part of a more sustainable and resilient food production system. *Food Security*. 2019: 1–14.
114. Sirakov, I. et al. The importance of microalgae for aquaculture industry. Review. *International Journal of Fisheries and Aquatic Studies*. 2015. 2(4): 81–84.
115. Milledge, J.J. Commercial application of microalgae other than as biofuels: a brief review. *Reviews in Environmental Science and Bio/Technology*. 2011. 10(1): 31–41.
116. Panis, G. and Carreon, J.R. Commercial astaxanthin production derived by green alga *Haematococcus pluvialis*: A microalgae process model and a techno-economic assessment all through production line. *Algal Research*. 2016. 18: 175–190.
117. Novoveská, L. et al. Microalgal carotenoids: A review of production, current markets, regulations, and future direction. *Marine Drugs*. 2019. 17(11): 640.
118. Ledesma-Amaro, R. and Nicaud, J.-M. Yarrowia lipolytica as a biotechnological chassis to produce usual and unusual fatty acids. *Progress in Lipid Research*. 2016. 61: 40–50.
119. Winwood, R.J. Recent developments in the commercial production of DHA and EPA rich oils from microalgae. *Ocl*. 2013. 20(6): D604.
120. Veramaris. https://www.veramaris.com/home.html.
121. Cohen, Z. and Ratledge, C. *Single Cell Oils: Microbial and Algal Oils*. 2015. Elsevier: Amsterdam.
122. Goldberg, I., *Single Cell Protein*. 2013. Springer: Berlin.
123. Sharma, S. et al. Microbial protein produced from brown seaweed and spruce wood as a feed ingredient. *Journal of Agricultural and Food Chemistry*. 2018. 66(31): 8328–8335.
124. Berge, G.M. et al. Bacterial protein grown on natural gas as protein source in diets for Atlantic salmon, *Salmo salar*, in saltwater. *Aquaculture*. 2005. 244(1–4): 233–240.
125. Storebakken, T. et al. Bacterial protein grown on natural gas in diets for Atlantic salmon, *Salmo salar*, in freshwater. *Aquaculture*. 2004. 241(1–4): 413–425.
126. Feedkind. http://www.feedkind.com/.
127. Øverland, M. et al. Evaluation of *Candida utilis*, *Kluyveromyces marxianus* and *Saccharomyces cerevisiae* yeasts as protein sources in diets for Atlantic salmon (*Salmo salar*). *Aquaculture*. 2013. 402: 1–7.
128. Xue, Z. et al. Production of omega-3 eicosapentaenoic acid by metabolic engineering of Yarrowia lipolytica. *Nature Biotechnology*. 2013. 31(8): 734.
129. Napier, J.A., Olsen, R.-E. and Tocher, D.R. Update on GM canola crops as novel sources of omega-3 fish oils. *Plant Biotechnology Journal*. 2019. 17(4): 703–705.
130. Ruiz-Lopez, N. et al. Successful high-level accumulation of fish oil omega-3 long-chain polyunsaturated fatty acids in a transgenic oilseed crop. *The Plant Journal*. 2014. 77(2): 198–208.

131. Napier, J.A. et al. Transgenic plants as a sustainable, terrestrial source of fish oils. *European Journal of Lipid Science and Technology*. 2015. 117(9): 1317–1324.
132. Nanton, D.A. et al. Effect of soy oil containing stearidonic acid on growth performance and fillet fatty acid composition of Atlantic salmon. *Aquaculture Nutrition*. 2012. 18(6): 640–650.
133. Betancor, M.B. et al. Oil from transgenic Camelina sativa containing over 25% n-3 long-chain PUFA as the major lipid source in feed for Atlantic salmon (*Salmo salar*). *British Journal of Nutrition*. 2018. 119(12): 1378–1392.
134. Costa-Pierce, B.A. Ecology as the paradigm for the future of aquaculture, in *Ecological Aquaculture: The Evolution of the Blue Revolution*. 2002. Blackwell Science: Oxford, pp. 339–372.
135. Crampton, V.O. et al. Demonstration of salmon farming as a net producer of fish protein and oil. *Aquaculture Nutrition*. 2010. 16(4): 437–446.
136. Ricker, W.E. *Growth rates and models. Fish Physiology*. 1979: 673–743.
137. Iwama, G.K. and Tautz, A.F. A simple growth model for salmonids in hatcheries. *Canadian Journal of Fisheries and Aquatic Sciences*. 1981. 38(6): 649–656.
138. Fôr Skretting – fôrkatalog. 2012.
139. Austreng, E., Storebakken, T. and Åsgård, T. Growth rate estimates for cultured Atlantic salmon and rainbow trout. *Aquaculture*. 1987. 60(2): 157–160.
140. Ewos Innovation. Ewos growth index. 2014.
141. Forsberg, O.I. Empirical investigations on growth of post-smolt Atlantic salmon (*Salmo salar* L.) in land-based farms. Evidence of a photoperiodic influence. *Aquaculture*. 1995. 133(3–4): 235–248.
142. Cho, C.Y. and Bureau, D.P. Development of bioenergetic models and the Fish-PrFEQ software to estimate production, feeding ration and waste output in aquaculture. *Aquatic Living Resources*. 1998. 11(4): 199–210.
143. Einen, O. and Mørkøre, T. *Fôringslære for akvakultur*. 1997. Landbruksforl.
144. Thorarensen, H. and Farrell, A.P. The biological requirements for post-smolt Atlantic salmon in closed-containment systems. *Aquaculture*. 2011. 312(1–4): 1–14.
145. Alanärä, A., Kadri, S. and Paspatis, M. Feeding management, in *Food Intake in Fish*. 2001. Wiley: Chichester, pp. 332–353.
146. Hancz, C. and Varga, D. Measuring fish metabolism–science and practice of development in fish feeding: A review. *Acta Agraria Kaposváriensis*. 2017. 21(1): 1–14.
147. Deslauriers, D. et al. Fish bioenergetics 4.0: an R-based modeling application. *Fisheries*. 2017. 42(11): 586–596.
148. Cho, C.Y. and Kaushik, S.J. Nutritional energetics in fish: energy and protein utilization in rainbow trout (*Salmo gairdneri*). *World Review of Nutrition and Dietetics*. 1990: 132–172.
149. Ali, M.A. Histophysiological studies on the juvenile Atlantic salmon (*Salmo salar*) retina: II. Responses to light intensities, wavelengths, temperatures, and continuous light or dark. *Canadian Journal of Zoology*. 1961. 39(4): 511–526.
150. Metcalfe, N.B., Huntingford, F.A. and Thorpe, J.E. Feeding intensity, growth rates, and the establishment of life-history patterns in juvenile Atlantic salmon *Salmo salar*. *The Journal of Animal Ecology*. 1988: 463–474.
151. Er dette ny vekstrekord for laks?, in *Merdkanten*. 2019. Skretting.
152. Wańkowski, J.W.J. and Thorpe, J.E. The role of food particle size in the growth of juvenile Atlantic salmon (*Salmo salar* L.). *Journal of Fish Biology*. 1979. 14(4): 351–370.
153. Skretting. Feed catalog (in Norwegian). 2012. Skretting.
154. Lekang, O.I. *Aquaculture Engineering*. 2020. Wiley: Chichester.

CHAPTER 11

Health, diseases and disorders in salmon farming

11.1 Introduction

Disease has always been a challenge for the salmon farming industry. New diseases are continually developing, seemingly as soon as one disease is controlled. Disease problems include interactions with wild fish. Diseases can be transferred from wild to farmed salmon and vice versa, but today it is typically the transfer of infectious diseases from farmed salmon to wild salmon that is significant. The number of farmed Atlantic salmon today is much higher, on a completely different scale, than the total number of wild Atlantic salmon in the world. There must be awareness when transporting farmed fish into new areas, as new diseases against which the wild salmon stocks have low immunity may be introduced.

Diseases in fish are significantly trickier to control than diseases in land animals. It is not so easy to observe the fish since they live in water, the individuals are often crowded together and the numbers of individuals are typically much larger.

Farmed salmon is more susceptible to disease than wild salmon. There are a number of reasons for this (section 11.2), but one major reason is the increased stocking density in farming that makes transfer of disease between individuals easier. When a fish in farming conditions is infected, they have to be removed from the population as fast as possible to avoid transfer of disease; in nature, however, natural predators quickly remove a sick fish.

Veterinarians with knowledge of this area must handle problems with disease; they are also legally permitted to write prescriptions for medications. This chapter will therefore only give a brief overview of the topic, and discuss the currently most significant diseases. Veterinary science uses specific terminology, often with a Latin or Greek origin [1], so some of the general terms used are explained in a simplified way. A great deal of specific literature is available regarding disease in fish and also more specifically for salmonids (for example [2–8]). However, the most important point when talking about disease is that it is more effective to prevent than to treat, and this will be focused on in the text. This is described as prophylactic treatment: treatment before a disease is detected, or preventative treatment. The term therapeutic treatment is used to describe treatment of detected disease.

There is a range of options for disease control. Good husbandry practice is, of course, of major importance: the environment must be optimal. Other measures include transport restriction, optimal feed (possibly with supplements of immunostimulants), vaccines, probiotics, disinfection, sectioning of the farm with separation of generations, fallowing of sea cage sites and general farm-level biosecurity. Breeding can also be used to selectively remove disease-sensitive individuals/families and those with low stress tolerance.

All farms therefore need to have good plans for:

- avoiding disease entry to the farm
- handling disease outbreaks on the farm.

In all salmon producing countries there are also government health regulations to ensure animal health and welfare and to prevent introduction and transfer of diseases. These include, for example, regular routine health checks by veterinarians, mortality recording by the farmers and biosecurity plans for the farm.

11.2 Infectious and non-infectious diseases

Diseases can be divided into viral, bacterial and fungal diseases. In addition, parasites can also cause health problems. All these can be grouped overall as infectious disease. A group of miscellaneous health problems is typically also included, called non-infectious disease. Infectious diseases are contagious, and normally possible to treat with medication. This is not the case for non-infectious diseases, which do not include pathogens, but are typically caused by a suboptimal environment or incorrect feed composition. They may therefore also be called production-related diseases.

11.2.1 Infectious diseases

There are enormous amounts of microorganisms, such as bacteria, viruses, fungi and parasites, present in water. A few of them are pathogenic, a term that describes organisms able to cause disease, and therefore these are of particular interest. The word pathogen originates from Greek, where pathos can be translated as 'suffer or experience', and -gen, from genes, as 'producer of'. When talking about a pathogen it is typically referred to as an infectious agent (within science, germ theory). It is important to remember that when talking about microorganisms, such as bacteria, viruses, fungi and parasites, only a few are pathogenic. However, pathogens are always present in the water, whether it be fresh- or saltwater, so it is important to create an environment that is suboptimal for these organisms, or where the 'good' microorganisms dominate. The fish must be in good condition so they can handle exposure to the pathogenic ones as well as possible themselves. If conditions are suitable for pathogenic growth, the pathogens can grow/multiply rapidly. The term virulence (from Latin = full of poison) is used to describe a pathogen's ability to cause disease in the fish: high virulence means they easily cause disease. The study of diseases is known as pathology.

An infection can be defined as when pathogenic microorganisms invade the body and start to reproduce. The process of an infection can be described as links in a chain, and all links must be present to result in a disease outbreak. The following links can be defined (based on [9]).

1. There must be enough pathogens with a high enough virulence to destroy normal tissue, a wound where they can invade, or a fish with reduced immune defences (a host for the pathogen).
2. The pathogen needs somewhere to develop and reproduce (multiply in the host) such as the body tissue.
3. The pathogen needs to leave the host after multiplication, for example from the intestine.
4. The pathogen needs to transfer from the host to another fish through a vector, such as water, or by direct contact.
5. The pathogen needs a portal to enter a new fish, such as open wounds or the intestinal tract.
6. The newly infected fish must not have adequate immune defences that can block the invading pathogen, and in this way disease will be spread.

When pathogens invade the body, it creates an infection locally or they can spread through the body liquids, such as the blood. The blood is normally sterile and when bacteria are present in the blood, it is named bacteraemia.

When an infection occurs in the fish's body, the body will react by producing antibodies and inflammation occurs. Inflammation can be described as a response to the infection, the mobilization of the defence mechanisms of the body. A self-healing process will start. The white blood cells (leucocytes) will try to eliminate the detected pathogens by releasing chemicals and substances toxic to the pathogens, the supply of red blood cells and plasma will usually also increase, and the infected area swells and becomes red.

Antibodies, also named immunoglobulins, are produced to resist the invading pathogens, which are also termed antigens. The immune response is defined as the response of the body to the introduction of an antigen. An antigen is described as a substance, typically harmful, such as a pathogenic bacterium, which results in production of antibodies within the fish. Antibodies are special y-shaped proteins produced by the white blood cells to fight the antigens. The immune response is temperature dependent as the metabolism increases with higher temperatures. An important aspect is that if the fish has completed this process once for a specific pathogen, it remembers this pathogen; the next time the same pathogen tries to invade, the antibody production system is already established and the invading pathogens will be attacked at once. The salmon will have developed what is called immunity. This is the basis of development of normal vaccines: a small dose of dead or weakened pathogens is administered to establish the immune defence system. The study of immunity is named immunology: the science of the immune system and how it fights antigens.

Immunity varies between individuals in a population; some are genetically coded to have a stronger response against specific pathogens. Selective breeding is therefore a method to improve disease tolerance in salmon farming, but it must be remembered that using breeding as a tool takes generations to achieve results.

Typically, the salmon will try to heal an infection itself, but if it cannot, a disease outbreak may be the result.

When a pathogen enters the body of the salmon there will be a certain incubation period before it eventually starts to proliferate and the disease symptoms develop. This period varies with each pathogen, the water temperature and the environmental conditions, and can range from days to months. Some diseases will be acute and develop very fast, while others can be sub-acute or chronic and take a longer time to develop. However, the pathogens may be latent in the body for a long time, showing no or minor symptoms, and may not develop into disease; in these cases, the fish can still host the pathogen and infect other fish. The pathogen may also be latent in the body and break out in situations when the immune defence of the salmon is reduced, for example, a situation creating stress in the fish.

11.2.1.1 Vertical and horizontal transmission

The transfer of pathogens can be visualized with vectors: a disease transfer vector is anything that can transfer a pathogen from one fish to another. The pathogen is transferred either vertically or horizontally, termed vertical or horizontal transmission (Figure 11.1). In vertical transmission the transfer is from generation to generation, meaning from parents to offspring. In salmon farming the pathogen can be transferred with the fertilized egg. Several viral and bacterial diseases can be transferred in this way, and some parasites. Horizontal transmission is transfer within the same generation, which could be due either to direct contact between sick individuals or to the disease agent being transferred through water, air or other hosts. Live organisms, called biological vectors, can host pathogens but will not develop disease themselves, and only host and transfer the pathogen. Non-biological objects may also transfer diseases, for example, a well boat that transfers pathogens. People can also transfer diseases when moving from farm to farm without proper disinfection or changing of clothes, so good biosecurity and hygiene is vital.

11.2.1.2 Epidemiology

The study of disease in a population is called epidemiology; a more accurate definition given in the *Oxford English Dictionary* is 'The branch of medicine which deals with the incidence, distribution, and possible control of diseases and other factors relating to health.' There are, however, a number of definitions of this term and the definitions are also changing [10].

One term used in epidemiology is prevalence, a statistical concept often used to describe the occurrence of a specific disease at a given time either in absolute numbers or as a proportion of

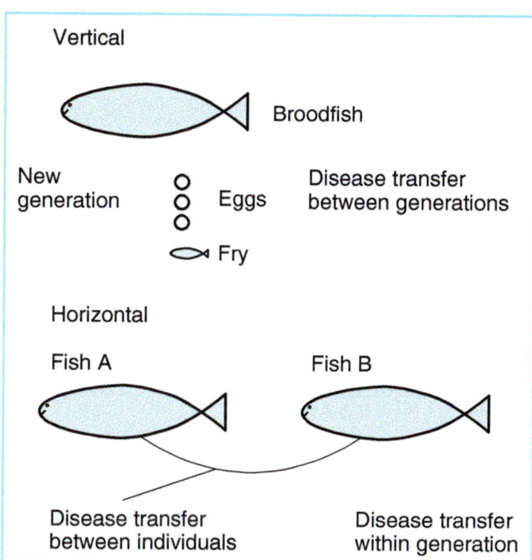

Figure 11.1 Pathogens can be vertically or horizontally transferred.

an overall number. The term incidence refers to the number of new outbreaks of a specific disease during a given time period. Chronic disease often has high prevalence but low incidence for a given period while acute disease can have a high incidence but low prevalence.

11.2.1.3 Notifiable diseases

Some diseases are notifiable; these are diseases that must be reported to the national authorities when symptoms are discovered. This can be either at national level or at international level (listed by the World Organisation for Animal Health [OIE]). Notifiable diseases are severe diseases where further development is undesirable. If such diseases are discovered on a farm immediate notification to the relevant authorities is required. This is, of course, essential in areas free of these diseases. Several of the diseases which occur in salmon farming are listed by OIE, such as the viral diseases infectious salmon anaemia (ISA), infectious haematopoietic necrosis (IHN) and viral haemorrhagic septicaemia (VHS) (described later), and the parasite *Gyrodactylus salaris* [11]. Reporting diseases to the authorities in this way is important so that national defensive strategies can be used. The reaction from the authorities varies with the disease: the worst outcome is immediate slaughter of the fish, or stamping out, which for some diseases will be the only possible course of action. Through this strategy, OIE is defending the animal and human population. A good example is the African swine fever outbreak in China, where the virus destroyed almost 25% of the world population of pigs. Everyone must understand that this step is not popular among the producers nor the country where it occurs, but it is a very effective way to prevent the much larger worldwide economic consequences. There is some discussion over whether producers and local authorities in China may have tried to cover up this outbreak [12].

11.2.1.4 Zoonotic disease

Some pathogens are active only in freshwater or only in seawater, but some can survive and cause outbreaks in both. Pathogens can either be specific for salmon or infect fish in general, and some may also infect other animals. Some fish pathogens may even lead to infection in humans. Pathogens that can be transferred from animals to humans and cause infections in humans are known as zoonotic [13–15]. Zoonotic disease is therefore, in our case, disease that can be transferred from salmon to humans, such as worms. However, these are not a problem in salmon farming, as salmon eat formulated feed that has completed heat treatment as part of the production process. Anisakid larvae have been found in the viscera of a farmed Atlantic salmon; this may result in the disease anisakiasis, but the risk is very, very low [16]. Care should also be taken regarding the rising seal populations in some areas as they may represent a vector [17].

11.2.2 Non-infectious diseases

As previously mentioned, there are a number of non-infectious diseases that can occur in salmon farming. They can be divided into categories related to a suboptimal environment and to suboptimal composition of the feed.

11.2.2.1 Environment-related disease

Included in the category of environment-related diseases is reduced water quality, which creates health problems mainly due to too high a concentration of certain substances in the farming water. Water quality can typically be divided into optimal, suboptimal and directly toxic. Suboptimal water quality will result in reduced growth and may also increase stress levels in the fish, thereby reducing their immune response and giving pathogens the opportunity to 'attack'; the result may be infectious disease. Suboptimal water quality over a period of time may result in a disease outbreak, and this can also occur if several water quality parameters are suboptimal at the same time. None of the parameters may be critical alone, but can cause problems together. The water quality parameters that are typically above optimal values include gases such as carbon dioxide, ammonia, hydrogen sulphide and dinitrogen, ions such as ammonium, nitrogen dioxide, nitrate, sulphate, and hydron (pH) and metal ions of iron and alumium. However, some values may also be too low, such as the oxygen gas concentration and the ionic concentration of the water; deionized freshwater is not suitable as fish need ions in the water. There can also be interactions between the ions in the water that create suboptimal or even toxic conditions, even if the individual ions are not at suboptimal concentrations; for example, pH resulting in interactions with metal ions [18]. In certain cases, substances such as chlorine, heavy metals, and sulphur may also present problems. If reusing water, as in RAS, there will be accumulation of all substances, and some substances which are normally not considered may create problems.

The water temperature is also a factor that is important to keep within certain limits; low temperatures close to freezing will reduce growth rate considerably, and too high a temperature will decrease the available oxygen in the water while increasing the oxygen consumption of the salmon. When the temperature is too high, it will result in 'breakdown' of the physiology of the salmon, and the inner organs responsible for metabolism will not function as they should. There have been examples of skeletal deformations when water temperatures on smolt farms are above 16°C. Incubation of salmon eggs above 8°C will also result in deformities.

Fish density and water current are other factors that can result in reduced welfare. Too high a fish density may result in wear on fins, creating areas for fungi to attack. The recommended density from an animal welfare perspective is an interesting topic: in open sea cages, regulations state 25 kg/m^3 but in tanks, densities of up to 50–80 kg/m^3 are used. Too high a water velocity may result in exhaustion of the fish leading to a reduced immune response.

Large quantities of algae and jellyfish may also be considered a non-infectious disease as they can clog the gills of the salmon, and some algae produce substances directly toxic to salmon.

11.2.2.2 Nutrition-related disease

It is of major importance that fish obtain the required nutrients for growth. Aquaculture is characterized by a requirement for individuals to maximize their performance, meaning the body is constantly 'pushed' to its maximum. The feed contains the necessary building blocks to run the body and to build new body mass. If there is a lack of essential nutrients, problems can develop; for example, if the feed contains insufficient amounts of phosphorus or vitamin C, salmon will suffer from improper bone formation, which leads to spinal deformities. Another example is if the raw materials in the fish feed are changed from being fish based to having a larger vegetable (soy) proportion, a lack of histidine can occur resulting in cataracts. Problems may also occur if there is too much of certain nutrients, which can lead to toxicity. The origin of the nutrients may also be important here, for example whether they are inorganic or organic. Too low an amount of feed in general compared to requirements may also result in disease due to lack of nutrients, but typically this will take time to develop because salmon are adapted to cope with periods of fasting.

Nutrition-related diseases are not very common in salmon farming today because of the increased knowledge about optimal feed composition. Dry formulated feed is delivered from large specialized feed-producing companies that have great expertise within this area. However, with the continuous push for increased growth and the alteration of main feed ingredients for optimal economy it is an ongoing challenge. In addition, novel ingredients are included in the feed and laws are updated to permit new additives. An example here is the regulation of the levels of selenium and zinc.

In rare cases disease can be related to genetics, such as skeletal deformation or abnormalities due to inbreeding or other breeding mistakes. However, with the current large expert breeding companies such problems are very rare.

11.3 Fish welfare

Ensuring good fish welfare is currently an important part of salmon farming and during recent years the consumer focus on fish welfare has increased. But what actually constitutes fish welfare? The definition is not exact and has been evolving over recent years [19–21]. There have also been different approaches to define animal welfare based on people's backgrounds [22]. One definition is that good health and good growth are synonymous with good welfare, another is that fish must have the opportunity of a natural life including a natural environment, while a third is based on feelings and states fish must not be exposed to fear or pain and must have the opportunity for positive experiences [22]. Merging all these approaches into one is very complex. Taking the current understanding of the brain function of salmon into consideration gives an even more complex system.

Here are two examples of definitions of animal welfare 'the quality of life as perceived by the animal itself' [23] and 'the protection of the health and well-being of animals' [24].

During the 1990s, the Five Freedoms of animal welfare were formulated [25] based on extensive work by the Brambell Committee in the 1960s [26], which was the starting point of the increased focus on animal welfare.

- Freedom from thirst, hunger and malnutrition
- Freedom from discomfort and exposure
- Freedom from pain, injury and disease
- Freedom from fear and distress
- Freedom to express normal behaviour.

The basis for the development of the Five Freedoms was animal livestock, but it is a general concept and can also be adapted to fish (Figure 11.2). Since the 1990s there has been an increase in knowledge and understanding of animal welfare and today it is possible to get a more in-depth understanding based on scientific knowledge. It can also be more species dependent, and the welfare needs of salmon can be divided into physical needs and behavioural needs [27]. The following factors can be included in the physical needs (direct quote [23]):

Figure 11.2 From an animal welfare perspective too high a density must be avoided.

- respiration – uptake of oxygen and release of carbon dioxide by pumping water over the gills
- osmotic balance – maintaining homeostasis of body cell fluids
- nutrition – intake of food containing the required energy, amino acids, minerals, vitamins and so on
- health – absence of disease, illness and malfunction
- thermal regulation – optimization of metabolism and temperature, including thermal comfort.

In the behavioural needs the following factors are included (direct quote [23]):

- behaviour control – ability of the fish to freely position themselves (including regulation of buoyancy) and respond to stimuli
- feeding – regular access to food
- safety – possibility to avoid perceived danger
- protection – possibility to keep the body undamaged from physical injury
- social contact – predictable interaction with conspecifics
- exploration – possibility to search for resources and information
- kinesis – being able to swim (physical activity)
- rest – possibility of reducing activity level or 'sleep'
- sexual behaviour – homeward migration, breeding behaviour, spawning, and so on
- body care – scratching, parasite cleaning, and so on.

How do we know if salmon really feel okay and have a good quality of life? They cannot talk to us. Therefore, it has been of major importance to identify some objective indicators, called welfare indicators. A welfare indicator can be defined as an observation or a measurement that gives

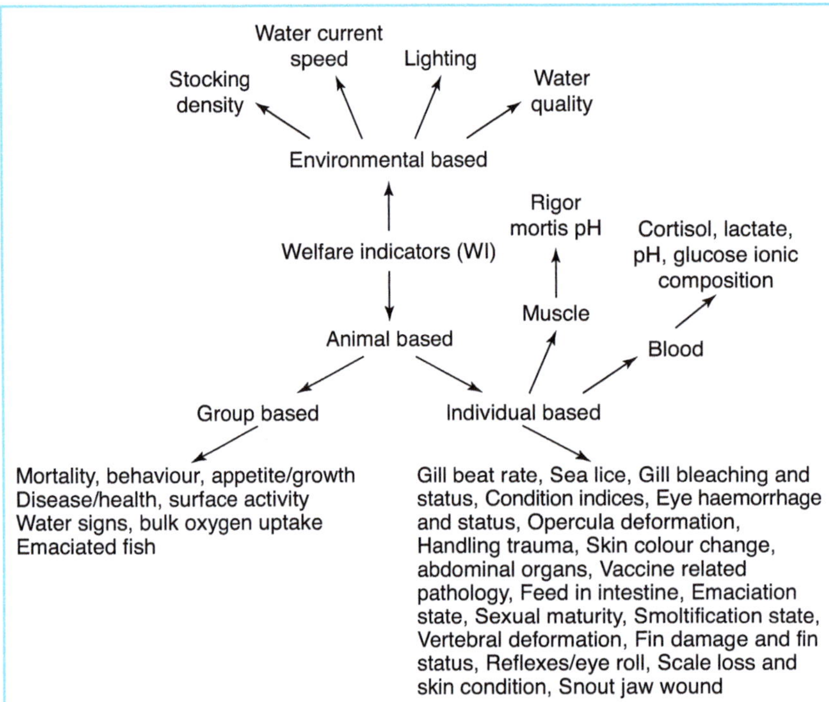

Figure 11.3 Welfare indicators can be separated into environment based or animal based; the animal based subset can again be defined as group or individual (based on [22]).

information about the welfare of the fish. In a newly published report [22] two different salmon welfare indicators were defined: indicators suitable for use under farming conditions (operational welfare indicators [OWI]) and indicators suitable for use under laboratory conditions (laboratory-based welfare indicators [LABWI]). The OWIs are further divided into salmon-based indicators, either at an individual level or at a group level, and environment-based indicators (Figure 11.3). These indicators can again be related to the different welfare needs of the salmon.

Welfare indicators can be used further to evaluate the welfare of salmon in different farming systems and when exposed to different farming routines/operations. Examples of production systems that can be evaluated are flow through systems and sea cages, while operations can be exemplified by crowding, pumping and grading.

A wider system for welfare of farmed fish in general has also been developed through the FishEthoBase, an open access database that provides information on fish welfare [28].

Animal welfare has become a large scientific field and much relevant literature is available; the aim of this text is only to give a brief overview of the area. For more information several textbooks are available, both related to animal welfare in general and specifically for fish (for example [29–32]).

11.4 Reasons for disease problems

Why are there problems with disease in aquaculture? To begin by considering infectious diseases, the pathogens responsible for diseases are generally not new; they are present in the natural environment even if they have not been transferred with the eggs (inside the eggs) or with the smolt taken to the ongrowing farms. What differentiates aquaculture from wild fish populations is the high biomass in a limited volume: a higher fish density. This creates a higher infection pressure, which means an increased risk of an infectious disease outbreak. The number of pathogens is higher and bacteria, for example, can produce thousands of offspring in a limited time, so the number per volume unit can increase rapidly if conditions are good for pathogen growth. Owing to the high fish density in salmon farming compared to that in the natural wild environment, the transfer from one individual to another is also much easier and faster.

The risk of disease outbreak will rapidly increase if the natural immunity of the salmon is reduced. In aquaculture conditions this will sometimes be the case as the salmon's stress levels may be higher. Stress can affect the immune system, increasing the risk for disease outbreak even more. Examples of factors that can result in higher stress levels, in addition to high fish density, are reduced water quality and the handling of fish.

In intensive aquaculture the performance of the individual is also pushed to the maximum, to achieve the best growth possible. This will also increase the risk of a disease outbreak.

11.5 Strategies for handling disease

There will always be discussion around how best to handle disease, as in one way or another most, probably all, intensive salmon farmers have contact with one or more diseases, especially in sea cage farming. As mentioned earlier, there are two important considerations to be borne in mind: (1) how to avoid diseases entering the farm; and (2) if there is a disease outbreak on the farm, how to mitigate its effects. Regardless of whether it is an infectious or non-infectious disease, it is important to remember that disease prevention will always be easier than treatment after an outbreak, so the focus should always be on actions to prevent disease. Here there is a great deal that can be done by farmers before veterinary expertise is needed. However, in a disease outbreak veterinary advice should be sought.

In medical terms, the word prophylactic is used. It originates from the Greek *phylax*, which can be translated as watcher or guard. In general, the term prophylactic therefore means to prevent, and in this case refers to methods for preventing disease. Examples of prophylactics in salmon farming are, for instance, the use of prophylactic antibiotics (which are undesirable) or the use of vaccines. These are further described later.

11.6 How to avoid disease problems

Whether infectious or non-infectious, a number of measures can be taken to prevent disease. Infectious or non-infectious diseases cannot be separated, as a non-infectious disease may result in outbreak of an infectious disease. There will always be a need for cost–benefit evaluation of the many preventative measures available, and even though all measures will help, the improvement will vary.

Here, primary and secondary disease problems should be mentioned. Primary disease can be illustrated by sea lice, and secondary disease represented by opportunistic bacteria that are present on the skin and take advantage of an immunological deficiency created by a sea lice infestation.

11.6.1 Optimal farming conditions and good fish welfare

Outbreaks of non-infectious disease should be avoided by having optimal feed composition, keeping water quality and water temperature within acceptable values and having a stable water quality. In open cages in seawater this will, of course, be difficult to control. This is why site selection is so important, and should include consideration of the history of the water quality of the sites and historical data over a long period: for example, whether there has been any algae or jellyfish blooming on the site. However, a start can be made by optimizing water quality in open sea cage farms by adding pure oxygen. The fish density must not be too high, nor the water flow too fast. The environment must be optimal regarding salmon welfare requirements, but with the knowledge platform of today this should not require control.

Avoiding infectious disease is a lot more difficult, but a range of measures can be taken. Of course, there must be optimal farming conditions as for non-infectious diseases. Suboptimal farming conditions may result in stress in the fish, reducing the general welfare and weakening the immune system which may be a trigger for the outbreak of diseases. The same is the case for situations that add a sudden stress response to the fish such as handling or crowding.

11.6.2 Preventing pathogens entering the farm

There are several measures that should be included to avoid disease entering a farm. Of course, the number of measures will depend on the farming system and the site selection. The term 'biosecurity' is used, which can be defined as (quotation from Wikipedia) 'Biosecurity, as originally conceptualized, is a set of preventative measures designed to reduce the risk of transmission of infectious diseases in crops and livestock, quarantined pests, invasive alien species, and living modified organisms.' In terms of salmon farming it can be simplified to preventing pathogens entering and spreading inside the farm. This is included on all farms, from broodstock farms through smolt farms to ongrowing farms. Biosecurity efforts start with measures that prevent pathogens from entering the farm. The possibilities for preventing entry are, of course, dependent on the production system used; it will be much easier for a closed farm compared to an open farm. This will also be reflected in the barriers that can be put in place. A biosecurity plan ought to be established for every farm, based on risk analysis and identification of critical control points.

All biological material entering the farm should have a health certificate documenting health status; the salmon stock must not be unwell, but documenting its history and origin will also be important. This will apply for eggs and smolt. The eggs should be surface disinfected when entering the farm, and care must be taken regarding handling of the packaging.

People entering must wash hands, disinfect or change shoes, and preferably change clothes. A sluice should be included in the farm design where possible (Figure 11.4). A log of people entering the farm should be kept (should a disease problem occur this can be used to track possible sources). Water, as explained later, can be disinfected on closed farms to reduce the risks of pathogens entering the farm with the water supply. Disinfection of equipment and other objects entering the farm is also a useful effort to reduce the risk of disease outbreak. Sectioning of the farm into clean and dirty zones should be included, see below.

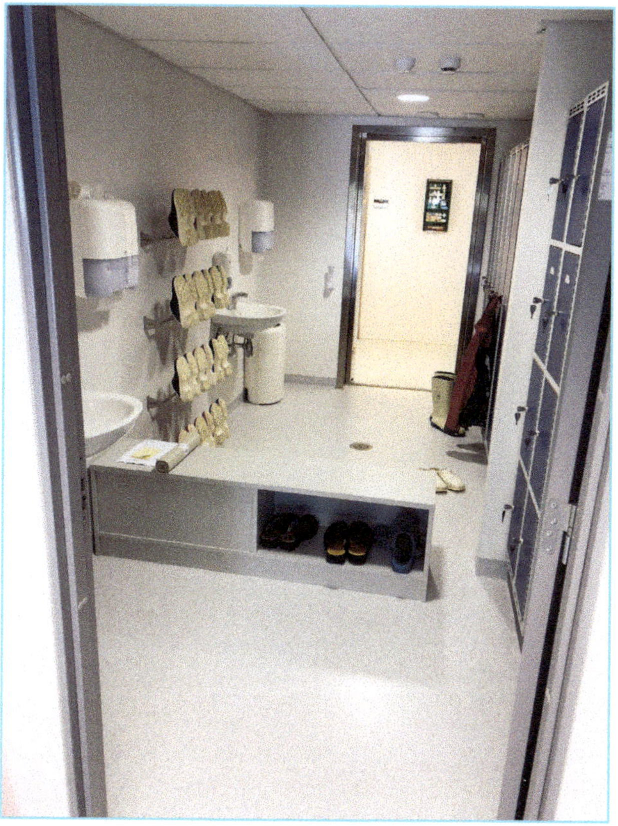

Figure 11.4 An infection sluice should be used when entering the farm to ensure diseases are not brought in or taken out of the farm with people.

11.6.3 Regular veterinary control

Even though veterinary control does not stop pathogens entering the farm, it is a means of discovering diseases at an early stage and increasing the likelihood of reducing the scale of any potential disease outbreak. This should therefore be an integral part of salmon production (Figure 11.5), with regular veterinary checks every 1 to 2 months. Many farmers employ their own veterinarian for most of the checks.

11.6.4 Sectioning of the farm

Sectioning of the farm is an important tool to prevent the spread of disease within the farm. On a smolt farm this can be achieved by dividing the farm into compartments. Typically, it is common to have separate hatchery, start feeding and ongrowing areas. A greater total number of individuals will increase the risk for a disease outbreak, so on large farms sectioning is a good solution, for instance, by having several ongrowing areas. On sea cage farms sectioning is not possible, so efforts here involve ensuring a certain distance between different farms, which is typically regulated by law. Another method is to have only one generation or intake per site and to fallow the sites between intakes. Only transferring smolt from one smolt farm to one sea cage site is also a way to reduce the risk, and decreases the risk of bringing in pathogens with the smolt.

Figure 11.5 Farm veterinarian checking fish for regular health control.

When a number of sea cage farms are located in the same area, a coordinated disease prevention programme should be implemented. This can be illustrated with sea lice treatment; the effect will be much better if a large geographical area carries out coordinated treatment.

However, all efforts to stop disease entering and spreading within the farm involve additional costs. It will therefore always be necessary to carry out an analysis of reduced risk in relation to the increased costs. Which measures to include will also depend on the production size, and large farms should put more efforts into this area.

11.6.5 Dead fish removal and disposal

Regular removal of dead fish and dead eggs in hatcheries is an absolute requirement for inhibiting pathogen proliferation. Both dead eggs and dead fish are an excellent substrate for microbial growth. On eggs this can easily be seen by the growth of the fungus *Saprolegnia*. When fish die, the cause of mortality is of interest: is there any disease? If this is found to be the case, removal of the fish must be done promptly to prevent transfer of disease to other individuals. Dead fish must be disposed of in a secure way. The most common method used today is to put the dead fish in an acidic solution (typically formic acid) that kills the pathogens. This is kept in a container on the farm (Figure 11.6). The container is regularly emptied by special companies handling such waste.

Figure 11.6 Equipment for handling dead fish should be installed.

11.6.6 Cleaning

All cleaning is a way to reduce the risk of pathogen growth. Biofilms growing on surfaces have been shown to be an especially good environment for hosting pathogens, creating a reservoir of potential infection. Regular removal of this is therefore advantageous. In closed units this will involve cleaning tanks and, if possible, other surfaces where growth occurs. Since this microbial growth is temperature dependent, exposing areas to periods of increased temperatures is a good control measure.

In sea cages, biofilm may become established on the nets, named fouling. This must also be regularly removed. Biofouling, in addition, reduces the water flow into the cage and thereby the oxygen supply and fish welfare and increases the forces on the mooring system.

11.6.7 Disinfection

Disinfection of water, equipment or people is a way of reducing the disease infection pressure further after cleaning. Water disinfection is only possible in closed units, either of the inlet or outlet water or in RAS. It is important to remember the difference between disinfection and sterilization. Sterilization will destroy or inactivate all life, while disinfection only reduces the levels; there will still be microorganisms in the water but at much lower levels. The typical disinfection degree is log 3 or log 4, meaning an inactivation degree of 99.9% and 99.99% of the initial number of microorganisms. Taking bacteria as an example of microorganisms, their initial numbers can be quite high and this presents a problem with log 3 or log 4 disinfection. A total colony count of 10^7 organisms per litre of water is not abnormal, so a log 3 disinfection means that 10,000 microorganisms per litre still remain. Those surviving are also the strongest ones and will act as 'broodstock' for the next generation of microorganism. However, only a very small proportion, if any, will be pathogenic.

Inlet water to a farm can be disinfected to reduce the risk of pathogens entering the farm. Several methods can be used for this, where the most common are UV light and ozone (Figure 11.7).

Figure 11.7 UV light may be used to disinfect inlet water.

If using seawater on a smolt farm, the seawater ought to be disinfected to avoid pathogen entry, which prevents the smolt that leave the farm from acting as pathogen carriers, or vectors, and transferring pathogens into new areas in the sea.

In RAS systems there is some discussion about whether or not to use disinfection, but most use some type of disinfection either on the main flow or on a part flow. The reason for the discussion is that disinfection is seldom very microorganism-specific, meaning that the overall number is reduced including both the 'good' and 'bad' organism. After the disinfection step, therefore, all microorganisms that are alive have the same potential for growth, so the microorganism community can change. Following disinfection, a lot of nutrients in the form of dead material become available for the remaining organisms, which provides the potential for good growth. Pathogenic bacteria can continue to grow in any biofilm within the system and thereby proliferate. Addition of a beneficial microorganism culture that is strong and can outcompete pathogenic species is therefore an alternative tool for controlling the microorganism community in a system. These are named probiotics, and is an area that is attracting more and more interest. It can be simplified as letting the 'good' microorganisms eliminate 'bad' (the pathogens), and is further described later. This principle has also been tested in the growing of marine larvae, which can be termed water maturation, where the incoming treated water is stored for a period so the microorganism culture in the water becomes stable and robust [33].

Disinfection should be carried out at the farm entrance, both for people and for vehicles. People can carry pathogens on shoes and clothes while vehicles can carry them on tyres. Disinfection is also a way to limit transfer of disease from one site to another. How difficult it is to kill/inactivate the microorganisms varies with the type of organism. In general, it is easier to destroy larger-sized microorganisms, while spores and eggs are typically the most difficult to kill. For each disinfectant and microorganism, a dose response trial will show the requirements for the dose of the disinfectant and the period of time to kill/inactivate each organism. Typically, the required dose will be reduced if the disinfectant is given a longer contact time.

Before disinfection of equipment or surfaces, general cleaning must be performed. This optimizes the disinfection. Considering total colony count, cleaning alone will provide a degree of disinfection because many microorganisms will be removed from surfaces. Organic matter on surfaces may also protect microorganisms from the action of the disinfectant. A thick biofilm, which can host pathogens, will have the same effect. It is important that there is no residue left after cleaning and disinfection, so proper rinsing is important. The methods and chemicals that can be used are dictated by the equipment that requires disinfection. For instance, soaking can be a good method for disinfection of nets and air stones [34]. One procedure for disinfection of equipment is as follows (after [35]):

- manually remove dirt and organic matter
- clean vigorously with detergent or soap and water
- rinse
- apply a disinfectant with appropriate contact time
- dry completely.

A range of disinfectants and methods are available, including physical methods such as heat and pressure, and the use of different chemicals. For surface disinfection, today chemicals are used almost exclusively. Typical chemicals used are iodine, chlorine and alcohol products, hydrogen peroxide, acids and bases (the last two involve pH regulation). Commercial products including these chemicals are available on the market under different brands; it is important to follow the recommendations from the supplier regarding concentration and contact time. In the past, formalin (formaldehyde) was used but this is not recommended any more due to its potential carcinogenicity, among other reasons.

Too much disinfectant use may, however, present a problem: just as for antibiotics, it promotes antimicrobial resistance. Over time, tougher microorganisms are bred and the result is that the necessary disinfection dose must be increased. Underdosing will also create problems, because only the weaker microorganisms are eliminated; the same is the case for antibiotics, as explained later. Therefore, proper dosage based on recommendations from the supplier must be used in order to maintain the efficacy for as long as possible. Use of chemical disinfectants also involves environmental considerations, as for pharmaceuticals, of the effect of releasing disinfectants into the environment [36].

11.6.8 Vaccines

Today it is common to vaccinate salmon smolt. Vaccination is prophylactic (preventative) and gives the salmon an improved natural defence against disease through stimulation of the immune system (see textbooks on the topic of fish vaccines for more in-depth knowledge [37–40]).

There are several advantages of vaccinating fish, and it has been one of the major factors in gaining control over the diseases found in salmon farming. It reduces the need for medications, including antibiotics, so contamination of the environment with medications will not be a concern, and there will be fewer problems with antibiotic resistance. Today diseases such as vibriosis and furunculosis no longer present any significant problem due to vaccines, similar to the success in human medicine with the disease smallpox, which has been eradicated due to vaccination. Herd immunity will also result: by protecting a certain number of individuals by vaccination, the overall risk of disease outbreak will be reduced. Even though the costs of vaccines can be high due to their high development cost, the overall economy will normally be positive because the total cost of a disease outbreak will most likely be higher. It is, of course, also a great benefit for the environment due to the reduction in medications used. Vaccines were first successfully developed for bacterial diseases but today there is also success in preventing and reducing viral diseases.

Once given the vaccine, the fish become immune to the disease and do not become sick. Antibodies (proteins, immunoglobulins), created in the white blood cells in millions, circulate in the body and are available wherever a pathogen may gain entry. The antibody can recognize a pathogenic microorganism and will try to inactivate it. When the body has been infected once, it will recognize the pathogen the next time it enters the body, and the response to inactivate it will start at once. This is the basis for vaccination. Based on the state of the microbe supplied, vaccines can be divided into

- live weakened bacteria or virus
- dead inactivated bacteria or virus.

The first group is most effective, because it stimulates the same level of response as the active pathogen. However, it is important that the dose is low enough to avoid development of the disease, and the natural immune system of the fish must not be reduced at the time of vaccination. Use of dead inactivated bacteria will also stimulate production of antibodies, but there will be a lower level of stimulation of the immune system.

To boost the immune response of a vaccine, an adjuvant is added. An immunological adjuvant is defined as (direct quotation from [41]) 'any substance that acts to accelerate, prolong, or enhance antigen-specific immune responses when used in combination with specific vaccine antigens'. The adjuvant results in increased antibody production and at the same time prolongs the effect of the vaccine in preventing disease, as vaccines typically provide protection only for a limited time. There is a range of possible adjuvants that can be included; oils (used in salmon vaccines) and aluminium salts being two possibilities. The adjuvants themselves will have no effect if used alone.

As mentioned, vaccines were originally used against bacterial diseases but in recent years several vaccines have been developed for viral diseases. However, the protection provided against viral diseases varies widely, although in general is improving. Development of viral vaccines is more demanding. Injection of weakened live viruses has been shown to be most effective. However, due to possible creation of new mutants and changes in the genome of the virus, there might be a risk of developing viruses that are even more pathogenic, and therefore care should be taken. Owing to this and other reasons, virus vaccine development is very time consuming and it takes a long time before a commercial product is available on the market.

It is also possible to inject not the entire microbe but only parts, named subunit vaccines, or second generation vaccines. Second generation vaccines are among others developed because of the risk of using live vaccines, especially related to viruses. Examples here are specific protein antigens or polysaccharides.

The vaccines of the future, also named third generation vaccines, are the DNA vaccines [42–44]. This is new technology (biotechnology) and it may have enormous possibility. It is the sequencing of the salmon genome that makes the development of such vaccines possible, enabling novel vaccines to be developed. Owing to the fact that genetically engineered DNA is injected there has been some discussion regarding approvals, but it seems that this has been resolved. The genetically engineered DNA (plasmid) that is injected is taken up by the cells producing an antigen that initiates a process that results in an immune response similar to that achieved by injecting a live bacteria or virus. DNA vaccines are specifically of interest for viruses where today's vaccines seem to be less effective, and where there have been problems finding effective vaccines. Good results have been achieved by using DNA vaccines for both pancreas disease (PD) and ISA. In the future this technology may also be the basis for the development of vaccines for fungi and parasites. As a digression, the same technology is being tested today to treat cancer in humans and is also used in some of the recent Covid-19 vaccines.

Vaccine development costs are very high and so a certain level of demand is necessary before development can start. The development costs are higher for more advanced vaccines, such as subunits and DNA vaccines.

Currently there are several vaccines in use against problematic salmon diseases. In some countries, there are even regulations that require smolt transferred to sea for further ongrowing in cages to be vaccinated against some major diseases where the vaccines have documented effects, for example, for bacterial diseases such as furunculosis, vibriosis and cold water vibriosis. It is also common to vaccinate for some viral diseases where there are available vaccines, such as winter ulcers and in some cases infectious pancreatic necrosis (IPN), PD, yersiniosis and ISA. For some of these diseases the vaccines may have limited effects but will at least give some increased protection.

The drawback of vaccines are the potential side effects, especially when including the adjuvants. The side effects of vaccines are normally not so severe for fish, but there have been examples of adhesions of internal organs. Vaccine side effects can be graded by the Speilberg scale, which is based on adherence of internal organs and melanin deposits in the belly. The scale ranges 1–6, with 1 denoting no damage in the abdominal cavity and 6 denoting extensive damage to all organs in the abdominal cavity, high amounts of melanin and adhesions between internal organs and fillet including a large amount of damage after gutting. Today, vaccine producers have side effects in mind, and they are continually reduced. Some growth reduction after injection may also be observed in comparison to unvaccinated fish, but only for a very limited period.

The fish must not be vaccinated if they have symptoms of any disease, and prior to vaccination must be starved to reduce the effects of handling stress and to have a better response to the anaesthetic used prior to the vaccination. The fish must be vaccinated in a disease-free area but must be exposed to a site of potential infection, such as the sea, within a certain time. However, the vaccine will not be effective at once; a certain time is required before it becomes effective. This is temperature dependent, but normally in the area of 1 month.

When the disease pressure is so high that the fish cannot defend themselves naturally, it might be said that the fish is 'manipulated' to be able to tolerate it, similar to the outcome of vaccination. Wild fish do not receive vaccines and will still be attacked by the same disease. But what does vaccination of salmon in farming do? The number of hosts that the pathogens can invade is reduced and this will, in general, reduce infection pressure and increase herd immunity. As mentioned, the smallpox virus was exterminated once a large proportion of the human population had been vaccinated and developed immunity.

Through the use of pharmaceuticals, disinfection and vaccination, it is probably possible to solve the problems for farmed fish, but not the transfer of disease to the wild fish population. Several examples are documented here, such as the history of *Gyrodactylus salaris* and furunculosis.

Today, sea lice pose the same problem. When wild salmon are migrating out of the rivers in the spring and out of fjords where aquaculture farms are sited, some of these farms will function as sea lice production factories, and the wild salmon have to run the gauntlet of possible infection. If the wild fish are infected, they are not going to be treated. Therefore, development of vaccines is important in relation to this because they will help wild salmon stocks too.

11.6.8.1 Vaccination methods

There are four ways to vaccinate fish: by injection, by immersion, by spray and orally. Most common today is by injection (Figure 11.8). The vaccine is injected into the abdominal cavity (peritoneum), except for DNA vaccines that are given intramuscularly. Prior to vaccination the fish are anaesthetized, and are vaccinated individually. Typical doses involve volumes of 0.05 to 0.1 ml. The needle is inserted into the abdominal cavity, just under the skin, approx. 0.5 cm deep, and directed forward at an angle of 45°. This is done either manually or by machines. Vaccines against different diseases, multicomponent vaccines, can be injected at the same time, and these are typically used today. Fish must be above a certain size, usually above 25–30 g for salmon, to be injected. Therefore, other vaccination methods must be used for smaller fish.

When performing immersion vaccination, the fish are dipped into a vaccine bath, so it is also called bath vaccination. Spray vaccination, where vaccines are sprayed over the body of the fish, is not common but can be used for large fish and functions in the same way as a bath. Bath vaccination is suitable for small fish and when large groups must be vaccinated effectively. When utilized, a number of fish are dipped in the vaccine bath at the same time for a given period (for example, 30 seconds for small fish (1–5 g) and significantly longer for larger fish). The drawback with bath vaccination is that the duration of immunity is short and repeated vaccination is needed. This method may be also be used for boosting injected vaccines to prolong the efficacy.

Oral vaccination is carried out by providing the vaccine with the feed. It can be mixed with the feed, coated on the feed surface or encapsulated within the feed, and is typically microencapsulated. Oral vaccination can also be carried out on large fish groups. However, dosage is a

Figure 11.8 Injection of vaccines can be done either manually or by machines.

problem, because it depends on the appetite of the individual fish at the time of vaccination, and the potency is sometimes not very good. The vaccine is released when the feed enters the gut of the fish. Oral vaccination may also be used for boosting a vaccine given previously.

Injection of the vaccine is the most effective method regarding both the efficacy and the duration. All fish receive the same vaccine dose, but the drawback is that it is both labour- and cost-intensive. The fish are stressed due to handling and side effects are possible. Injection of vaccines can be performed manually by humans operating a syringe, but vaccination machines are becoming increasingly common.

Oral vaccines do not stress fish and save labour, but the protection and duration are highly reduced compared to injection. Mass vaccination is possible as for immersion vaccination, but the amount of vaccine required is higher. On larger smolt farms at least, it seems that the use of injectable vaccines, preferably with machines, is the best solution.

11.6.9 Probiotics, prebiotics and immunostimulants

In response to the problems with disease in aquaculture, and specifically to address the problem with possible resistance to antibiotics generally, there is fast-growing interest in finding alternatives to traditional pharmaceuticals (chemotherapeutics) for fighting disease.

The use of probiotics, prebiotics and immunostimulants is included here, and in addition phage therapy is described in section 11.6.10. The term probiotics is derived from two Greek words: pro = for, and bios = life [45]. There is a range of definitions of probiotics which are evolving due to the increased knowledge of microbiology. Probiotics can be defined as live microorganisms, or their products, that provide health benefits, boosting the natural defences of the host when consumed in adequate doses. However, a wider definition includes microorganisms that help to maintain an improved water quality and control the microbiota of the water. Stimulation of appetite and fish growth may also be included. Use of probiotics in aquaculture may therefore have a number of advantages including increased growth, inhibition of pathogens, improved digestion of nutrients, increased stress tolerance, and improved reproduction, in addition to improved water quality [46–50]. The following microorganism groups are proposed as effective probiotics in aquaculture: Gram-negative and Gram-positive bacteria, bacteriophages, yeasts and unicellular algae [51].

The term prebiotics can also be derived from Greek (pre = before, and bios = life). The typical definition of prebiotics is non-digestible food ingredients that selectively stimulate the growth and/or the metabolism of health-promoting bacteria in the intestinal tract [48] improving the microorganism balance in the gut of the fish [52]. In relation to probiotics, prebiotics are not living microorganisms, but typically a type of fibre promoting bacterial growth in the gut. Prebiotics are currently widely used in aquaculture [53, 54] and are supplied with feed, which is termed functional feed (a feed with additional properties).

Immunostimulants may be defined as substances that stimulate the immune system, but as a group may also include several other substances. Immunostimulants are used in aquaculture and may also be defined as being within the group of prebiotics [55,56]. Some also include immunostimulants within the group of probiotics [57] because of the use of whole or parts of microbial cells to stimulate the immune system, such as glucan from yeast cell walls [57]. A mix of probiotics, prebiotics and immunostimulants may be used. In literature, the effects of probiotics, prebiotics and immunostimulants are often listed together [48,58,59].

In fish, prebiotics and probiotics affect the intestinal balance in a way that achieves improved digestion and growth. An improved microbiota in the gut may also inhibit pathogens from attacking, and can be compared to improving the mucus layers on the surface of the fish which strengthen it against pathogen attack. One type of bacteria that has been used as a probiotic is lactic acid bacteria, which tolerate low pH and bile salts [60]. Feed with added prebiotics and/or probiotics may also be included in the group of functional feed.

By adding probiotics to water, the growth of beneficial microbes may increase and at the same time the growth of pathogenic microbes is reduced. In this way the microbiological water quality will be improved, and the fish will be healthier and more robust. This can therefore be a method to combat diseases without the use of traditional chemotherapeutics [48], and a way of reducing the use of antimicrobial compounds such as antibiotics; interest in the knowledge and use of probiotics is therefore increasing.

Probiotics can, as mentioned, be added to the feed, or provided as a bacterial culture added directly to the water as a powder or liquid. However, they can also be administered to the fish in a bath, or injected. In the last case there might be some confusion with vaccines. Several commercial probiotic products are available on the market, which are typically a mix of bacteria cultures. Probiotics have been introduced in salmon farming, both included in feed and added to the water. This is carried out both on smolt and ongrowing farms and trials are also being conducted in RAS. Different types of prebiotics are typically used in functional feed for salmon as a feed additive.

11.6.10 Bacteriophages

A bacteriophage can be described as a virus that enters a bacterium and kills it, and treatment can be visualized as using specific viruses that target pathogenic bacteria (general description textbooks, general description review articles, aquaculture specific articles [61–66]). Bacteriophages can therefore be an alternative to the use of antibiotics [67,68], so general interest in their use is increasing, especially in relation to antimicrobial resistance (AMR). The active use of bacteriophages (called phages for short) against diseases is called phage therapy.

A simplified explanation of the process is that the virus (phage) injects its DNA into a bacterium. The DNA in the bacterium may start to reproduce, resulting in the cell 'exploding' (cell lysis). However, the virus can also live in synergy in the bacterium with no consequences, or can change the DNA of the bacterium giving it higher virulence. The latter, of course, is not wanted because it can result in even more problematic pathogens. Therefore, the choice of virus strain is of great importance, including proper documentation of the effect of the phage.

Compared to antibiotics, bacteriophages are highly specific, which means they target the pathogens but leave the rest of the microbiota intact. This sets the requirements when developing bacteriophages for specific bacteria, but it is also possible to add a concentrated cocktail of phages. The phages used are natural parts of our ecosystem, so nothing new is supplied to the body system, only an increased concentration of some specific types. In natural ecosystems it is believed that phages have a role in controlling the bacterial community [69,70].

Phages were discovered early in the 20th century and soon after this trials with therapeutic treatment started [68,71]. But the development was slow compared to antibiotics, which were developed at the same time, treatment was much more target-specific and not as broad as that provided by antibiotics [72]. Therefore, the focus changed to the use of antibiotics. However, during recent years, the interest in bacteriophages has boomed in all medical treatment and, as mentioned, as an alternative for antibiotics. An improved effect has also been achieved combining antibiotics with bacteriophages: phage-antibiotic synergy [73,74]. They may also have a synergetic effect in combination with disinfectants. Bacteria may develop strategies to protect themselves against bacteriophages, but it seems that this is not as simple as the strategies for antibiotic resistance; however, more knowledge of this area is required [75]. By using cocktails of phages it will be more difficult for bacteria to develop resistance, and it seems comparatively easier to find other phages against which resistance has not developed than it is when bacteria become antibiotic resistant [76].

Phages can be administered in a bath solution, by spray or supplied with the feed via a coating on the pellet. It is also possible to add the culture as a liquid to the water, or to give it by injection: the same delivery methods as for probiotics. Proper dosing of the phage is very important for the result [77]. Phage therapy is typically used prophylactically, but may also be used therapeutically for direct treatment of an infection.

Bacteriophages are produced/concentrated in laboratories with fermentation technology like that used for antibiotics, and involve production in a bioreactor [78–80]. However, establishing a good, secure method for supplying phages is still a challenge [77].

Today there are commercial phage products on the market targeted for the aquaculture industry, and also specifically for salmonids, for instance, against *Yersinia ruckeri*, *Pseudomonas* spp. and *Aeromonas* spp. infections [71,77,81,82].

11.7 Handling disease outbreaks on the farm

If disease problems occur on the farm action must be taken, but what form this takes depends on the type of disease. If it is non-infectious, action must be taken to improve the environment or the feed, but this type of disease should be avoided on salmon farms with good husbandry standards. If it is an infectious disease, the pathogen must be identified. This is a job for veterinarians, and typically samples are taken from the fish and are sent to laboratories for analysis. It must then be decided how to treat the disease, a decision which again must be taken by a veterinarian, especially if it includes treatment with pharmaceuticals.

It is important to monitor the behaviour of the fish on a regular basis, in order to identify a disease outbreak as early as possible. This allows for early treatment and hopefully the extent of the outbreak can be reduced. A typical change in behaviour is a drop in appetite, observed during feeding. Another behavioural change is the salmon crowding together at the surface, which is unusual for salmon. Technology for monitoring behaviour has been developed and is being developed further. Among other methods, it is possible to use dynamic feeding systems that can provide an early warning system for the detection of possible diseases [33,83].

When a pathogen is positively identified there are different options depending on the type of disease that is present. For some severe diseases, notification is a requirement. The type of disease also dictates which action to take: for some severe diseases there will be a requirement to immediately destroy the entire fish group on the farm. Effective drugs and pharmaceuticals are only available for some diseases. If the disease is not severe and the fish can live with it, it may be possible to continue to feed the fish until they can be slaughtered.

There are a number of drugs to choose between for different diseases. This brings us to the term pharmacology, a word originating from Greek (*pharmakos* = drug, and *logos* = study), and which can be defined as follows (direct quotation from https://biologydictionary.net/pharmacology/) 'Pharmacology is the study of drugs including their origins, history, uses, and properties.' This is a large scientific area with much literature available [84–86], including specific literature for fish [87]. This area will not be described further in this book because pharmaceutical treatment should be handled by veterinarians with prescribing rights. As discussed, the chosen drug will depend on the disease. For some ectoparasites, chemicals can be used and farmers can perform this treatment themselves after positive identification. This includes, for example, bath treatment with hydrogen peroxide.

11.7.1 Use antibiotics with caution

For an infectious disease caused by bacteria, a quick fix can be to perform treatment of the sick fish with antibiotics. The word antibiotics originates, again, from Greek (*anti* = against, and *bios* = life). They are common and important medicines used to treat bacterial diseases [88]. They either kill the bacteria (by damaging their cell walls) or inhibit multiplication. There are about 100 different antibiotics; penicillin, the most widely known, was the first discovered. Based on functionality, they can be separated into narrow-spectrum and broad-spectrum antibiotics. The first group targets specific bacteria while the second acts against a larger group of different bacteria. The antibiotics used in aquaculture are not aquaculture specific, but are the same as for other animals and humans. Which antibiotics are permitted to be used varies based on each country's legislation.

Today there are three major concerns about the use of antibiotics:

- residues in the salmon for human consumption
- contamination of the environment
- resistance and spread of bacteria and genes.

Aquaculture may provide a reservoir of transferable resistance genes [89]. Direct transfer of resistant pathogenic bacteria from animals to humans is also possible [90].

Some humans have allergic reactions to antibiotic residues that can be quite severe. Antibiotics may also accumulate in the human body if low doses are consumed over a long time, and this can cause organic lesions [90]. Some antibiotics may also be metabolized in the human body and become more toxic [90].

Supplying antibiotics to the fish can be carried out with feed containing antibiotics (medicated feed), but here the first problem occurs. When fish are suffering from disease they typically have a reduced appetite, so there will be unwanted loss of medicated feed. Antibiotics can leach into the surrounding environment, which of course is not desirable. Another solution is to dissolve antibiotics in water, then immerse the fish in this antibiotic bath. This results in crowding of the fish, creating even more stress for fish that are already sick, which of course is a bad combination.

Antibiotics may further be used as preventative, or prophylactic, treatment. This involves adding antibiotics to the feed on a regular basis, so the fish always have a degree of antibiotic intake and become more resistant to disease. Even though antibiotics are only useful for bacteria, they improve general tolerance, and bacteria are usually involved to some degree in several disease outbreaks.

But does this represent a good strategy? In a number of markets it is not permitted to sell salmon directly for human consumption if they have recently been treated with antibiotics. The fish must be withheld from the consumer market for a specified time period after the last antibiotic treatment. However, the main problem is that the bacteria can develop resistance/immunity against antibiotics, so after a while, treatment with antibiotics will have no effect.

The typical way that this develops is that first a narrow-spectrum antibiotic is used and resistance occurs. A new type of antibiotic is then used and the same happens again, and so on. When this continues, the bacteria might develop multi-resistance, which means resistance to more than one antibiotic. Bacteria that are resistant to all known antibiotics are named pan-resistant bacteria. Resistance to antibiotics is one of the most serious threats to human health in the world today. If antibiotics are no longer effective, the last tool for fighting bacterial disease is ineffectual, and mortality may occur from common bacterial diseases. This is also a reason for not stopping an antibiotic course too early, because some of the bacteria, the hardest ones to kill, will still be alive and will then become the breeding stock for the new generation of bacteria.

However, the regulations, understanding and public opinion regarding the use of antibiotics varies a lot throughout the world. Despite the dark clouds of the multi-resistant bacteria, the overall use of antibiotics in human and animal healthcare is increasing. The use of antibiotics in salmon production varies greatly between the countries involved. In Norway the consumption is almost zero on production farms today, while a peak was reached in 1987 with almost 50 tonnes of active substances. A major reason for the reduction, and today's very low consumption, is related to the development of effective vaccines for certain bacterial diseases. However, in Chile the numbers are still high, above 500 g per tonne of salmon produced, and it has increased during recent years [91]. A major reason for this is the prevalence of the bacterial disease salmonid rickettsial septicaemia (SRS) (later described), for which no effective vaccines have yet been developed.

Regulations regarding use and antibiotic residues vary greatly throughout the world; regulations in Europe and the USA are stringent, while in other countries, for instance in Asia, they are not so strict [92]. What is certain is that salmon containing residues of antimicrobials have been found, and such fish may also contain antimicrobial bacteria [93,94].

11.8 A brief overview of diseases in salmon aquaculture

As discussed, disease problems can be divided into bacterial, viral and fungal diseases. In addition, parasites can create health problems for fish. A group of miscellaneous health problems can also be included. The diseases vary somewhat between the salmon producing countries, but below some of the major infectious diseases and diseases prevalent in salmon farming are briefly described. These diseases also represent the largest economic problems for the salmon farming industry today. In addition to the descriptions in a number of textbooks [7–8] there are also several web pages describing salmon diseases, including observable symptoms. These are the basis for this short overview of specific salmon diseases and are where additional information can be found [95–101].

There are also websites available that document the current disease situation in the fish farming industry (for example [102]).

11.8.1 Bacterial diseases

Bacteria are small single cell organisms, size from 0.001 mm to 0.01 mm. They can live on their own, have their own metabolism and multiply by splitting. They grow fast with short generation intervals, depending on temperature. For example, the generation interval for *E. coli* bacteria is around 20 minutes. There are tremendous numbers of bacteria and they can live in all types of natural environments. However, only a few are pathogenic and result in disease outbreaks. Bacteria produce toxic substances and enzymes. Most bacterial diseases can be treated with antibiotics, but care should be taken regarding resistance as mentioned earlier.

Bacterial diseases occur in both the fresh- and seawater phases of salmon farming. In the past, a number of bacterial diseases, including furunculosis, vibriosis and cold water vibriosis, created large problems for farmers, resulting in high losses. With the development of good vaccine protection, these diseases have almost been eliminated from the salmon farming industry. Below some of the major bacterial diseases of significance today are described briefly. Bacterial diseases do not represent the major overall disease problem in salmon farming today and often antibiotics can be used, at least to some degree, for treatment. For individual farms that get these diseases, however, the consequences can be significant. The exception here is the fact that the bacterial disease SRS is a large problem in Chilean salmon farming. Examples of general literature on bacterial disease include [103,104].

11.8.1.1 Winter ulcer

The winter ulcer is a bacterial disease that occurs in the seawater phase and is associated with infection with the bacteria *Moritella viscosa*. It is multifactorial and other bacteria are typically also found in the ulcers, which are usually located on the head and side of the fish and lead to reduced slaughter quality. The ulcers result in osmotic problems for the fish. Transfer of smolt to seawater at low temperatures seems to increase the risk of outbreak. Smolt farms using seawater may also get the disease, and rough handling can also result in development of winter ulcers. The disease can be treated with antibiotics, but vaccines have been developed that give protection.

11.8.1.2 Bacterial kidney disease

Bacterial kidney disease (BKD) is a particularly undesirable chronic disease caused by the bacteria *Renibacterium salmoninarum*. Symptoms are swollen kidneys with white spots, and the disease results in wounds and bleeding from the kidneys and liver. The fish may lose appetite, and some swim uncontrollably at the water surface. High mortality will occur in severe outbreaks. The disease is transferred horizontally (contact between fish) and vertically with eggs, therefore control of the broodstock is important. The disease occurs in both fresh- and seawater. There is no actual treatment, and neither vaccines nor antibiotics are used with high success. It is of major importance to ensure that eggs and smolt are free of the pathogen when transferred.

11.8.1.3 Yersiniosis

Yersiniosis, also named enteric redmouth disease (ERM), is caused by the bacteria *Yersinia ruckeri*. The disease occurs most commonly in the freshwater phase and some months after the smolt have been transferred to the sea. However, it has also been found on salmon ready for slaughter. It is transferred both horizontally (easily) and vertically (more difficult) so it is important to carry out checks before sending out eggs and smolt to other sites. The disease can be observed by abdominal bleeding and protruding eyes, and inside the fish there will be bleeding from internal organs. Behavioural changes can be observed with abnormal swimming and the mortality will increase. The outbreak is often related to some kind of stress. In RAS plants the bacteria can be in hosted in the biofilm and results in repeated problems [105,106]. The disease can be treated with antibiotics and a vaccine for prophylactic treatment has also been developed.

11.8.1.4 Salmonid rickettsial septicaemia syndrome

The bacterial disease SRS is resulting in high economic losses, approximately US$500 million yearly, for the salmon industry in Chile [107]. Outbreaks of this bacterial disease have also occurred in other salmon producing countries but not on the same scale or with the same mortality. It started by causing mortality of coho salmon, and was called coho salmon septicaemia initially, but later Atlantic salmon have also suffered from mortalities. The disease is caused by the bacteria *Piscirickettsia salmonis*, and results in increased mortality. The gills become pale and the belly swollen, and in addition salmon appear lethargic and dark [107–109]. The disease can be treated with antibiotics and is the disease responsible for the large consumption of antibiotics in Chile.

11.8.2 Viral diseases

Viruses are the smallest of the microorganisms; the size is typically much smaller than bacteria. Most are sub-microscopic size, between 5 nm and 300 nm (10^{-9} m), so an electron microscope must be used for identification. Viruses do not have their own metabolism and can be categorized as a life form between living and non-living. Viruses cannot live on their own but need a host cell to gain access to metabolism. They invade the host cell and use the cell's components for themselves and for multiplication. They multiply rapidly and take control of the cell, damaging it. They may also release their genetic material, so the invaded cell starts to multiply uncontrollably. Viruses can be spread via body liquids, faeces, water and air (aerosol). They are hard to protect against, and cannot be treated with antibiotics. Only some viruses can be inhibited with medication, which either improves the immune system or prevents the virus from multiplying. The major defence against viral disease is to have a good natural immune system, and not to expose the salmon to stressful situations that lower the immune defences.

Viral diseases seem in general to present the greatest problems in salmon farming today. There is a great effort to try to develop vaccines against these diseases. Some vaccines have been developed but the efficacy varies from very good to less effective.

11.8.2.1 Pancreas disease

Pancreas disease (PD) is a serious contagious (transmissible, by direct or indirect contact) viral disease and in Norway, as an example, it represents the greatest problem among viral diseases today. PD is caused by the salmonid alphavirus (SAV) which may also be called salmon pancreas disease virus (SPDV). It results in damage to the pancreas, with a subsequent reduction in the production of digestive enzymes, together with damage to the heart muscle. Weight loss and mortality in the seawater ongrowing phase (especially the first year) is seen, and the economic losses can be considerable. Potential disease outbreaks can be observed with a sudden reduction in appetite, increased swimming activity and the fish crowding together at the water surface. There is no treatment, but a vaccine that has a limited effect can be used. The disease is spread between sick fish both inside the farm and to other fish farms in the surrounding area by the water. Infection may also be transferred when sick fish are transported along the coast in well boats with open well valves. The disease can be latent in the farm but not apparent, but when the fish are stressed it may strike. It can also be latent in the farm for several weeks before disease develops. In Norway there is a local authority regulation to prevent PD in salmon farming. Today it is possible to obtain eggs with increased PD resistance compared to normal eggs (QTL-eggs, see also under IPN).

11.8.2.2 Infectious salmon anaemia

ISA is a serious viral disease which may result in significant mortality. It was first detected in Norway in 1984 and later in Canada, Scotland and Chile, almost eliminating the Chilean salmon farming industry in 2012. The ISA virus is an orthomyxovirus. It is similar to the human flu virus and it easily spreads between the cages inside the farm. The disease attacks the veins in the fish and blood loss is one problem, resulting in anaemia. Mortality occurs due to failure of blood circulation and therefore lack of oxygen. Loss of appetite, pale gills and abdominal bleeding are other signs. The fish may move close to the water surface snapping air, but there may be no odd behaviour and sudden death can be the only sign. If 'opening' the fish, bleeding from veins and internal organs can be observed. The virus is transferred from fish to fish and can be transferred from farm to farm through natural water currents if the distance between them is short. It can also be transferred with humans and equipment and therefore zoning is important. There is no treatment for the disease, therefore it is of major importance to have good husbandry. Early slaughter of sick fish may reduce the overall outbreak. A vaccine is available for ISA, but gives limited protection. If a farm gets ISA the mortality will vary. ISA is a notifiable disease.

11.8.2.3 Infectious pancreatic necrosis

IPN is a viral disease that has been responsible for high mortality rates in salmon farming. It is caused by the aquabirnavirus which is a very tough virus, able to survive both freezing and drying for periods of time. It occurs both in the fresh- and seawater phases and can be carried by smolt. The mortality caused by IPN varies; often it is not too high but 100% mortality may occur both in fresh- and in seawater in rare cases. It seems most easy to contract the disease in freshwater and therefore it is most critical here. In a number of smolt farms there are 'in house' IPN strains. The disease can be latent in an individual and stress may result in disease, for instance, when smolt are transferred to the sea. The disease can be observed by odd behaviour, moving the mouth against the surface and having an erratic swimming pattern. Internal bleeding and a light coloured, or pale, liver may be observed on cutting open a fish. The disease is transferred both horizontally and vertically, including inside the eggs. There is no treatment but today it is possible to obtain IPN-free eggs (QTL-eggs), where the broodstock have been screened to ensure they are free of IPN. Vaccines for IPN have also been developed and give protection.

11.8.2.4 Heart and skeletal muscle inflammation

Heart and skeletal muscle inflammation (HSMI) is a common viral disease in the seawater phase and typically outbreaks occur some months after smolt transfer. The disease is transferred horizontally between cages in a farm. It is also believed that there can be vertical transfer. The virus is named piscine orthoreovirus (PRV) and it is common in fish, but may not necessarily result in a disease outbreak. Stress seems to be an important factor in an outbreak. The virus attacks the heart and inhibits blood circulation. Inflammation of the heart is typical. Before mortality occurs, the fish usually become thin, having a low condition factor. There is no protection today, but vaccines are under development, and mortality is typically related to some kind of stress. Mortality is highest in the seawater phase and generally ranges from almost 0 to up to 20% in a cage.

11.8.2.5 Cardiomyopathy syndrome

Cardiomyopathy syndrome (CMS), also known as broken heart syndrome, is a viral disease common in salmon farming. The virus infects the heart and results in inflammation. This may result in rupture of the heart if the fish is stressed. If cutting open the fish, clotted blood can be found close to the heart due to broken veins in the heart. Mortality may occur in large fat fish, especially during feeding. The disease may be misidentified as HSMI or PD as it has some of the same external characteristics/symptoms. There is no protection today, but vaccines are under development, and CMS-QTL smolt are available. Mortality typically occurs late in the seawater phase, since the disease develops slowly, and even if mortality is not particularly high the economic losses can be high due to the large fish size. Again stress, for instance when handling, is an important factor for an outbreak.

There are several other viral diseases that are present in salmon farming, such as viral haemorrhagic septicaemia (VHS), infectious haematopoietic necrosis (IHN) and salmon pox, but the diseases discussed above are the most significant currently.

11.8.3 Fungal diseases

Fungus is a microorganism that grows on organic matter; it is called a heterotrophic organism. It takes energy for its own metabolism from oxidizing the organic matter, which is used as the nutrient. A fungal disease is also called a mycotic disease and is caused by a fungal invasion. Fungi reproduce by producing spores that spread. The spores are very hard to destroy, requiring heavy disinfection, so it is difficult to remove them and they may survive for a long time and under harsh conditions. Fungal diseases in aquaculture include, for example, saprolegniasis, branchiomycosis and ichthyophonus. The most common fungal problem in salmon farming is caused by *Saprolegnia* spp. Infection occurs in the entire freshwater phase, starting with the broodstock, on to the eggs and continuing until smolt. *Saprolegnia* spp. spores are present in most freshwater sources and accompany the intake water into the farm. It spreads through mobile spores, which are difficult to destroy with traditional water disinfection methods, such as UV or ozone, and high doses are therefore required. The fungi are easily recognizable as a white fluffy layer and therefore it is also named cotton wool fungus. However, fungi will not grow on healthy fish skin or live eggs. Dead eggs are a naturally good surface for the fungi to start to grow. Fish that are missing scales, or have wounds, will also be more exposed to fungal attack. It is generally when the immune response of the fish is reduced that fungal problems occur, and then infection occurs in areas where there is a lack of scales, such as the fin base. The fungi destroy the skin (dermis) and osmotic problems will occur for the fish. *Saprolegnia* can be treated with fungicides such as formalin baths. However, good hygiene and careful handling of the fish is of major importance both to keep the level of spores low and to avoid wounds on the fish.

There is a range of antifungals that can be used. In the past, a highly potent antifungal, malachite green, was commonly used. It has been banned since the early 2000s due to its mutagenic and carcinogenic effects. In use today are, for instance, hydrogen peroxide, formalin and bronopol. Bathing the fish in a bath with an increased salt concentration for short periods may also be used.

11.8.4 Parasites

Parasites, like viruses, need to live on another organism, named a host, to get nourishment, to be able to reproduce and in many cases for shelter. Parasites can be separated into ectoparasites and endoparasites. The first live on external surfaces on their host, while the second live inside their host. Typically, there are some parasites on fish, but the numbers are low and they do not create any problems: the parasites need the host to survive so do not normally kill the host. Problems occur as the number of parasites increases. In nature, parasites are usually not a problem because natural parasite predators keep the levels low. In a sea cage, however, there are no natural predators but many parasite hosts so it is easy for the parasites to multiply. The most important parasite in relation to costs to today's salmon farming industry is the salmon louse, but there are also a number of other parasites, mostly ectoparasites, that are problematic both in the fresh- and seawater phases. Some endoparasites, however, may also be a problem; in wild salmon these include worms, and because of this it is not recommended to consume wild salmon. The worms infect the salmon through the food, so there is no problem in farmed salmon because the production process of salmon feed involves heat, which parasites will not survive.

11.8.4.1 Sea lice

Sea lice represent the largest disease problem facing the salmon farming industry today (Figure 11.9). Sea lice are ectoparasitic copepods that are members of the family Caligidae, where the major species are members of the *Lepeophtheirus* and *Caligus* genera. For salmon farming a major problem is presented by the salmon louse (*Lepeophtheirus salmonis*). This species is only found in the northern hemisphere, and therefore in salmon producing countries in this region. Problems with another louse, *Caligus elongatus*, seem to increase in the north. In Chile the louse *C. rogercresseyi* presents the problem [110,111].

Sea lice can be observed on salmon when they reach the large, adult life stages. If there are problems with sea lice, salmon increase their jumping activity [112] to try to get rid of the sea lice.

A salmon louse has different life stages, and it is only in the last stages that they attach to the salmon. It is when the lice reach the preadult stages that the problems really start, because the lice start to 'eat' the salmon. However, a stress response may occur at earlier attachment stages [113], although the effects here are not severe even with high numbers [114]. Preadult stages have a number of negative effects on salmon including: increased stress, reduced appetite, reduced feed conversion efficiency, damage to surfaces (such as skin and gills), osmotic problems and reduced

Figure 11.9 Sea louse attached to a salmon.

swimming performance [115–119]. How severe the effects are depends on the number of sea lice on the salmon. This can be expressed as the degree of infection and the following limits are proposed [27,120]:

- short-term effect infection degree 0.01 lice/cm^2
- long-term effect infection degree 0.05 lice/cm^2
- mortality occurs with infection degree >0.12 lice/cm^2

As shown, this degree of infection is expressed as an area, meaning that a large salmon tolerates higher numbers of salmon lice than a smaller one due to its larger surface area. A new model proposed to estimate the surface area of Atlantic salmon based on the weight is shown below for fish slightly above 1 kg [121]:

$$SA = 13.9\ W^{0.61}$$

There are also a number of other proposed estimations, but they give quite similar results.

For *C. elongatus* the problems have been less severe until now. *C. elongatus* is not a species-specific louse like the salmon louse, and is found on several marine fish species. Wild cod may carry them into the salmon farm. It also seems that wild cleaner fish may host this type of louse and therefore could bring it into salmon farms. Unlike the salmon louse, *C. elongatus* may jump from fish to fish: after attachment to one individual, the louse may detach, jump and attach to another.

Typically, *C. elongatus* does not create open wounds but lives in the slime and the epithelial tissue [22], but there are examples of the opposite. The problem with this species of lice is quite new and exactly what and how numerous the harms it can cause is still not well known.

The other *Caligus* species, *C. rogercresseyi*, also has a lower host specificity and, like *C. elongatus*, can be found on wild fish [110].

A problem with these lice, even though they do not directly create open wounds, is that the risk of secondary infections from viruses and bacteria increases due to a reduction in the mucus layer protecting the fish. Another problem occurring with all sea lice is that if there are high amounts of salmon lice in the sea cages, lice can be transferred to wild salmon stocks in the area and this can be very critical.

The sea louse develops in stages. The early stages are free swimming pelagic stages while the last and problematic stages are when the lice attach to the salmon and parasitize the host. The life cycle starts when an adult female louse releases fertilized eggs in the water column, where they hatch. There are then two free swimming pelagic nauplius stages, where the length is 0.5–0.6 mm. The next stage is the copepodid stage, and at the end of this stage they attach to the host, the salmon, and enter the chalimus stages, of which there are two for the *L. salmonis* [122] and five for *C. rogercresseyi* [123]. This is followed by two preadult stages, where the parasites now start causing damage and develop into male and female lice. The last step is the adult stage where the females develop two egg strings. The length of a salmon louse is 8–18 mm for the female and 5–6 mm for the male [124].

Sea lice have a high reproductive potential; for *L. salmonis*, which is the most productive, one egg string may contain up to 183 eggs and each louse has two of them. The salmon lice can produce up to 10–11 generations of eggs during its life cycle that can last up to 210 days (7.2°C) [125].

The generation interval for the salmon louse is around 8–9 weeks at 6°C and 4 weeks at 18°C, or around 400°D [124].

The nauplius stage is where the louse is a pelagic swimmer and this stage lasts from 2–10 days, depending on water temperature. The first part of the copepodid stage is also pelagic. They mostly drift on the current and can be transported tens of kilometres during this period [126–128]. This is, of course, of major importance when designing strategies for delousing and for performing regional delousing. Salmon lice have a vertical migration due to their positively phototactic nature, meaning they rise during daylight. The survival of the pelagic stages varies with the water: when temperature drops below 4.2°C the lice will not survive [129]. Low salinity will also reduce survival; below 1.5% salinity the nauplius will not survive and below 2–2.5% reduced survival is seen [124,126,130,131]. In some cases, however, sea lice have adapted to reduced salinity [130]. Higher fish densities and more individuals on a site will contribute to higher densities of copepodites in the area [132]. The pelagic sea lice stages moving vertically in the water column has been practically observed in closed sea cages with a water inlet at 25 m deep, where sea lice have been discovered in the water inlet into the cage. This is, however, very site dependent and based on the vertical water circulation on the site, and has not been reported as a problem on other sites.

There are a number of ways to treat salmon infected with sea lice, including physical, medical, chemical and biological (described in Chapter 8). The problem is achieving 100% effective treatment, and treatment must be coordinated within an area to gain full effect. Another alarming point here is resistance against the common treatment methods. Typically, some type of medication has been used for antiparasitic treatment, but it is common for lice to develop immunity against the active agents.

The search for new delousing methods is therefore intense. Breeding has a documented effect, but this is a long-term aim. There are hopes of developing vaccines against sea lice with a number of ongoing vaccine trials, and a number of methods are under patent.

11.8.4.2 Amoebic gill disease

Amoebic gill disease (AGD) is a gill disease caused by the marine amoeba *Neoparamoeba perurans*. The result of infection is gill inflammation. White areas on the gills can be observed and the lamellae of the gills appear clogged together. The fish develop respiration problems; the frequency of movement of the gill covers increases and the fish gasp. The mucus production of the gills increases and the behaviour of the fish changes to slower movement and crowding at the surface, with a decrease in appetite. The parasite is similar to salmon lice in that is has stages where it lives in the free water mass (pelagic), and stages where it is parasitic on the fish. During the free stages it can be transported with the water for days and infect farms in the surrounding area. The parasite prefers full salinity and warmer temperatures, so there are fewer problems in areas with lower temperatures and brackish water. Treatment of AGD can be performed by bathing the salmon in freshwater. A chemical method is to bathe the salmon in hydrogen peroxide for a given period. Treatment may not be 100% effective and the best result is achieved when treating the early stages, just after attachment to the gills. Stress is a factor that can enable an outbreak. If no treatment is carried out, mass mortality may occur [133]. AGD is a problem in all salmon producing countries today [134], after beginning in Tasmania.

11.8.4.3 Gyrodactylus salaris

Gyrodactylus salaris (Gs) is a small freshwater parasite, 0.5–1 mm long, and is in the *Gyrodactylus* genus. Its distribution is restricted to Europe and it attacks Atlantic salmon. The consequences of infection on the wild salmon stocks are catastrophic, with large mortalities as the result. Several natural stocks in rivers have almost been eliminated. It is a notifiable disease for this reason. The parasite infects the gills, skin and fins of salmon, and multiplies rapidly causing wounds due to the excretion of protein-dissolving enzymes. Fry are most susceptible, with larger fish being able to cope much better with infection. However, if the parasite invades a salmon river it can have a drastic effect on the salmon population because no fry reach maturity. Several other salmonids and trout may also host the parasite, but with a much lower impact. The problem is that these species may function as horizontal vectors in spreading it to Atlantic salmon in some areas. The parasite can easily be spread by contaminated fish, by humans and by equipment. Owing to the consequences it is very important to carry out proper disinfection; equipment may be disinfected by traditional disinfection solutions. It is important to have proper veterinary controls before sending fish to other areas or farms.

The main problem with the parasite does not occur in salmon farming, where traditional antiparasitic treatment can be performed to control the disease, and stamping out may also be a solution. The parasite may be present within the biofilm in a farm, where it is rather difficult to get rid of, similar to biofilm in pipes. The main problem is, however, related to the spread of the parasite to wild salmon stocks.

Therefore, it is of major importance to disinfect all clothes and equipment including angling equipment and water sport equipment used in salmon rivers. The equipment must be cleaned and treated to kill the parasite, for instance, by one of the following methods [135]:

- drying at a minimum temperature of at least 20°C for at least 2 days
- heating for at least 1 hour at temperatures of above 60°C
- deep freezing for at least 1 day
- immersion in a solution* suitable for killing the parasite for a minimum of 10 minutes. (*Virkon ® Aquatic (1%), Wescodyne (1%), sodium chloride (3%), and sodium hydroxide (0.2%).)

If salmon rivers have become infected by the parasite, the consequences can, as stated, be catastrophic because juvenile salmon will not survive. To restock the river, the entire river must be treated with an antiparasitic to kill or eliminate the parasite. Today rotenone is used with quite good results, even though it is a very large and costly operation to perform. The river is closed and rotenone, a toxin, is pumped or sprayed so all parts of the river receive the necessary dose of toxin, which will kill all life in the river. After a period, life will gradually regenerate, and the river can be reopened for salmon migration.

11.8.4.4 Other

There are also a number of other parasites that can create problems for salmon farming. One example is parvicapsulosis caused by the parasite *Parvicapsula pseudobranchicola*. The disease occurs specifically after transferring smolt to cage farms. The fish typically become thin, and bleeding in the eye can be observed [105,106]. There is no known treatment but generally the best chance of avoiding problems is for the fish to be in good condition and to have a good health status.

Ichthyobodo spp. are another parasite group found in salmon farming. Different parasites within the group occur in both fresh- and seawater and can damage both skin and gills. In freshwater *Costia* (*Ichthyobodo necator*) is known for resulting in gill problems. *Trichodina* is another ectoparasite that may create gill problems in the freshwater phase.

Paranucleospora theridion is an intracellular microsporidian parasite. It is also known for causing autumn disease. It is recognized by pale swollen gills, a drop in appetite and increased swimming activity.

In freshwater, treatments can include dipping the fish in a bath with chemicals such as hydrogen peroxide, a common general treatment for removing ectoparasites.

References

1. Etymonline. Etymology dictionary. https://www.etymonline.com/.
2. Buchmann, K. et al. *Fish Diseases: An Introduction*. 2009. Biofolia.
3. Smith, S.A. *Fish Diseases and Medicine*. 2019. CRC Press: Baton Rouge, FL.
4. Woo, P.T.K. and Gregory, D.W.B. *Diseases and Disorders of Finfish in Cage Culture, 2nd Edition*. 2014. CABI: Wallingford.
5. Jeney, G. *Fish Diseases: Prevention and Control Strategies*. 2017. Elsevier: Amsterdam.
6. Aoki, T. *Fish Diseases*. 2016. EOLSS Publications.
7. Roberts, R.J. and Shepherd, C.J. *Handbook of Trout and Salmon Diseases*. 1997. Wiley: Chichester.
8. Bruno, D.W., Noguera, P.A. and Poppe, T.T. *A Colour Atlas of Salmonid Diseases*. 2016. Springer: Amsterdam.

9. Medical Dictionary. Infection. http://medical-dictionary.thefreedictionary.com/infection.
10. Frérot, M. et al. What is epidemiology? Changing definitions of epidemiology 1978–2017. *PLOS One*. 2018. 13(12): e0208442.
11. Animal Health in the World. http://www.oie.int/animal-health-in-the-world/oie-listed-diseases-2019/.
12. Patton, D. Special Report: Before coronavirus, China bungled swine epidemic with secrecy. 2020. Reuters.
13. dos Santos, C.A.L. and Howgate, P. Fishborne zoonotic parasites and aquaculture: a review. *Aquaculture*. 2011. 318(3–4): 253–261.
14. Beran, G.W. *Handbook of Zoonoses, Second Edition: Viral Zoonoses*. 1994. Taylor and Francis: Abingdon.
15. Murrell, K.D. and Fried, B. *Food-Borne Parasitic Zoonoses: Fish and Plant-Borne Parasites*. 2007. Springer: Dordrecht.
16. Marty, G.D. Anisakid larva in the viscera of a farmed Atlantic salmon (*Salmo salar*). *Aquaculture*. 2008. 279(1–4): 209–210.
17. Buchmann, K. and Mehrdana, F. Effects of anisakid nematodes *Anisakis simplex* (s.l.), *Pseudoterranova decipiens* (s.l.) and *Contracaecum osculatum* (s.l.) on fish and consumer health. *Food and Waterborne Parasitology*. 2016. 4: 13–22.
18. Lekang, O.I. *Aquaculture Engineering*. 2020. Wiley: Chichester.
19. Mancuso, M. Fish welfare in aquaculture. *J. Aquac. Res. Development*. 2013. 4: e107.
20. Hewson, C.J. What is animal welfare? Common definitions and their practical consequences. *The Canadian veterinary journal = La revue veterinaire canadienne*. 2003. 44(6): 496–499.
21. Carenzi, C. and Verga, M. Animal welfare: review of the scientific concept and definition. *Italian Journal of Animal Science*. 2009. 8(sup1): 21–30.
22. Noble, C. et al. Welfare Indicators for farmed Atlantic salmon: tools for assessing fish welfare. 2018. https://dspace.stir.ac.uk/handle/1893/28435.
23. Stien, L.H. et al. Salmon Welfare Index Model (SWIM 1.0): a semantic model for overall welfare assessment of caged Atlantic salmon: review of the selected welfare indicators and model presentation. *Reviews in Aquaculture*. 2013. 5(1): 33–57.
24. Collins Dictionary. Animal welfare. ttps://www.collinsdictionary.com/dictionary/english/animal-welfare.
25. Webster, J. Assessment of animal welfare, the five freedoms, in *Animal Welfare, A Cool Eye Towards Eden*. 1995. Blackwell: Oxford.
26. Brambell, F.W.R. *Report of the Technical Committee to Enquire into the Welfare of Animals kept under Intensive Livestock Husbandry Systems*. 1965. Her Majesty's Stationery Office: London.
27. Stien, L.H. et al. Salmon Welfare Index Model (SWIM 1.0): a semantic model for overall welfare assessment of caged Atlantic salmon: review of the selected welfare indicators and model presentation. *Reviews in Aquaculture*. 2013. 5(1): 33–57.
28. Saraiva, J.L. et al. A global assessment of welfare in farmed fishes. *The FishEthoBase*. 2019. 4(2): 30.
29. Grandin, T., *Improving Animal Welfare, 2 Edition: A Practical Approach*. 2015. CABI: Wallingford.
30. Haynes, R.P. *Animal Welfare: Competing Conceptions And Their Ethical Implications*. 2008. Springer: Dordrecht.
31. Branson, E.J. *Fish Welfare*. 2008. Wiley: Chichester.
32. Butterworth, A. *Animal Welfare in a Changing World*. 2018. CABI: Wallingford.
33. Lekang, O.-I. Aquaculture hatchery water supply and treatment systems, in *Advances in Aquaculture Hatchery Technology*. 2013. Elsevier: Amsterdam, pp. 3–22.

34. Sadler, J. and Goodwin, A. *Disease Prevention on Fish Farms*. 2007. Southern Regional Aquaculture Center.
35. Yanong, R.P. and Erlacher-Reid, C.J.S.P. Biosecurity in aquaculture, part 1: an overview. 2012. http://fisheries.tamu.edu/files/2013/09/SRAC-Publication-No.-4707-Biosecurity-in-Aquaculture-Part-1-An-Overview.pdf.
36. Burridge, L. et al. Chemical use in salmon aquaculture: A review of current practices and possible environmental effects. *Aquaculture*. 2010. 306(1): 7–23.
37. Gudding, R.J.F. Vaccination as a preventive measure. 2014. 1: 12–21.
38. Adams, A. *Fish Vaccines*. 2016. Springer: Basel.
39. Assefa, A. and Abunna, F.J.V. Maintenance of fish health in aquaculture: review of epidemiological approaches for prevention and control of infectious disease of fish. 2018.
40. Muktar, Y., Tesfaye, S. and Tesfaye, B. Present status and future prospects of fish vaccination: a review. *J Veterinar Sci Technol*. 2016. 7(2).
41. Immunological adjuvant. https://en.wikipedia.org/wiki/Immunologic_adjuvant.
42. Dalmo, R.A. DNA vaccines for fish: Review and perspectives on correlates of protection. *Journal of Fish Diseases*. 2018. 41(1): 1–9.
43. Koprowski, H. and Weiner, D.B. *DNA Vaccination/Genetic Vaccination*. 2012. Springer: Berlin.
44. Hølvold, L.B., Myhr, A.I. and Dalmo, R.A. Strategies and hurdles using DNA vaccines to fish. *Veterinary Research*. 2014. 45(1): 21–21.
45. Gismondo, M., Drago, L. and Lombardi, A. Review of probiotics available to modify gastrointestinal flora. *International Journal of Antimicrobial Agents*. 1999. 12(4): 287–292.
46. Martínez Cruz, P. et al. Use of probiotics in aquaculture. *ISRN Microbiology*. 2012. 2012: 916845.
47. Banerjee, G. and Ray, A.K. The advancement of probiotics research and its application in fish farming industries. *Research in Veterinary Science*. 2017. 115: 66–77.
48. Gatlin III, D. and Peredo, A.M. Prebiotics and probiotics: definitions and applications. SRAC Publication, 2012(4711).
49. Hoseinifar, S.H. et al. Probiotics as means of diseases control in aquaculture, a review of current knowledge and future perspectives. *Frontiers in Microbiology*. 2018. 9: 2429–2429.
50. Michael, E.T., Amos, S.O. and Hussaini, L.T. A review on probiotics application in aquaculture. *Fisheries and Aquaculture Journal*. 2014. 5(4).
51. Irianto, A. and Austin, B.J.J Probiotics in aquaculture. *Journal of Fish Diseases*. 2002. 25(11): 633–642.
52. Gibson, G.R. and Roberfroid, M.B.J.T.J. Dietary modulation of the human colonic microbiota: introducing the concept of prebiotics. *Journal of Nutrition*. 1995. 125(6): 1401–1412.
53. Ringø, E. et al. Prebiotics in aquaculture: a review. *Aquaculture Nutrition*. 2010. 16(2): 117–136.
54. Guerreiro, I., Oliva-Teles, A. and Enes, P. Prebiotics as functional ingredients: focus on Mediterranean fish aquaculture. *Reviews in Aquaculture*. 2018. 10(4): 800–832.
55. Song, S.K. et al. Prebiotics as immunostimulants in aquaculture: a review. *Fish & Shellfish Immunology*. 2014. 40(1): 40–48.
56. Dawood, M.A., Koshio, S. and Esteban, A. Beneficial roles of feed additives as immunostimulants in aquaculture: a review. *Reviews in Aquaculture*. 2018. 10(4): 950–974.
57. Salminen, S. et al. Probiotics: how should they be defined? *Trends in Food Science & Technology*. 1999. 10(3): 107–110.
58. Akhter, N. et al. Probiotics and prebiotics associated with aquaculture: a review. *Journal of Fish Shellfish Immunology*. 2015. 45(2): 733–741.
59. Merrifield, D.L. and Ringo, E. *Aquaculture Nutrition: Gut Health, Probiotics and Prebiotics*. 2014. Wiley: Chichester.

60. Bentzon-Tilia, M., Sonnenschein, E.C. and Gram, L. Monitoring and managing microbes in aquaculture–Towards a sustainable industry. *Microbial Biotechnology*. 2016. 9(5): 576–584.
61. Abedon, S.T. et al. *Phage Therapy: Past, Present and Future*. 2017. Frontiers Media SA.
62. Borysowski, R.M.J. *Phage Therapy*. 2014. Caister Academic Press.
63. Doss, J. et al. A review of phage therapy against bacterial pathogens of aquatic and terrestrial organisms. *Viruses*. 2017. 9(3): 50.
64. Dy, R.L., Rigano, L.A. and Fineran, C. Phage-based biocontrol strategies and their application in agriculture and aquaculture. *Chem Soc Trans*. 2018. 46(6): 1605–1613.
65. Culot, A., Grosset, N. and Gautier, M.J.A. Overcoming the challenges of phage therapy for industrial aquaculture: A review. *Aquaculture*. 2019: 734423.
66. Kalatzis, P. Phage Therapy Applications in Aquaculture. 2019. Capsid and Tail.
67. Lin, D.M. et al. Phage therapy: An alternative to antibiotics in the age of multi-drug resistance. *World J Gastrointest Pharmacol Ther*. 2017. 8(3): 162.
68. Wittebole, X., De Roock, S. and Opal, S.M.J.V. A historical overview of bacteriophage therapy as an alternative to antibiotics for the treatment of bacterial pathogens. *Virulence*. 2014. 5(1): 226–235.
69. Carlton, R.M. Phage therapy: past history and future prospects. *Archivum immunologiae et therapiae expermentaltalis*. 1999. 47: 267–274.
70. Imbeault, S. et al. Using bacteriophages to prevent furunculosis caused by Aeromonas salmonicida in farmed brook trout. *Journal of Aquatic Animal Health*. 2006. 18(3): 203–214.
71. Vincent, A.T. et al. Would Bacteriophages Be a New Old Complement to Antibiotics in Aquaculture?, in *Microbial Communities in Aquaculture Ecosystems: Improving Productivity and Sustainability*, N. Derome (ed.). 2019. Springer: Dordrecht, pp. 51–68.
72. Summers, W.C.J.B. The strange history of phage therapy. *Bacteriophage*. 2012. 2(2): 130–133.
73. Torres-Barceló, C. and Hochberg, M.E. Evolutionary rationale for phages as complements of antibiotics. *Trends in Microbiology*. 2016. 24(4): 249–256.
74. Comeau, A.M. et al. Phage-antibiotic synergy (PAS): β-lactam and quinolone antibiotics stimulate virulent phage growth. *PLOS One*. 2007. 2(8): e799.
75. Oechslin, F. Resistance development to bacteriophages occurring during bacteriophage therapy. *Viruses*. 2018. 10(7): 351.
76. Matsuzaki, S. et al. Bacteriophage therapy: a revitalized therapy against bacterial infectious diseases. *J Infect Chemother*. 2005. 11(5): 211–219.
77. Gon Choudhury, T. et al. Advances in bacteriophage research for bacterial disease control in aquaculture. *Reviews in Fisheries Science and Aquaculture*. 2017. 25(2): 113–125.
78. García, R. et al. Bacteriophage production models: an overview. *Frontiers in Microbiology*. 2019. 10.
79. Jurač, K. et al. Bacteriophage production processes. *Appl Microbiol Biotechnol*. 2019. 103(2): 685–694.
80. Agboluaje, M. and Sauvageau, D. Bacteriophage production in bioreactors, in *Bacteriophage Therapy*. 2018. Springer: Dordrecht, pp. 173–193.
81. Schulz, P. et al. Influence of bacteriophages cocktail on European eel (Anguilla anguilla) immunity and survival after experimental challenge. *Fish and Shellfish Immunology*. 2019. 84: 28–37.
82. Fletcher, R. A targeted alternative to antibiotics?. *The Fish Site*. 2019.
83. Lekang, O.-I., *Aquaculture Engineering 3rd edn*. 2020.Wiley Blackwell: Oxford.
84. Hacker, M., Messer, W.S. and Bachmann, K.A. *Pharmacology: Principles and Practice*. 2009. Elsevier Science: Amsterdam.
85. Huang, S.M. et al. *Principles of Clinical Pharmacology*. 2012. Elsevier Science: Amsterdam.
86. Wanamaker, B.P. and Massey, K.L. *Applied Pharmacology for Veterinary Technicians*. 2009. Saunders.
87. Treves-Brown, K.M., *Applied Fish Pharmacology*. 2013. Springer: Dordrecht.
88. Gallagher, J. and MacDougall, C. *Antibiotics Simplified*. 2011. Jones and Bartlett Learning.

89. Santos, L. and Ramos, F. Analytical strategies for the detection and quantification of antibiotic residues in aquaculture fishes: A review. *Trends in Food Science and Technology*. 2016. 52: 16–30.
90. Liu, X., Steele, J.C. and Meng, X.-Z. Usage, residue, and human health risk of antibiotics in Chinese aquaculture: a review. *Environ Pollut*. 2017. 223: 161–169.
91. Miranda, C.D., Godoy, F.A. and Lee, M.R. Current status of the use of antibiotics and the antimicrobial resistance in the Chilean salmon farms. *Frontiers in Microbiology*. 2018. 9: 1284.
92. Liu, X., Steele, J.C. and Meng, X.-Z. Usage, residue, and human health risk of antibiotics in Chinese aquaculture: a review. *J Environmental Pollution*. 2017. 223: 161–169.
93. Cabello, F.C. and Godfrey, H.P.J.A. Salmon aquaculture, Piscirickettsia salmonis virulence, and one health: Dealing with harmful synergies between heavy antimicrobial use and piscine and human health. 2019.
94. Elbashir, S. et al. Seafood pathogens and information on antimicrobial resistance: A review. 2018. 70: 85–93.
95. Thefishsite. Disease guide. https://thefishsite.com/disease-guide.
96. Scottish Government. Diseases (Finfish, Molluscs and Crustaceans).
97. Norwegian Veterinary Insitute. Sykdom og agens (in Norwegian). from: https://www.vetinst.no/sykdom-og-agens.
98. Marin Helse. Fiskesykdommer (in Norwegian). https://www.marinhelse.no/fiskesykdommer/.
99. Norwegian Veterinary Institute. An overview of emerging diseases in the salmonid farming industry – Technical report. 2019. https://www.vetinst.no/rapporter-og-publikasjoner/rapporter/2019/an-overview-of-emerging-diseases-in-the-salmonid-farming-industry-technical-report.
100. FAO. Salmo salar (Linnaus, 1758). http://www.fao.org/fishery/culturedspecies/Salmo_salar/en#tcNA00B1.
101. CABI. Invasive species compendium. https://www.cabi.org/isc/search/index?q=Salmo%20salar.
102. Barents Watch. Norwegian fish health. 2020. https://www.barentswatch.no/en/fishhealth/.
103. Austin, B. and Austin, D.A. *Bacterial Fish Pathogens: Disease of Farmed and Wild Fish*. 2010. Springer: Dordrecht.
104. Toranzo, A.E., Magariños, B. and Romalde, J.L. A review of the main bacterial fish diseases in mariculture systems. *Aquaculture*. 2005. 246(1): 37–61.
105. Bornø, G. and Lie, L.M. Fish health report 2014. 2015.
106. Hjeltnes, B. et al. The health situation in Norwegian aquaculture 2016. 2017: 127.
107. Cabello, F.C. and Godfrey, H.P. Salmon aquaculture, Piscirickettsia salmonis virulence, and One Health: Dealing with harmful synergies between heavy antimicrobial use and piscine and human health. *Aquaculture*. 2019. 507: 451–456.
108. Evensen, Ø., Immunization strategies against piscirickettsia salmonis infections: review of vaccination approaches and modalities and their associated immune response profiles. *Frontiers in Immunology*. 2016. 7: 482–482.
109. Health Management Program. SRS background. https://comitepgsa.cl/srs-background/.
110. Gallardo-Escárate, C. et al. The race between host and sea lice in the Chilean salmon farming: a genomic approach. *Reviews in Aquaculture*. 2019. 11(2): 325–339.
111. Health Management Program. Caligidosis background. https://comitepgsa.cl/caligidosis-background/.
112. Furevik, D.M. et al. Surface activity of Atlantic salmon (Salmo salar L.) in net pens. *Aquaculture*. 1993. 110(2): 119–128.
113. Finstad, B. et al. The effect of sea lice on Atlantic salmon and other salmonid species, in *Atlantic Salmon Ecology*, O. Aas, S. Einum, A. Klemetsen and J. Skurdal (eds). 2011. Wiley: Chichester, pp. 253–276.

114. Finstad, B. et al. Laboratory and field investigations of salmon lice [Lepeophtheirus salmonis (Krøyer)] infestation on Atlantic salmon (Salmo salar L.) post-smolts. *Aquaculture Research*. 2000. 31(11): 795–803.
115. Bui, S. et al. Sea lice infestation level alters salmon swimming depth in sea-cages. *Aquaculture Environment Interactions*. 2016. 8: 429–435.
116. Boxaspen, K.J.I.J. A review of the biology and genetics of sea lice. 2006. 63(7): 1304–1316.
117. Skugor, S. et al. Local and systemic gene expression responses of Atlantic salmon (Salmo salar L.) to infection with the salmon louse (Lepeophtheirus salmonis). 2008. 9(1): 498.
118. Bowers, J. et al. The physiological response of Atlantic salmon, Salmo salar L., to a single experimental challenge with sea lice, Lepeophtheirus salmonis. 2000. 23(3): 165–172.
119. González, M.P., Marín, S.L. and Vargas-Chacoff, L. Effects of Caligus rogercresseyi (Boxshall and Bravo, 2000) infestation on physiological response of host Salmo salar (Linnaeus 1758): establishing physiological thresholds. *Aquaculture*, 2015. 438: 47–54.
120. Nolan, D. et al. Infection with low numbers of the sea louse Lepeophtheirus salmonis induces stress-related effects in postsmolt Atlantic salmon (Salmo salar). 1999. 56(6): 947–959.
121. Frederick, C., Brady, D.C. and Bricknell, I. Landing strips: Model development for estimating body surface area of farmed Atlantic salmon (Salmo salar). *Aquaculture*. 2017. 473: 299–302.
122. Hamre, L.A. et al. The salmon louse Lepeophtheirus salmonis (Copepoda: Caligidae) life cycle has only two chalimus stages. 2013. 8(9): e73539.
123. González, L. and Carvajal, J.J.A. Life cycle of Caligus rogercresseyi,(Copepoda: Caligidae) parasite of Chilean reared salmonids. 2003. 220(1–4): 101–117.
124. Marine Institute Ireland. Life cycle of the salmon louse. https://www.marine.ie/Home/site-area/areas-activity/aquaculture/sea-lice/life-cycle-salmon-louse.
125. Heuch, P., Nordhagen, J. and Schram, T. Egg production in the salmon louse [Lepeophtheirus salmonis (Krøyer)] in relation to origin and water temperature. *Aquaculture Research*. 2000. 31(11): 805–814.
126. Costello, M.J. Ecology of sea lice parasitic on farmed and wild fish. *Trends in Parasitology*. 2006. 22(10): 475–483.
127. McKibben, M.A. and Hay, D.W. Distributions of planktonic sea lice larvae Lepeophtheirus salmonis in the inter-tidal zone in Loch Torridon, Western Scotland in relation to salmon farm production cycles. *Aquaculture Research*. 2004. 35(8): 742–750.
128. Kristoffersen, A.B. et al. Large-scale modelling of salmon lice (Lepeophtheirus salmonis) infection pressure based on lice monitoring data from Norwegian salmonid farms. 2014. 9: 31–39.
129. Bravo, S. The reproductive output of sea lice Caligus rogercresseyi under controlled conditions. *Exp Parasitol*. 2010. 125(1): 51–4.
130. Bravo, S., Pozo, V. and Silva, M. The tolerance of Caligus rogercresseyi to salinity reduced in southern Chile. *Bulletin of the European Association of Fish Pathologists*. 2008. 28(5): 198–206.
131. Brooks, K.M. The effects of water temperature, salinity, and currents on the survival and distribution of the infective copepodid stage of sea lice (Lepeophtheirus salmonis) originating on Atlantic salmon farms in the Broughton Archipelago of British Columbia, Canada. *Reviews in Fisheries Science*. 2005. 13(3): 177–204.
132. Penston, M.J. and Davies, I.M. An assessment of salmon farms and wild salmonids as sources of Lepeophtheirus salmonis (Krøyer) copepodids in the water column in Loch Torridon, Scotland. *Journal of Fish Diseases*. 2009. 32(1): 75–88.
133. Steinum, T. et al. First cases of amoebic gill disease (AGD) in Norwegian seawater farmed Atlantic salmon, Salmo salar L., and phylogeny of the causative amoeba using 18 S cDNA sequences. *Journal of Fish Diseases*. 2008. 31(3): 205–214.

134. Oldham, T., Rodger, H. and Nowak, B.F. Incidence and distribution of amoebic gill disease (AGD)—An epidemiological review. *Aquaculture*. 2016. 457: 35–42.
135. Scottish Government. Gyrodactylus salaris. https://www2.gov.scot/Topics/marine/Fish-Shellfish/aquaculture/diseases/notifiableDisease/g-salaris.

CHAPTER 12

Breeding

12.1 Introduction

In every population there is a variation in the traits of the individuals. For instance, some salmon grow fast and some grow slowly. The reason for this is a combination of environment and genetics (genotype). The genotype of the individual is based on heredity from the parents, so by choosing parents with good performance in a desired trait, normally the offspring will have improved performance. If this is done in a controlled way, the performance of the broodstock population will gradually improve. Breeding has been of major importance in the domestication of wild salmon in order to keep them under aquaculture conditions and to increase the performance of Atlantic salmon in aquaculture.

Selective breeding can be described as controlled mating of parents so the offspring inherit desired traits. This is done by selecting and mating parents with these desired traits to achieve transfer of genes from generation to generation. This can be compared to what happens under natural conditions over generations. The individuals best suited for the environmental conditions will survive, grow fastest and outcompete other individuals, and will most probably have the highest breeding success; it will be the genes of these individuals that are transferred to the next generation. In this way the adaptation to the environment of their habitat will be improved. This illustrates why it is so important to use local strains when stocking local lakes and streams.

Under aquaculture conditions, however, the environment is controlled and optimized so the majority of the individuals will survive to adults. Selective breeding can then be carried out for traits other than survival and adaptation to the environmental conditions. Breeding programmes (section 12.12) are designed to achieve desired goals, and by optimal programme design the goals can be reached in a relatively short time compared to natural breeding in the wild. The effect of selective breeding is cumulative, meaning the breeding gain will increase from generation to generation.

However, to be able to perform selective breeding for fish species in aquaculture some basic requirements must be fulfilled. First of all, the entire life cycle must be known and it must be possible to control and manipulate the life cycle. This is the case for Atlantic salmon where the reproductive biology and methods to artificially control the life cycle are well known.

A great advantage for salmon compared to traditional animal livestock is that fertilization occurs externally. Milt can be collected from the males and eggs from the females; these are then mixed, and fertilization takes place. This makes it easy to control which parents produce offspring.

To perform selective breeding with success, there has to be a certain genetic variation within the population, and the larger the variation, the larger the genetic progress will be. The population must also be of a minimum size to avoid problems with inbreeding depression. Inbreeding may potentially be a problem in aquaculture due to the high number of siblings and keeping control over the individuals may be a problem. The fish used for selective breeding ought to be individually marked to be able to identify their pedigree and avoid the crossing of near relatives.

Selective breeding has nothing to do with genetic modification. When genetically modifying a salmon, part of the process is carried out in a laboratory where either new foreign genes are introduced into the salmon or the salmon's own genes are edited/manipulated. Which processes are included in genetic modification differs slightly between definitions. The AquAdvantage salmon is an example of a genetically modified salmon (see Chapter 5). Manipulating genes that are already within the salmon is known as gene editing and there are several techniques for doing this. There has been fast development in the sector and it has become easier to perform with the development of new techniques such as Crisper-Cas9. By editing genes mutants can be created, but these can also occur in nature even though most of them will be extremely rare.

A very simplified description of selective breeding is as follows: decide on the goal for selective breeding, for instance, growth; find the best individual of both sexes for this trait; crossbreed them and then statistically the offspring should be improved (in this case, grow faster). But as described later, the breeding programmes used today are much more sophisticated and include breeding for a number of traits and utilizing family information, genomics and statistics to increase the probability of success. Therefore, performing selective breeding now requires a great deal of knowledge and is carried out by skilled specialists in different areas, including genetics, statistics and computer science. Biotechnology, molecular genetics and genomics have become increasingly important parts of selective breeding.

Selective breeding has been a great success for the Atlantic salmon farming industry for two major reasons: salmon produce a high number of offspring, which enables a strong selection of the best individuals (selection intensity, described more later); and the degree of heritability of the desired traits is quite high which results in good progress in genetic gain (more later).

The purpose of this chapter is only to give a brief overview of selective breeding in order to provide more understanding of this very important topic for the success of the Atlantic salmon farming industry. Genetic improvement has been of major importance in reducing the production cost of Atlantic salmon. Important basic information regarding breeding is also given in Chapter 4 on biology, and specifically in Chapter 5 on broodstock, in the sub-chapter on controlling maturation.

A lot of information is available about genetics and selective breeding through textbooks, review articles and web pages (like for instance [1–6]): both in general, for traditional animal livestock, and specifically for aquaculture.

12.2 The commercial success of selective breeding in Atlantic salmon

The selective breeding programmes for Atlantic salmon with the aim of increasing yield in aquaculture started in the early 1970s and have, as stated earlier, been a great success bringing importance development in the industry [3]. The documented achievements include better growth rate, better feed utilization, increased age and fish size at maturation, reduced mortality from disease, improved slaughter quality (as meat colour) and in general a better suitability for aquaculture conditions [7].

12.3 Breeding stations

Selective breeding is typically performed by specialist companies that have extensive knowledge in the area and are able to set up breeding programmes, perform the necessary statistical analyses and, increasingly, also have expertise in genomics. These companies then support the rest of the industry with eyed eggs bred with the best traits possible, which improve from generation to generation. There is a handful of companies worldwide that have specialized in selective breeding of Atlantic salmon. Some of the largest salmon producing companies have completely or partly in-house selective breeding programmes. Companies that carry out selective breeding and sell eyed eggs have their own broodstock farms, either one or several, and their own breeding specialists. Much information about breeding is therefore also found in Chapter 5. Broodstock farms are also often named breeding stations. Breeding stations are characterized by having a number of tanks to keep the eggs and fry from different families separately before they reach a size where they can be individually marked and pooled together in larger tanks (Figure 12.1).

12.4 Some basic genetics

Genetics is part of biology and can be described as the study of genes and how genes are transferred between generations. Genes are part of the overall DNA of the fish included in the chromosome and carry the genetic information from parents to offspring (see Chapter 5). Different genes have specific locations on the chromosome. A quantitative trait locus (loci – plural) refers to a specific location on the DNA that controls a specific quantitative trait (see later). When genes are transferred from parents to offspring, genes are transferred from both mother and father. The offspring will therefore receive genetic input from both mother and father, with half coming from each. During fertilization, sperm enters the egg; both egg and sperm contain unique genetic information from the parents, each representing a haploid cell (with half the number, or a single set, of chromosomes). When the egg and sperm merge, the fertilized egg becomes a diploid cell, a zygote, that contains all the genetic information (more in section 12.11.2). Each gene can come in different forms named alleles, and the differences are seen in the code for the bases in the backbone of the helix. (see chapter 5.13.3). This gives the variation between individuals because the alleles may differ. If the alleles of the genes from both mother and father are the same, they

Figure 12.1 A breeding station with a large number of individual production units.

are called homozygous. If they differ, they are called heterozygous. An allele of a gene from one parent may be dominant, and the allele of the gene from the other parent is then of no relevance. If both mother and father carry a recessive gene, this will be transferred to and expressed by the offspring. The total DNA, the salmon genome, includes all the different genes of the salmon and will be unique for each individual based on the base pairs (the code), because the overall combination of alleles will differ. Therefore, two full siblings will never be exactly equal.

12.5 Variation is the basis for breeding success

The basic principle behind breeding is that there is a difference in traits in a population. Normally it is expected that there is a normal distribution in a population regarding how it scores for a quantitative trait. This can be described by the mean value of the population ($\bar{x}_n$) and the variation in the population can be described by the variance or standard deviation (σ^2 or σ) (Figure 12.2). The variance can further be described as the sum of the square from each individual from the mean value of all individuals in the population divided by the number of individuals (n) or described by the following mathematical equations.

Empirical variance (estimated):

$$\sigma_e^2 = \frac{1}{n-1}\sum_{i=1}^{n}(x_i - \bar{x}_n)^2$$

Real variance (knowing all values):

$$\sigma_r^2 = \frac{1}{n}\sum_{i=1}^{n}(x_i - \bar{x}_n)^2$$

The standard deviation can then be calculated by taking the square root of the variance:

$$\sigma = \sqrt{\sigma^2}$$

Another important parameter in relation to breeding is whether there is correlation between traits or not, or the covariance. A correlation between two traits can be positive or negative, and varies between 1 and –1; 1 means full correlation between two traits, 0 means no correlation while –1 means full opposite correlation (Figure 12.3). If there is a negative correlation between two traits it is more difficult to breed for both of them, because as one increases, the other decreases. For several of the important traits a positive correlation has been observed, such as body weight and survival and body weight and flesh colour.

The covariance (measure of relationship, Cov_{xy}) between two traits x and y can be described as follows:

$$Cov_{xy} = \frac{1}{n-1}\sum_{i=1}^{n}(x_i - \bar{x}_n)(y_i - \bar{y}_n)$$

The correlation (Cor_{xy}) between the traits may then be calculated by the following equation

$$Cor_{xy} = r_{xy} = \frac{Cov_{xy}}{\sigma_x \cdot \sigma_y}$$

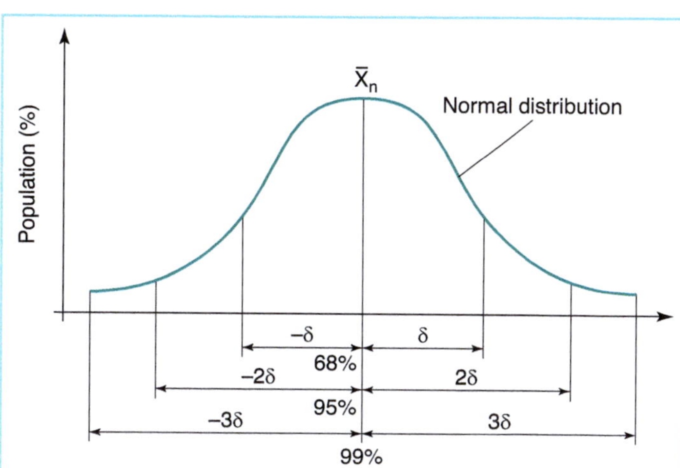

Figure 12.2 A biological population is expected to have a normal distribution which can be described with the average value and standard deviation. Here 1.2 and 3 standard deviations are described with the amount of the population that is within this range.

This is an important value which has to be taken into account when breeding for different traits. If it is positive, which is normal for most traits, there is no problem but when it is negative it will be more difficult to breed for both traits (Table 12.1). However, there might be negative correlation between some diseases [8,9].

When selecting individuals or families for breeding, it is individuals that best exhibit the desired trait that are taken out of the group for breeding use. If expecting a normal distribution, the right part of the distribution is selected. The selection differential, S, gives information about the withdrawal and represents the difference between the mean value of the entire group and the mean value of the individuals selected for breeding (the withdrawal) (Figure 12.4).

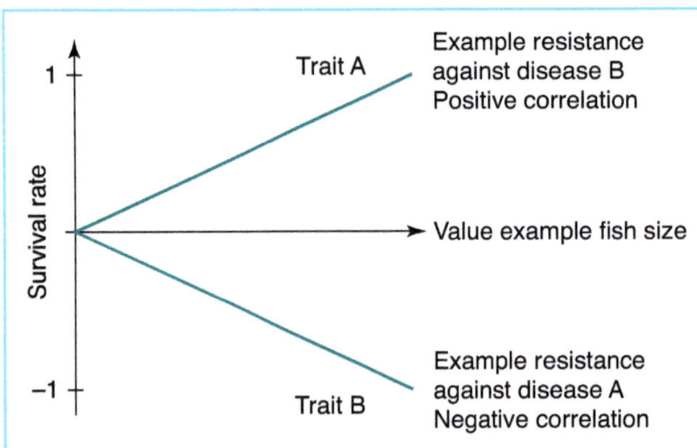

Figure 12.3 Correlation between traits may be positive or negative. In the illustration there is a positive correlation between fish size and disease A and a negative correlation between fish size and disease B.

Table 12.1 Genetic correlation between some selected traits (after [1]).

Trait	Genetic correlation
Body weight versus survival	+ 0.25
Body weight versus fat %	+ 0.23
Body weight versus flesh colour	+ 0.25
Fat % versus flesh colour	+ 0.10

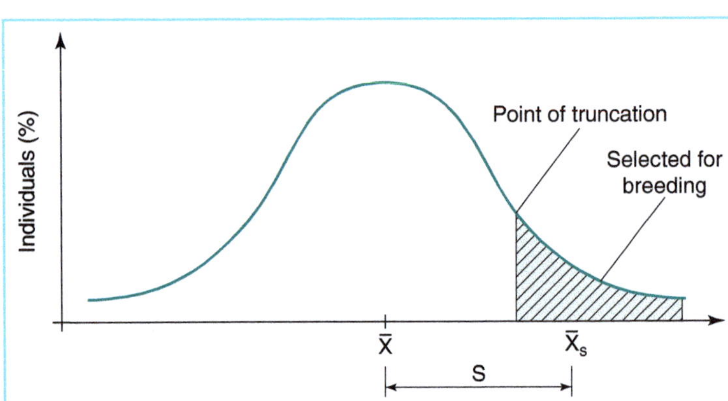

Figure 12.4 The best individuals of the population are selected for further breeding. The selection differential (S) represents the difference between the average value for the entire population and the average value of the selected individuals.

The lowest point of the size withdrawn may be called the point of truncation. By dividing the S value by the standard deviation of the individuals (σ_p) in the group, the intensity of the selection can be found (i).

$$i = \frac{S}{\sigma_P}$$

The selection intensity can be calculated or found in selection intensity tables (see Table 12.2) [10,11]. The given selection intensity is then based on the proportion of the individuals in the population that are taken out for breeding; for example, if the entire population is used the intensity will be 0, if 50% is used it will be 0.80 and if 1% is used it will be 2.67.

12.6 Phenotype and genotype

The observable or measurable traits of an individual at a given point in time represents what is named the phenotype (P) of the individual. The number of phenotypes will therefore equal the number of traits that can be observed or measured.

The phenotype is the sum of genotype (G) and environment (E). The genotype is based on the genes of the individual and represents the heritage from the parents. The environment represents the conditions that the individual is exposed to, such as type or amount of food. The phenotype may change based on the environment and life stage of the individual while the genotype is fixed based on heritage.

Heritability may therefore be described as how much of a trait is represented by the genes, or alternatively how much of the phenotype is represented by the genotype. The following equation can be used:

$$P = G + E$$

Taking growth rate as an example to illustrate this, it will be affected by both G and E. If the environment for the fish is suboptimal, the growth rate will be reduced even if the genes are good. If the G is good, selective breeding will be more favourable than if it is low.

This can also be described as variance in a population; there will be a variation in a trait and this variation can be based on the genetics or the environment. An example can be how much of the variance in growth in a group is based on genetics and how much is based on environment. Typical values for salmon are that 20–30% is based on genetics and the rest is based on environmental conditions, but this is highly dependent on trait and is described more later in terms of heritability. The variance in phenotype can be described as follows:

$$V_P = V_G + V_E$$

Owing to the interaction between genotype and environment, it is important to create an optimal environment for the fish to reduce the influence of the environment on the overall phenotype when monitoring the effects of breeding. As an example, water temperature can be included here. A suboptimal environment will increase the variance in the environmental effects that may camouflage any genetic effects.

The genetic variance (V_G) can further be described by a sum of the variance of the following three components:

$$V_G = V_A + V_D + V_I$$

where V_A represents the additive effects of alleles, V_D represents the dominance effects of alleles and V_I represents interactions among loci. The first is additive genetic variance while the other two are non-additive genetic variance and are based on interactions between genes.

12.7 Heritability and genetic gain

In a population there will, as discussed, be variation in the traits of the individuals, for instance growth rate. This variation can be caused not only by the environment but also by the genes, so raises the question of whether the growth rate is genetically high or low. The heritability (h^2) will be 1 if all the variation is related to genes and 0 if all is based on the environment. If the heritability for one trait, such as growth, is high, it will be very effective to only select the fastest growing fish as broodfish for the next generation because then the offspring will have an increased growth rate.

Heritability can be described as broad sense heritability, h_B^2, or narrow sense heritability, h_N^2. Broad sense heritability can be described as follows:

$$h_B^2 = V_G / V_P$$

while narrow sense heritability can be described as follows:

$$h_N^2 = V_A / V_P$$

Narrow sense heritability may also be described as the ratio between observed phenotypic correlation (r_{obs}) between individuals and true genetic correlation (r_{exp}) between the same individuals, as 0.5 between siblings

$$h_N^2 = \frac{r_{obs}}{r_{exp}}$$

The genetic gain (ΔG) is a term that can be used to describe the genetic progress from generation to generation for a desired trait (Figure 12.5). Mathematically it can be described by the following equation:

$$\Delta G = S \cdot h^2 = i \cdot \sigma_p \cdot h^2$$

where S = selection differential, h^2 = heritability, σ_p = standard deviation in the population and i = intensity of the selection, how large a proportion of the population is used for breeding, which can be calculated or found in tables [10,11] (Table 12.2).

If the genetic gain per year is to be found it must be divided by the generation interval in years (L):

$$\Delta G_y = \frac{i \cdot \sigma_p \cdot h^2}{L}$$

The genetic gain attained by the breeding programmes of salmon has been large for a number of traits. For example, the genetic gain per generation of increased growth has been up to 15% [4,12,13].

As shown from the equation, the genetic gain depends on the heritability. In general, the heritability for aquaculture species is quite high, which can be seen in the achieved genetic gain. The heritability varies between traits, typically from close to 0 to almost 0.6 (Table 12.3). During recent years, disease resistance has been an important part of the breeding goals and here the heritability is quite good, though it varies between diseases. For ISA, values between 0.13 and 0.40 have been reported and for *Aeromonas*, reported values are between 0.34 and 0.6 [14].

Table 12.2 The selection intensity will be based on how large a proportion of the salmon is used for breeding (based on [10,11]).

Percent of individuals selected	Intensity
100	0
50	0.8
20	1.4
10	1.76
1	2.67
0.1	3.37

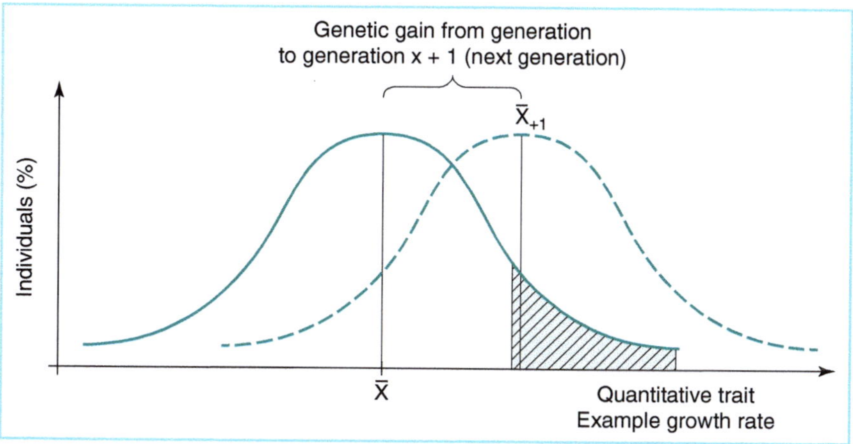

Figure 12.5 The genetic gain describes the genetic progress from generation to generation.

Table 12.3 Some measured heritability for specific traits of salmonids (based on [13–15]).

Trait	Heritability (h^2)
Growth	0.3–0.6
Age at maturation	0.1–0.4
Carcass quality	0.1–0.4
Fat content of flesh	0.1–0.2
Flesh colour	0.1–0.2
Disease resistance	0.1–0.5
A. *salmonica*	0.3–0.6
IPNV	0.4–0.6
ISAV	0.1–0.4
Caligus	0.1–0.4
Processing yield	0.2–0.5

The heritability may be measured in the individual or in siblings depending on the trait; the latter includes traits where the individual must be killed or attacked by disease (a challenge test) before measurement.

One simple way to visualize the breeding progress is through a breeding stair. The steps represent the generation interval, which is 3 years for salmon today, while the rise is represented by the genetic gain per generation (Figure 12.6) [1]. A steep stair is preferable and this can, as seen, be achieved with a short generation interval, a high genetic gain or preferably both. Since

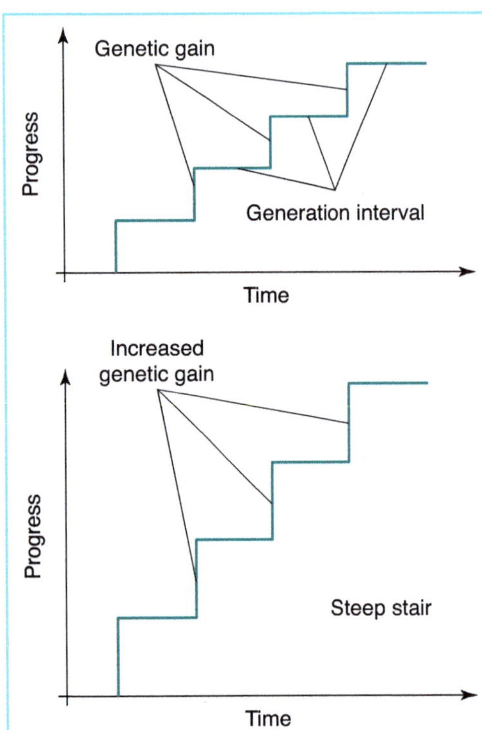

Figure 12.6 In a breeding stair the rise represents the genetic gain and the steps the generation interval. An increased genetic gain will increase the steepness of the stair.

the generation interval is fairly fixed, the genetic gain will be the important factor affecting the steepness of the stair. During the last decade, the growth of salmon has increased so much it has been possible to go from a generation interval of 4 years to 3 years, which of course has had a major impact on the yearly genetic gain.

Variation in the population is, as shown, of major importance to gain improvement through selective breeding (Figure 12.7). However, it is to be expected that the natural variation will decrease after a period due to only a part of the population being selected each time. Here Atlantic salmon and aquaculture species in general have an advantage compared to traditional animal livestock. Traditional animal livestock have been exposed to domestication and selective breeding (directly or indirectly) for more than 10,000 years while Atlantic salmon is quite newly domesticated [12,16]. However, selective breeding is still very important in traditional animal livestock and there is still a good genetic gain per generation, which shows future possibilities for Atlantic salmon breeding.

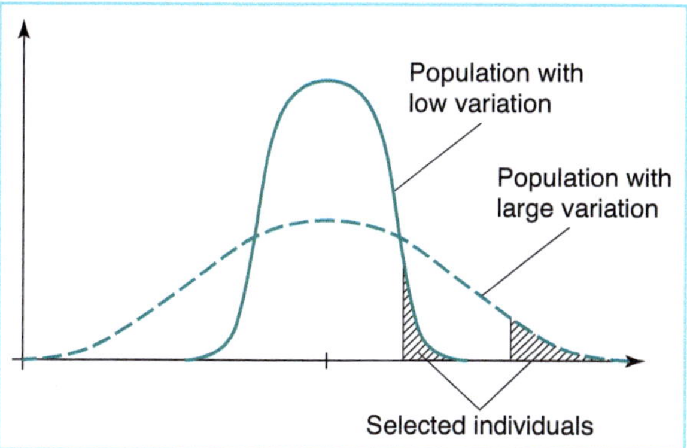

Figure 12.7 It is the variation between traits in a population that ensures genetic progress from generation to generation; increased variation increases the possible genetic gain.

Box 12.1

The yearly genetic gain in a population is to be calculated. The heritability is set to 0.3, the average weight of the fish is set to 6 kg and the coefficient of variation (CV) is set to 25%. For breeding, 3% of the females and 1% of the males from the population are used. The generation interval is set to 3 years.

Calculate first the standard deviation σ:

$$CV = \frac{\sigma \times 100}{Avg}$$

$$\sigma = \frac{CV \times Avg}{100} = \frac{25 \times 6}{100} = 1.5$$

Continue to find the intensity i. Using the table, for females it is found to be 2.27 and for males 2.67, which gives an average of 2.47.

Then the genetic gain can be found:

$$\Delta G_y = \frac{i \times \sigma_p \times h^2}{L}$$

$$\Delta G_y = \frac{2.47 \times 1.5 \times 0.3}{3} = 0.37$$

But this is the progress, and this has to be related to the weight to calculate the percentage genetic gain in growth per year.

$$\frac{6 \times x}{100} = 0.37$$

$$x = \frac{0.37 \times 100}{6} = 6.17$$

This means that the genetic gain per year is 6.17%.

If setting the CV to 5%, the genetic gain in growth per generation will be reduced to 1.24% which also shows the importance of variance in the fish group to achieve progress.

If only using 0.1% of the population for breeding, the intensity will increase to 3.37 and the genetic gain in growth per generation will be increased to 8.43%, but now care must be taken to avoid inbreeding depression.

12.8 Inbreeding

Inbreeding occurs if relatives are mated, and the degree of inbreeding is based on how closely related the individuals are. This is the coefficient of relationship, or inbreeding coefficient, which is defined as the probability of two alleles having an identical origin [10]. The offspring are 50% related to both the father and the mother, since 50% of the genes come from the mother and 50% from the father. Identical twins (clones) are 100% related, full siblings are 50% related, half-siblings are 25% related, first cousins 12.5% and so on. If the inbreeding coefficient between the parents ($F_{parents}$) is known, the inbreeding coefficient of the offspring can be calculated with the following equation:

$$F_{offspring} = 0.5 \times F_{parents}$$

When inbreeding, the number of different alleles of genes will be reduced, because the chances of transferring the same alleles from both parents will increase. This means that the variance in the population (genetic pool) will be reduced.

The performance of individuals in relation to the desired traits can be reduced due to inbreeding, which is termed inbreeding depression. The same applies to the viability of individuals.

The result of inbreeding is that the offspring perform worse than the parents for a number of traits and thereby reduce the selection response. Recessive genes may be transferred to offspring. An example to illustrate this is that if all individuals are susceptible to the same disease, then it is easy for the entire population to be wiped out. The variation in the population will be reduced, meaning individuals are more similar to each other for the desired traits. Therefore, the population, the genetic pool, needs to be a certain size to start selective breeding. In general, inbreeding problems will be more noticeable when inbreeding levels rise above 10% [2], and very approximately there will be a 2–20% performance decrease for each 10% increase in inbreeding coefficient [2]. It is recommended to keep inbreeding below 1–2% per generation [15,17].

There will always be a negative correlation between the selection intensity and the inbreeding problem, because the fewer individuals that are used for breeding, the higher the risk for inbreeding will be. Therefore, there is a requirement to use a minimum population size. The effective population size, Ne, represents the sum of males (N_m) and females (N_f) used as brood fish. An Ne of 100 is considered the minimum to ensure an inbreeding rate of less than 1%, so if the number of males and females are equal there will be at least 50 of each sex. If one of the sexes is overrepresented, the following equation can be used to calculate the Ne [18]:

$$Ne = \frac{4N_m N_f}{N_m + N_f}$$

Based on the effective breeding population size (Ne), the degree of inbreeding (F) or inbreeding coefficient can be calculated by the following equation per generation [10,18]:

$$F = \frac{1}{2Ne}$$

As can be seen from the equation, a large effective breeding population will reduce the average inbreeding degree.

To calculate the inbreeding per generation (F) for each generation (t) the equation will be as follows [1]:

$$F_t = \frac{1}{2Ne} + \left(1 - \frac{1}{2Ne}\right) F_{t-1}$$

Inbreeding is, as stated, a problem when relatives are bred together, which can be a problem in salmon breeding. Salmon have a large number of offspring and there are no simple natural identification marks on the individuals. Mating of siblings can therefore occur when the heritage is unknown. The way to avoid inbreeding is to keep families apart or to perform genetic testing. In breeding stations (broodstock farms), this is achieved by keeping the families in different tanks or marking them individually. Today, individual marking is usually done by electronic marking using PIT tags. The fry are marked/tagged when they reach an acceptable size, and after this they can be pooled. This enables later division of the different families (Figure 12.8).

The main negative effects of inbreeding may therefore be divided in two, where the short-term effect will be inbreeding depression and the long-term effect will be reduced genetic variance in

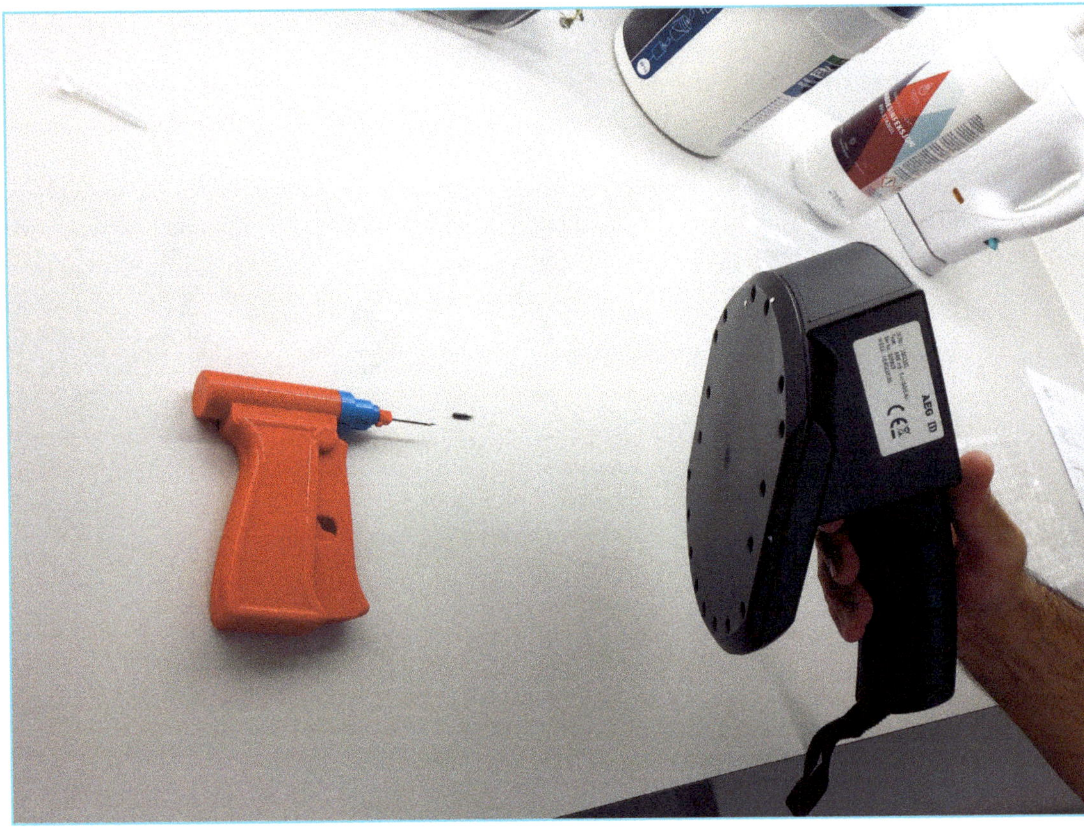

Figure 12.8 Electronic tagging of fish to keep track of the individuals. Figure shows the PIT tag, the unit for implanting the PIT tag into the belly of the salmon and the electronic reading unit.

the population. In optimally designed breeding programmes, it is therefore of major importance to restrict the degree of inbreeding.

Negative side effects of breeding may also occur when only focusing on high production efficiency. This can, for instance, be observed with traditional animal livestock where the animals grow fast, but the strength of the legs is not sufficient to support the fast growth. When performing selective breeding for high efficiency, it must therefore be combined with breeding for animal welfare and disease resistance. Any negative correlation between traits must also be addressed.

12.9 Quantitative and qualitative traits

A trait represents something that can be observed or measured in the individual. Inherited traits can be divided into quantitative and qualitative traits. Qualitative traits can be defined in categories and may also be named discontinuous traits or simply inherited traits. Quantitative traits vary continuously over a range and may also be called continuous traits. An example of a qualitative trait is sex, while growth rate is an example of a quantitative trait. Quantitative traits are typically controlled by a number of genes while qualitative traits are usually controlled by few genes or just one.

Traits can be directly production related, such as growth and survival, but may also be related to the fish quality as observed by the consumer, for instance fillet colour. A trait with ever-increasing focus is disease resistance.

The traits may also be differentiated by whether it is possible to measure them in the individual or if the effect can be measured in siblings, so it is not necessary to kill the individual. An example of the first includes growth, while the second may include fillet quality.

12.10 Varying importance of genes

The importance of different genes varies. Some may have high importance and qualitative properties, and then the environment will not be so important; other genes may have an important involvement in quantitative properties, but these traits may also be influenced by environmental conditions. A group of genes or an interaction of genes may have the same result.

12.11 Classic selective breeding and the use of molecular genetics

Today both traditional selective breeding and new molecular techniques are used in breeding programmes. The latter includes the use of advanced biotechnological methods to gather genetic information about individuals, and includes molecular genetics and genomics.

12.11.1 Individual mass selection

The simplest form of traditional breeding involves what is called individual mass selection, where the individuals with the best traits are used as brood fish for the next generation (Figure 12.9). The phenotype of the individuals forms the basis of selection, and only traits that can be easily monitored in live fish can be used as breeding goals. No information from earlier generations of the individual is used. However, there may be problems with inbreeding due to

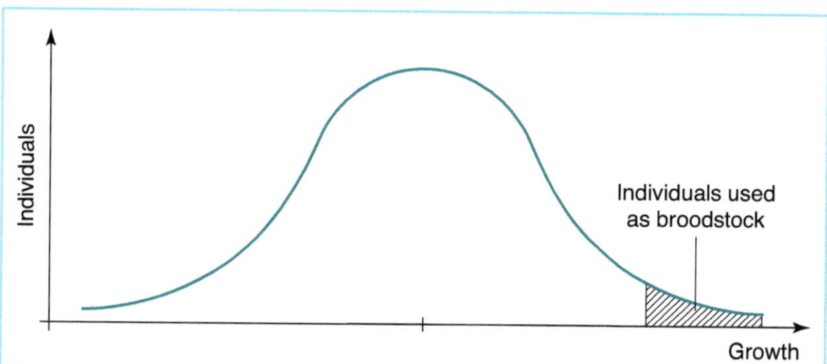

Figure 12.9 In individual mass selection the best individuals in a population are used as brood fish, for instance the individuals that show the best growth.

the high number of siblings if the individuals are not marked and families are not kept apart. By only selecting the best individuals in a population there will always be a risk that they are from the same family or share a genetic origin, because the offspring of the best parents will typically also be the better individuals. Therefore, in real selective breeding programmes family selection (information) is used as a basis, and in addition molecular techniques have been used more and more in the last decade. Although starting up a selective breeding programme based on family selection involves high investment and running costs, the overall pay off for the salmon industry has been very good. It is therefore important to remember that when starting a breeding programme, the yield of the investment will be in the long term.

12.11.2 Family selection

Family selection has successfully formed the basis for the development of selective breeding programmes for Atlantic salmon since 1970. Since salmon have a large number of siblings (>10,000) all closely related by heritage, the average presentation of a quantitative trait will be similar for the entire family. Therefore, families can be graded in relation to how well they present a trait on average. This makes it possible to categorize families as good or bad for particular traits based on the average values attained for these traits. It is also possible to calculate a breeding value for the family. This is the basis for selective breeding, where the aim is to breed males and females from different top-rated families regarding the desired traits.

All information from siblings can also be used when testing for performance for different traits. Traits that cannot be measured in live fish, such as slaughter quality, may also be evaluated, because siblings can be used in further breeding. The same is the case for disease challenge tests that can be performed on some selected individuals within the family. Also, for traits with low heritability, it is an advantage to include information from siblings which can be combined with statistics; higher accuracy and breeding progress can be achieved when using family information.

In the broodstock farms there is therefore a core of families with a high performance in one or more traits that are used in the selective breeding programme, usually more than 100 families. The families are kept separately in individual production units or marked individually. The broodstock farms that perform selective breeding therefore use a number of incubators and start-feeding units to keep the offspring of the mated families apart after fertilization. When the offspring reach a certain size, it is possible to mark them with individual tags, and then the offspring from different families can be pooled in larger production units (tanks). Here individuals from different families may be mixed and later divided again based on reading the tags. Generally, as mentioned, implanted electronic tags are used so it is easy to separate the fish according to family, but in addition, genetic marking is becoming more and more commonly used. When fry from different families grow, the performance of the traits for each family are monitored, and the best families are then used for the next generation. In this way there will be a continuous improvement from generation to generation in the performance of the selected trait. When the broodstock farms deliver eyed eggs to smolt farms, they are therefore from the best available breeding material. The families may be ranked on how well they present the specific traits, and by using all available information from all siblings and relatives it is possible to differentiate environmental influence from genetic influence (Figure 12.10). This makes it possible to calculate the

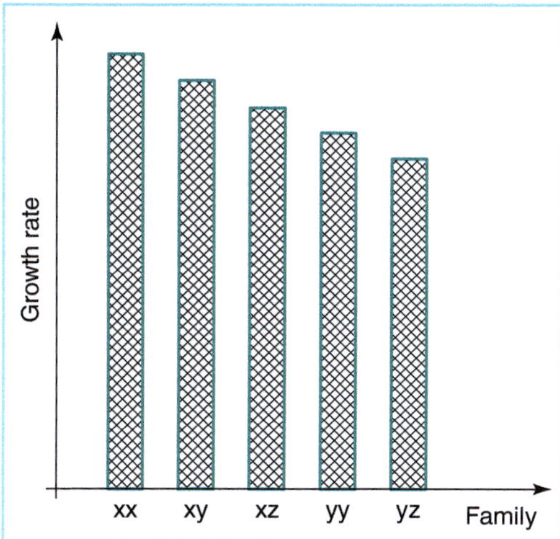

Figure 12.10 Families can be ranked based on how well they present on average regarding a trait.

breeding values (based on genetics) for the different families and thereby rank them. The families can be ranked in relation to one trait and also for the sum of traits in relation to an overall breeding goal. When calculating the breeding value, different traits may also be weighted. Some can be focused on more than others, for instance growth in seawater compared to growth in freshwater. Early maturation is a limited problem today due to genetic progress, and therefore the focus on this trait can be toned down.

Compared to individual mass selection, family selection involves a much higher cost because a number of families must be kept and must be kept apart, so a number of production units are required. However, the payback of the selective breeding will be on another economic scale. The improvement in traits will be much higher, even though it takes longer to see a return on the investments. Therefore, it is highly economically beneficial for the entire salmon farming industry to continue with family based breeding programmes for Atlantic salmon.

Combined family and individual selection may also be used for some traits. For instance, regarding growth the best individual in the best family may be used as brood fish, but here also the use of molecular techniques will become increasingly important.

12.11.3 New molecular genetic tools

During the last few decades, new techniques have been introduced into the selective breeding programmes to make them more accurate. This includes the use of marker-assisted selection and genomic selection to gain additional information about the genetics of individuals. Within each family there will, of course, be individual variation and, if taking growth as an example, some will grow better than others. The best individuals in a family can be selected by such techniques, or individuals that carry or do not carry specific gene frequencies. These new techniques are of major importance for improving knowledge of salmon biology [19].

Marker-assisted selection (MAS), finding a marker that can be related to a trait, has been developed since the nineties. It can be based on morphological signs, on biochemical substances or on genetics. A genetic marker is a piece of the DNA which has been focused on to try to link a trait to a variation on the chromosome [20,21]. MAS is therefore an indirect method: the actual trait is not measured but the marker that is linked to the trait. MAS is used with success for traits that are difficult to measure and where the heritability is low, but where a variation in the measured marker is linked to the trait.

A QTL represents a specific area on the genome/DNA that is important for a trait. As described earlier, the locus important for higher IPN resistance has been identified, and by selecting individuals with this variation, salmon with high resistance to IPN can be produced. However, there is only a limited number of traits that are controlled by a single locus on the DNA. For other traits, such as resistance to other diseases, it can be tricky to locate the locus and they are typically controlled by several parts of the DNA, not just one single locus. Different biotechnological techniques can be used to identify QTL, for example low-density markers such as microsatellites [22].

By utilizing MAS, progress in breeding can be sped up, because individuals that have or do not have these markers can be selected or ruled out for further use as a brood fish. This is done by taking a sample from the individuals to detect if it is carrying the genetic marker or not.

12.11.4 Genomic selection

When utilizing genomic selection, it means that more detailed information of the genome is used; it may also be called a form of MAS but with higher resolution [23]. The genome is divided into smaller elements that are scanned for differences [24] to look for genetic markers, segments of the DNA. With the identification of the overall salmon genome (see chapter on broodstock fish), new possibilities have opened for increased accuracy. By going further and more accurately into the genome, and going from low-density markers (microsatellites) to high-density genetic markers (SNP, see Chapter 5, broodstock), much more information can be gained and used in selective breeding. It can then be attempted to link the observed genetic differences to traits desired for breeding. However, as a number of genes and parts (SNP) of the DNA are usually responsible for a trait, this is not simple. This is also still quite an expensive technique even though the prices are dropping, but in the future will be an increasingly important tool for breeding progress.

The use of molecular genetics therefore makes it possible to select breeding stock not only based on family achievement but also based on information from individuals. By combining traditional selective breeding with molecular genetics it is possible to improve the results of breeding with faster progress. Molecular genetics will, in addition, be an improved tool to avoid inbreeding as more information is gathered. However, the cost involved in development of breeding will increase, and knowledge of molecular genetics and facilities for performing such techniques will also be required.

12.12 Breeding programmes and breeding goals

To set up a family based breeding programme requires a high knowledge of breeding [25], especially when including genomic selection [26,27], but the long-term economic benefits of a good well-designed breeding programme are high [28].

When setting up a family based breeding programme the goals of selective breeding are important [29]. At the start of the salmon farming industry, growth and age at sexual maturation were the two major goals of breeding, but the number of breeding goals has since increased in number; for example, breeding companies have over 20 traits in their breeding programmes. By adding more traits, breeding for the goals becomes more and more advanced and the requirement for knowledge in genetics and statistics increases. The cost of the breeding programme will also be directly related to the number of traits included, and economy has to be considered as an important factor for inclusion of each new trait. It is also recommended to monitor important non-selected traits, to avoid negative correlation with the selected traits. The goals can be separated into directly production related and related to disease and animal welfare, even though of course they interact highly. Examples of production-related goals can include growth rate, feed utilization, age at sexual maturation and a number of parameters related to slaughter quality. Disease- and welfare-related goals can include resistance against different diseases and how families handle stress. Those goals are important for achieving a healthy and robust salmon suited to farming conditions.

The different traits may also be weighted, meaning some are more important than others in the overall breeding programme. This also makes it possible to produce tailor-made roe based on how the different traits are weighted.

It is also important to remember that when selecting breeding goals and setting up a breeding programme it forms the basis of a strategy to give long-term results, not just for one generation. For example, the first established Atlantic salmon breeding programme is now on its 14th generation of breeding, and genetic progress is attained in each generation.

A simplified summary of selective breeding is as follows. First, good individuals from good families are mated. The fertilized eggs from the mating are kept in individual incubators on the breeding station, which is also the case when the eggs hatch and the fry is start fed. The fry continue to grow in separate tanks until they reach a size where they can be individually marked and can be pooled with other fry from other families. Then the fish grow, and during growth, traits related to the breeding goals are monitored. The families and individuals that perform best will then be used as new broodfish for the next generation and in this way genetic gain is ensured.

Several individuals per family are kept for breeding, and in addition to keeping some on the breeding station some are sent out to different producers which function as test stations. This includes sea cage farms where they are grown to slaughter size and desired traits are measured, including harvesting and measuring of slaughter quality. By utilizing several test stations, the fish will be exposed to different environments to see how they display the desired traits. This also provides the opportunity to see how environment affects the phenotype, and is a security measure to ensure that fish are available when they start to mature and are collected for use as brood fish (if they perform well according to the traits). If for some reason the fish die during ongrowing, brood fish can be collected from another farm.

The breeding values for the individuals or how well they perform according to the breeding goals can be estimated based on all types of available information, both for individual presentation, family presentation, heritage and statistics. The breeding value can be given as a BLUP index (best linear unbiased prediction – a statistical method) [30]. This is used in family based breeding, but can also include marker-assisted BLUP. Genomic breeding values can also be included to gain a more accurate value [31]. There is software available for calculation of breeding values.

References

1. Gjedrem, T. and Baranski, M. *Selective breeding in aquaculture: an introduction*. Vol. 10. 2010, Springer: Dordrecht.
2. Oldenbroek, K. and van der Waaij, L. *Textbook Animal Breeding : Animal Breeding and Genetics for BSc Students*. 2014≥ Wageningen University and Research Centre.: Centre for Genetic Resources and Animal Breeding and Genomics Group.
3. Gjedrem, T., Robinson, N. and Rye, M. The importance of selective breeding in aquaculture to meet future demands for animal protein: a review. *Aquaculture*. 2012. 350: 117–129.
4. Gjedrem, T. and Rye, M. Selection response in fish and shellfish: a review. *Reviews in Aquaculture*. 2018. 10(1): p. 168–179.
5. Janssen, K. et al. Impact of selective breeding on European aquaculture. *Aquaculture*. 2017. 472: 8–16.
6. Thodesen, J. and Gjedrem, T. Breeding programmes on Atlantic salmon in Norway: lessons learned, in *Development of Aquatic Animal Genetic Improvement and Dissemination Programs: Current Status and Action Plans*, R.W. Ponzoni, B.O. Acosta and A.G. Ponniah (eds). 2006. WorldFish Center Conference Proceedings 73. Penang, Malaysia, pp. 22–26.
7. De Verdal, H. et al. Improving feed efficiency in fish using selective breeding: a review. *Reviews in Aquaculture*. 2018. 10(4): 833–851.
8. Yáñez, J.M. et al. Negative genetic correlation between resistance against Piscirickettsia salmonis and harvest weight in coho salmon (Oncorhynchus kisutch). *Aquaculture*. 2016. 459: 8–13.
9. Gjøen, H.M. et al. Genetic correlations between survival of Atlantic salmon in challenge and field tests. *Aquaculture*. 1997. 158(3–4): 277–288.
10. Falconer, D.S. and Mackay, T.F.C. *Introduction to Quantitative Genetics*. 1996. Pearson: Harlow.
11. Mackay, I. Selection intensity, manual, in *Breeding Optimization Lead, CGIAR Excellence in Breeding Platform*, G.E. Covarrubias-Pazaran (ed.). 2020.
12. Houston, R.D. and Macqueen, D.J. Atlantic salmon (Salmo salar L.) genetics in the 21st century: taking leaps forward in aquaculture and biological understanding. *Animal Genetics*. 2019. 50(1): 3–14.
13. Lhorente, J.P., et al., Advances in genetic improvement for salmon and trout aquaculture: the Chilean situation and prospects. *Reviews in Aquaculture*. 2019. 11(2): 340–353.
14. Yáñez, J.M., Houston, R.D. and Newman, S. Genetics and genomics of disease resistance in salmonid species. *Frontiers in Genetics*. 2014. 5: 415.
15. Duncan, N., Sonesson, A. and Chavanne, H. Principles of finfish broodstock management in aquaculture: control of reproduction and genetic improvement, in *Advances in Aquaculture Hatchery Technology*. 2013. Elsevier: Amsterdam, pp. 23–75.
16. Mignon-Grasteau, S. et al. Genetics of adaptation and domestication in livestock. *Livestock Production Science*. 2005. 93(1): 3–14.
17. Pante, M.J.R., Gjerde, B. and McMillan, I. Inbreeding levels in selected populations of rainbow trout, *Oncorhynchus mykiss*. *Aquaculture*. 2001. 192(2–4): 213–224.
18. Tave, D. *Inbreeding and Broodstock Management*. 1999. FAO: Rome.
19. Yáñez, J.M., Houston, R. and Newman, S. *Genomics in Aquaculture to Better Understand Species Biology and Accelerate Genetic Progress*. 2016. Frontiers Media SA.
20. Wakchaure, R. et al. Marker-assisted selection (MAS) in animal breeding: a review. *J. Drug. Metab. Toxicol.* 2015. 6(5): e127.

21. Yue, G.H. Recent advances of genome mapping and marker-assisted selection in aquaculture. *Fish and Fisheries*. 2014. 15(3): 376–396.
22. Abdurakhmonov, I.Y. Introduction to microsatellites: basics, trends and highlights. *Microsatellite Markers*. 2016. 1: 13.
23. Goddard, M. and Hayes, D. Genomic selection. *Journal of Animal Breeding and Genetics*. 2007. 124(6): 323–330.
24. Meuwissen, T.H., Hayes, B.J. and Goddard, M.E. Prediction of total genetic value using genome-wide dense marker maps. *Genetics*. 2001. 157(4): 1819–1829.
25. Gjerde, B. Design of breeding programmes, in *Selection and Breeding Programmes in Aquaculture*. 2005. Springer: Dordrecht, pp. 173–195.
26. Khatkar, M.S. Genomic selection in aquaculture breeding programmes. *Bioinformatics in Aquaculture: Principles and Methods*. 2017: 380–391.
27. Sonesson, A.K. and Meuwissen, T.H. Testing strategies for genomic selection in aquaculture breeding programmes. *Genetics Selection Evolution*. 2009. 41(1): 1–9.
28. Janssen, K., Saatkamp, H. and Komen, H. Cost-benefit analysis of aquaculture breeding programmes. *Genetics Selection Evolution*. 2018. 50(1): 1–16.
29. Gjedrem, T. *Selection and Breeding Programmes in Aquaculture*. 2005. Springer: Dordrecht.
30. Robinson, G.K. That BLUP is a good thing: the estimation of random effects. *Statistical Science*. 1991. 6(1): 15–32.
31. Nielsen, H.M. et al. Comparison of accuracy of genome-wide and BLUP breeding value estimates in sib based aquaculture breeding schemes. Aquaculture. 2009. 289(3–4): 259–264.

CHAPTER 13

Future trends – salmon farming of tomorrow

13.1 Introduction

Although there is no certainty as to how the salmon farming industry will develop in the future, trends can be observed and qualified guesses are possible. The worldwide growth in farming of Atlantic salmon will certainly continue. It is a species that is well known in all markets and it can be used in a number ways: cooked or raw (such as sashimi), or processed by smoking, curing or marinating. The world population is growing, and so is the proportion of the population with reasonable income. The health aspects of food are also more in focus. Therefore, there is no reason why the market should not continue to grow, and if production does not increase to meet demand, prices will increase and thereby regulate the market. This has already been documented with very high Atlantic salmon prices in periods during recent years, due to a lack of available salmon on the market (Figure 13.1). This general increase in the demand for salmon during recent years, with increased prices, has provided an opening for new production methods that give the possibility of salmon farming in new areas. These methods for ongrowing salmon potentially have increased investment and production costs and include closed cages, offshore cages and land-based production systems, including both flow through and RAS.

It is not only the production methods that will be developed in the future, even though this is the focus of this chapter. In addition, change must occur with the feed and feed ingredients used, because this is a major factor in the production cost of salmon. If new high-volume, low-cost ingredients for use in feed are not found, the feed costs will continue to increase due to limitations in supply (also discussed in Chapter 10). Disease problems must also be addressed, especially sea lice. However, there are currently great efforts being made, which will certainly continue. Some of the new production methods may also handle these problems more effectively.

Robotics and computer science will also affect how the salmon industry develops. More working operations will be handled by robots and artificial intelligence will replace the human brain in a number of operations, such as feeding and feed control. However, this will require the development and use of more and more sensors.

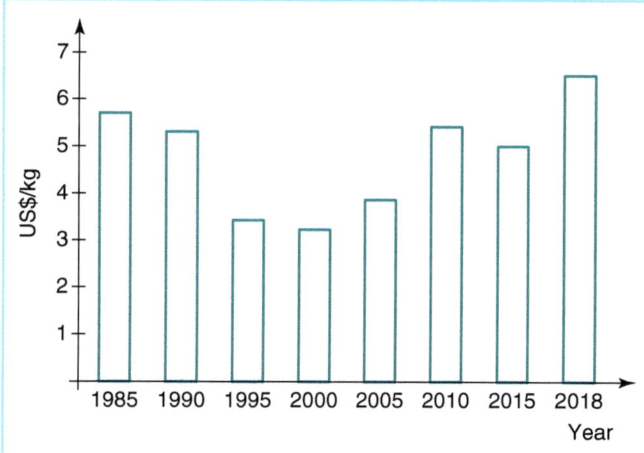

Figure 13.1 Development of the prices for Atlantic salmon during recent years (values from [1]).

13.2 High prices make new production methods economically beneficial

Common to all the new methods is, as stated, a higher production cost compared to farming in the traditional open sea cages; however, this is only the case when no problems occur in the traditional sea cage farming of Atlantic salmon. Today this is not the case; there are reasonable problems, at least in some areas, with the traditional salmon farming methods which make growth in salmon production difficult. The problems are represented by disease, especially sea lice, but in addition there is an increased focus on methods for avoiding fish escape and methods preventing accumulation of waste close to the open sea cages. To avoid these problems increasing, new regulations are continually set for traditional sea cage farming, regarding both the requirements for the technology used and also how the farms must be operated. The latter may, for instance, involve programmes to measure and document accumulation of waste at the site. All new requirements are costly to meet, meaning that the production costs for salmon farmed by traditional methods is increasing. Countries starting to introduce environmental taxes will, of course, also have a negative effect on the production costs.

There is also a limitation in the availability of suitable production sites for use of the traditional sea cage technology: either insufficient shelter, which leads to problems with high waves and strong water currents, or too great a degree of shelter, which results in water currents too low to ensure adequate oxygen supply and effective dilution of metabolic waste products.

13.3 Increased industrialization

There is no reason for the ongoing industrialization not to continue, nor the increasing size of the companies. If no political action is taken, the small enterprises will be bought up by the large companies, at least if the current high profitability continues. In a number of areas there are large-scale advantages, meaning larger companies can produce salmon more cheaply than smaller ones. For instance, when buying feed or equipment, larger companies have greater economies of

Figure 13.2 A salmon smolt farm under development that has capacity for delivering over 30 million smolt a year.

scale and purchasing power in the market. Several of the new production methods also require enormous investment costs, not only for the equipment but also for the establishment of biomass before harvesting can start (Figure 13.2). Investments of over €100 million are not rare; this represents a financial risk that only large companies can take.

However, special care must be taken regarding disease when production units become large. A greater biomass in a restricted area will increase the risk for outbreak and transfer of diseases, including parasites, so this has to be a special focus. When diseases occur, the economic consequences will also be higher, as will the cost of any other failure. This is a heavy economic burden to carry.

13.4 Atlantic salmon farming in new areas

Until today, salmon farming has been restricted to areas where Atlantic salmon has its natural distribution or where the natural conditions are suited to the species. For farming in sea cages, the seawater temperature must suit the requirements of Atlantic salmon, being neither too high nor too low. The waves cannot be too high, nor the water current. Too low a water exchange on the site will also limit ongrowing production. Overall, this restricts traditional sea cage farming to some limited areas worldwide, as can been seen by where salmon farming is carried out today. Through the introduction of new technology and production methods, this area can be greatly expanded. The new methods include offshore cages, closed cages and land-based farms using either flow through technology or RAS (section 13.6.1).

13.5 Timing the introduction of new methods

Regarding the new and partly new production methods that are now under development, there is of course a great advantage in starting the development and production of these methods when salmon prices are high. This makes it possible to gain at least a small positive cash flow from the start. The period of high prices will also give the potential to achieve increased practical knowledge with these new production methods and to optimize the production systems. The result is that the production costs can be reduced, which will also make these systems more compatible with traditional sea cage farming if salmon prices drop, as will be the result when production volume increases.

13.6 New industry structure

The structure of the industry will probably also change due to the development of new technology and production methods. Until recently there has there been a rigid structure: broodstock farms deliver eyed eggs to smolt farms, which in turn deliver smolt to ongrowing farms to produce salmon for harvesting in traditional sea cages located in quite sheltered areas. In the future this structure will be challenged by a variety of new models.

What already has been observed is the interest in producing large smolt up to 250 g, and also to produce what are called post-smolt, up to 1.5 kg, before transfer to traditional sea cages. This reduces the period the fish are in the sea when they are exposed to the risks of sea lice and disease. In open sea cages it is impossible to control the water flowing into the cages and whatever microbiota it carries, so disease risk increases; the particular problem today is, of course, sea lice.

13.6.1 A three-step model

By introducing RAS on smolt farms, the production capacity of the available freshwater supply can be increased. This capacity increase can be realized with a higher number of fish, a larger fish size or both. RAS also allows for higher water temperatures, so the growth rate will increase providing the opportunity to reach a higher weight of smolt in a shorter time. It might be possible to produce a salmon ready for harvesting before it reaches the age of 2 years.

There is a significant interest in the introduction of a three-step model (Figure 13.3), where the traditional smolt farms deliver their smolt to a new intermediate farm that uses seawater and produces post-smolt in a closed production system. The size of the fish produced in this farm can be up to around 1 kg, before the fish are delivered to traditional sea cage farms. The production system used in the post-smolt farm may be either land-based or closed sea cages (closed

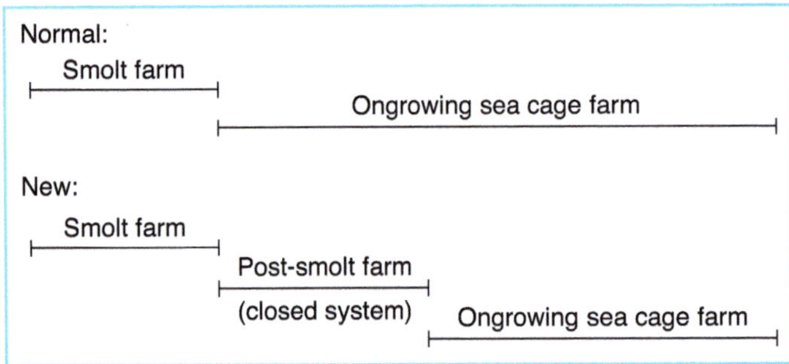

Figure 13.3 Introduction of a three-step model, smolt-post-smolt-ongrowing.

contaminant systems). The fish in the farms have smoltified and seawater is required, so the farms are equipped with pumping stations that supply seawater to the production units.

Most of the land-based farms with such production, either already in operation today or that are under construction, use RAS technology, but flow through is also considered. By using RAS technology, a higher water temperature can be used, thereby increasing growth rate and turnover.

Various designs of closed sea cages are used for this production, but they commonly use direct flow through. Most supply pure oxygen to the inlet water and have the potential to purify the effluent water.

The obvious advantage of such a model is that the fish reach a larger size before transfer to open sea cages, therefore the required time in the sea cages before harvesting will be reduced. If a disease breaks out in closed units it is also easier to control the outbreak and limit the extent, and disease treatment will be easier. The problems with smoltification when fish enter the open sea cages will also be eliminated because the fish have already been exposed to seawater in the post-smolt farm.

13.7 Ongrowing salmon farms of tomorrow

Methods for ongrowing salmon other than the traditional open cages have already been used to a limited degree, but the interest in alternatives is increasing and there is much further development and optimization of methods in progress. The methods can be divided into offshore cages, closed cages and land-based farms. The common factor for all is that more areas can be used for salmon farming.

13.7.1 Offshore cages

Owing to the limited number of suitable sites and problems with local pollution at over-sheltered sites, the interest in offshore cages is increasing. However, when discussing offshore cages a definition is required and even then, the differences are fluid. The traditional cages typically tolerate waves up to 3–5 m (significant wave height, the average of the highest one-third of the waves), in addition to the current. If exposed to higher waves, problems with the construction will occur. Therefore, the sea cages, mooring systems and important additional equipment must be certified for a given wave height and water current to avoid breakdown of construction or construction components resulting in escape of salmon.

Cages or similar constructions that tolerate higher waves have been developed and a large number of new constructions are either under development or under testing. These are constructions that tolerate significant wave heights of up to 10–15 m. The great advantage of using such constructions is, of course, that new areas can be used for ongrowing salmon. Some shelter is still required but the trend is that areas with higher and higher waves could be used for salmon farming. Even fully offshore farming in the open ocean may be a possibility in the near future. But even if such constructions can be built to tolerate the harsh weather and environmental conditions, it must still also be possible to perform the regular operations on the farm such as feeding control and observation of fish behaviour. It must also be possible to harvest the fish on a regular basis, and for boats to supply the farms with operators/crew and other necessities.

The size of some of the offshore constructions built is very large; this results in very large salmon production at each site, with biomasses above 10,000 tonnes. This means that when such cage constructions are used, significant quantities of salmon can be produced for the market in addition to the production levels today.

The method of production in such cages will be the same as production in traditional cages, with the difference that the distance from the shore will be larger and transport will take longer. The operators/crew will also most probably stay at the farm.

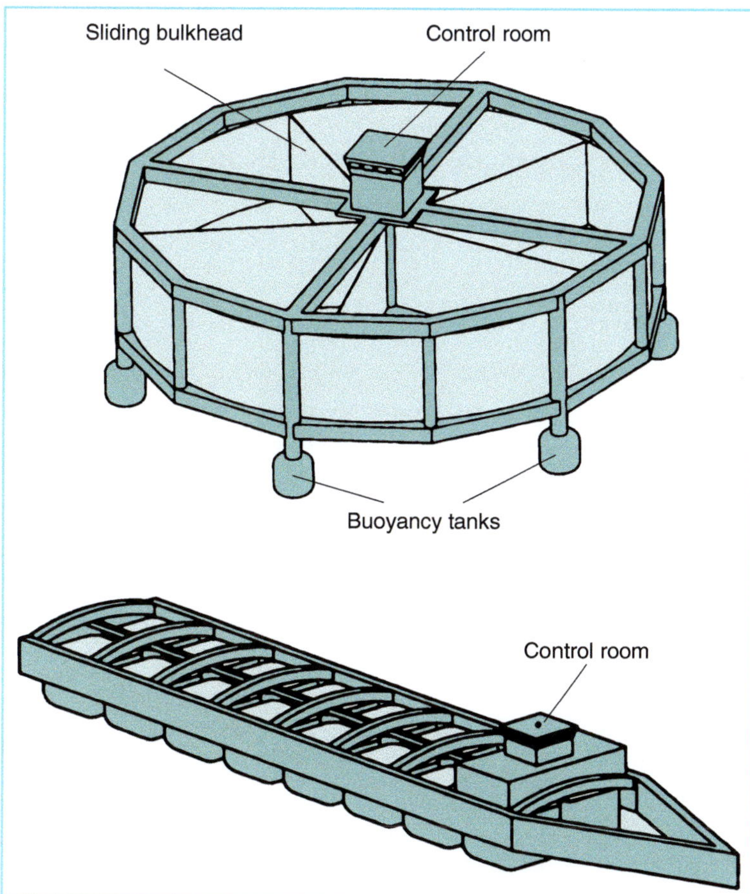

Figure 13.4 Offshore cages tolerate higher wave heights [2].

13.7.2 Farming in closed cages, closed contaminant systems

In a closed cage the net is exchanged for a dense material. Various materials such as tarpaulin, stiff PE or fibreglass are used instead of a net bag. Various constructions and designs are also used, ranging from a simple substitution of the net bag with a tarpaulin bag to far more advanced structures such as large domes, ovals or horizontal floating pipes. Some also include a roof, where one such example is a floating egg-shaped construction.

By changing the net to a dense material, the enclosed area must be filled with water, and therefore the closed systems are supported by one or several pumps moving the water. Here there is an advantage to utilizing floating systems compared to traditional land-based farms because the height that the water must be lifted into the production volume is much lower. The height required is only that necessary to defeat the friction loss to enable water to flow though the inlet pipe, through the production unit at the desired velocity and out of the production unit. The other advantage is that no land area is used. Land-based farms often require much excavation and/or land filling in coastal areas. In many coastal areas there is also competition for use of land, and different regulations must also be met.

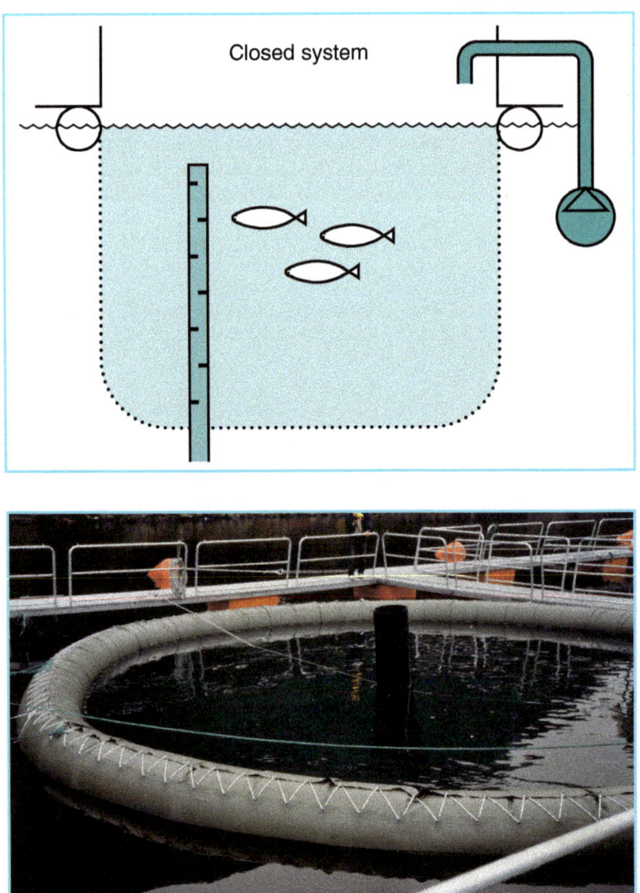

Figure 13.5 Closed cage [2].

One great advantage of closed cages compared to open cages is that it is possible to clean the outlet water because all effluent water can be collected. This makes it possible to use suboptimal sites close to the shore and sheltered areas with limited water exchange, greatly increasing the number of potential sites for salmon farming. They will, however, require a sheltered coast because the currently available designs do not tolerate high waves and current.

Production will be similar to that performed in traditional sea cages; however, the temperature will be different because the water is taken from some depth, normally below 15–20 m. At this depth the temperature will be higher during winter and lower during summer compared to closer to the water surface. This will affect the growth rate. The reason for taking water from lower depths is to avoid taking in sea lice and to reduce the number of microorganisms taken into the cage, because the amount of microorganisms reduces with the depth.

13.7.3 Land-based ongrowing production

With the introduction of land-based technology, the farming of Atlantic salmon can be carried out in areas where a sheltered coastline with suitable temperatures for traditional sea cages is not available. By using RAS even more areas can be used, both for juveniles and for ongrowing to harvesting. Farms with a high degree of reuse will have very low requirements for the supply of new water and can be built further away from large water sources. Ongrowing can also be carried out away from the coast, and the low new water requirement means freshwater with the addition of salt to produce the required salinity can be used (technology is required for this).

When using land-based systems for ongrowing, the smolt are taken directly from the smolt farm to the land-based ongrowing farm, which may actually be the same design as a traditional smolt farm but using seawater or brackish water, as for a post-smolt farm. Typically, however, the production unit size will be larger.

An interest in production on land-based farms was observed in the 1990s, when a number of farms (>10) of different sizes were established in Norway and Iceland, among other locations. However, they suffered from poor economy and either closed down or went bankrupt. During those days the production costs for salmon in traditional sea cages were lower than today, as were salmon prices. Prices for salmon were in the area of NOK 25/kg and production prices in sea cages were NOK 20–25/kg; the land-based farms had problems getting production prices down to NOK 30/kg. The land-based farms in this period were built in different ways with different materials, and all were flow through systems so seawater was pumped from the sea through the farm and back to the sea. All used additional oxygen to reduce the amount of water to pump. None of the farms at this time used RAS technology.

Land-based salmon production through to harvesting size should be grouped into flow through and RAS systems because the production regime differs. However, when operating land-based farms, the management requirement is increased both for flow through and for RAS. Both the investment and running costs of the production volume (cost for investment and operation of 1 m^3 farming volume) are higher than for sea cages. The result is that the yearly salmon production per cubic metre of farming volume must be considerably higher, and the higher the investment and running costs of the volume, the higher the yearly production per cubic metre must be to ensure the required profitability. This is solved by keeping a higher fish density to increase the growth in biomass per cubic metre, but this also requires regular intake, internal transport and harvesting. Therefore, the logistics from intake to harvesting must be well designed and planned. However, with higher fish densities animal welfare will also be affected, and this must be in high focus when developing these new production methods.

13.7.3.1 Flow through

A flow through farm is, very simplified, just a replacement of the cages with onshore tanks. Water is pumped from the sea, through the fish tanks and back out to the sea, including a particle filtration plant on the effluent water. To reduce the amount of water required to be pumped through the fish tanks, water is supersaturated with pure oxygen gas. Deep water is taken in, as for closed sea cages, and taking water below the thermocline can be an advantage as the water temperature will be more stable, in addition to the advantages regarding disease and parasites. From the point of production planning, a stable predictable temperature will be an advantage. The farms can also be covered with a roof, which makes it possible to control light and use 24-hour light which can increase the overall growth rate. However, this could result in problems with early maturation because there is no direct trigger for maturation. Covering the farm will also prevent any problems with birds, and no leaves or other pollutants from above will drop into the tanks.

The location must be at the coast, at no distance from the sea. It is an advantage for the coastline not to be too exposed to weather and waves both because of the inlet pipe and because a pumping station must be installed close to the water; otherwise special designs must be used. The fish production tanks must not be placed too high above the sea level to avoid increasing the pumping height which leads to increased production costs. However, by using more advanced technology the potential areas where Atlantic salmon can be grown to harvesting size are increased, and there will not be a requirement for a sheltered coastline.

13.7.3.2 RAS

With the use of RAS, the geographical areas where Atlantic salmon can be farmed are hugely expanded. This trend can already be seen with the building of land-based ongrowing farms in central Europe, the southern USA and China. RAS and technology that enables almost zero water exchange remove the requirement for farms to be located by the sea. Owing to the low requirement for new water with the high degree of recycling, salt can be added to freshwater to achieve the desired salinity, where around 15–20 per mille may be the most beneficial level to use. This allows production of salmon in areas with a warmer climate and that are far away from any coastline, so a large number of new nations can begin Atlantic salmon production.

The advantage of such farming concepts is that the fish can be produced close to the large consumer markets avoiding lengthy transport and thereby prolonging shelf life. There are, of course, also disadvantages, the main one being the investment and production costs, of which the long-term values are not yet actually known. What has also been seen is that if using RAS for salmon in warmer climates, the cost of chilling the water can be considerable. Handling the effluent in a secure way will also represent an additional cost.

Regarding production control and production time, RAS has the advantages of the water temperature being easier to control and the possibility of using a higher temperature, resulting in an increased growth rate and thereby a shorter production time from eyed egg to harvesting. Because the farms are roof-covered, the light can be controlled which also provides an advantage regarding the production and production planning.

Figure 13.6 Land-based production farm with flow through system.

However, there are some problems related to the production of harvest-size salmon in RAS, especially when a high degree of water reuse is utilized. One is the need for purging before sending the salmon to consumers, because the fish take up substances from the water which affects the flavour of the meat ('off-flavour'). Today this is usually done by keeping the salmon in clean water without feeding for a period (14 days) prior to slaughter so the flavour is released from the fish meat, but several trials are ongoing to evaluate other methods.

Other problems can be caused by accumulation of substances in the water. One of these substances is hydrogen sulphide, which can be produced in areas with accumulation of waste in anaerobic zones, and will be directly toxic for the fish. However, all sorts of substances, including hormones, can accumulate and may have an effect. The knowledge of water maturation and the microbiological community in RAS systems is not fully understood and the unexpected can occur.

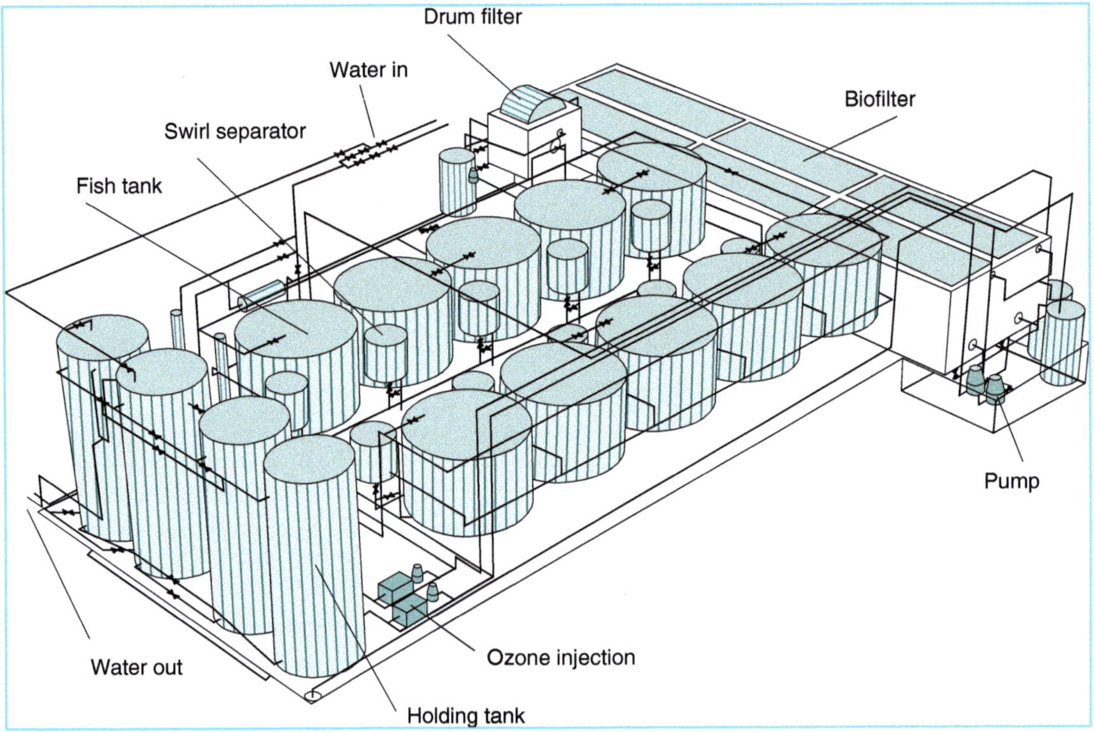

Figure 13.7 Land-based ongrowing farm using RAS [2].

13.8 Full production farms

With the introduction of land-based farms, preferably using RAS, the overall concept of salmon farming can be changed. It will be possible to perform the entire production cycle, from egg to harvesting, at the same site: a full production farm. It is, of course, an advantage to have the entire production on one site because the smolt will not need transporting. There are also advantages regarding production planning. When controlling the water temperature, it will be possible to control the entire production cycle and to make more accurate prognoses of when to deliver fish and at what size. This is important for the market.

13.9 Sustainability

In the future there will be an increasing focus on sustainability. The UN have defined sustainability goals, and a number of them could directly affect the production methods for Atlantic salmon. The answer to which production methods are most sustainable, however, is not clear. All methods have advantages and disadvantages and it is a matter of how the different sustainability goals are weighted. The obvious advantage with the traditional sea cages is the simple construction with low material input. When up and running there is no input needed apart from feed, and the water exchange is assured by nature. A land-based RAS plant is an advanced construction with high material input. When up and running it is possible to collect all effluent, so the environmental impact can be almost fully controlled. It does, however, require a continuous supply of energy for many of the processes, such as pumping, oxygenation and temperature regulation. There are also questions related to the use of land for the land-based farms; the land used has to be planned, and blasting and land filling can be required. Therefore, it is not straightforward to perform sustainability analyses to evaluate which method is most sustainable.

13.10 New knowledge required

When new production methods and new technology are used, skilled operators are vital. Today there is a lack of people with the skills needed by RAS technology. This results in a lack of optimization of production, and failure due to poor knowledge can occur, both of which negatively affect the production cost. The same will also be the case when offshore cages on weather-exposed sites become more common. Farm operation under the onslaught of 10 m high waves is far removed from the lesser challenges of 2 m high waves. Here human safety also constitutes an important issue.

For the new production methods to be fully and economically utilized more and better trained operators will be required.

References

1. FAO. Online query panels. http://www.fao.org/fishery/topic/16140/en.
2. Lekang, O.I. *Aquaculture Engineering*. 2020. Wiley: Chichester.

Index